To Patti
With best personal regards
Dale

Natural Resources Management in Agriculture

Methods for Assessing Economic and Environmental Impacts

Natural Resources Management in Agriculture

Methods for Assessing Economic and Environmental Impacts

Edited by

B. Shiferaw, H.A. Freeman
ICRISAT, and ILRI, Nairobi, Kenya

and

S.M. Swinton
Michigan State University, East Lansing, Michigan, USA

CABI Publishing

CABI Publishing is a division of CAB International

CABI Publishing
CAB International
Wallingford
Oxfordshire OX10 8DE
UK

Tel: +44 (0)1491 832111
Fax: +44 (0)1491 833508
E-mail: cabi@cabi.org
Web site: www.cabi-publishing.org

CABI Publishing
875 Massachusetts Avenue
7th Floor
Cambridge, MA 02139
USA

Tel: +1 617 395 4056
Fax: +1 617 354 6875
E-mail: cabi-nao@cabi.org

Published in association with:

The International Crops Research Institute for the Semi-Arid Tropics (ICRISAT)
Patancheru 502 324
Andhra Pradesh, India

The International Crops Research Institute for the Semi-Arid Tropics (ICRISAT) is a non-profit, non-political organisation for science-based agricultural development. ICRISAT conducts research on sorghum, pearl millet, chickpea, pigeonpea and groundnut – crops that support the livelihoods of the poorest of the poor in the semi-arid tropics, encompassing 48 countries. ICRISAT also shares information and knowledge through capacity building, publications and information communication technologies. Established in 1972, it is one of 15 centres supported by the Consultative Group on International Agricultural Research (CGIAR).

A catalogue record for this book is available from the British Library, London, UK.

Library of Congress Cataloging-in-Publication Data

Natural Resource management in agriculture : methods for assessing economic and environmental impacts / edited by B. Shiferaw, H.A. Freeman, and S.M. Swinton.
p.cm.
Includes bibliographical references and index.
ISBN 0-85199-828-3 (alk. paper)
1. Agriculture--Environmental aspects--Congresses. 2. Agriculture--Economic aspects--Congresses. 3. Agricultural resources--Management-Congresses. I. Shiferaw Bekele. II. Freeman, H.A. III. Swinton, Scott M. IV. Title
S589.75.N368 2005
338.1--dc22

200402093

ISBN 0 85199 828 3

Printed and bound in the UK by Cromwell Press, Wiltshire, from copy supplied by the editors

Contents

Preface

In response to increasing concerns about degradation of natural resources and the sustainability of agricultural production potentials in many poor regions of the world, many national and international organisations have initiated research and development programmes for natural resource management (NRM). Efforts in this direction include the design and development of low-cost technological options for integrated management of soil and water resources, the development of ecologically sound cropping systems, and options for the conservation and management of agro-biodiversity and forestry resources. Among others, the Consultative Group for International Agricultural Research (CGIAR) has substantially increased its research investments in the area of NRM. Development agencies in developing countries also invest substantially in measures to sustain productivity and conserve both the agricultural resource base and the environment. Donors, policy makers, development agents, and researchers are all anxious to evaluate the potential social benefits and environmental outcomes resulting from such investments through the adoption of new resource conserving and/or productivity enhancing technologies. Although methods for evaluating the impacts of crop improvement technologies are well developed and widely applied, there is a dearth of methods to evaluate the impacts of NRM interventions. This is partly due to the methodological difficulties encountered in assessing the impacts from NRM research, including those arising from inter-relationships among natural resources, spatial and temporal dimensions of impact, and the valuation of direct and indirect environmental benefits and costs.

Despite a handful of attempts to assess the impacts from NRM research, until now researchers from a range of disciplines and institutional backgrounds have not joined to critically address the challenges and develop methods for NRM impact assessment. This book is an effort towards filling this gap. Its objective is to examine methodological difficulties and present practical methods that can be used to assess the economic and environmental impacts of NRM technology and policy interventions. It synthesises recent methodological advances and results from frontier research in this field. The methodological and conceptual chapters are enriched and illustrated with case studies and examples. Several chapters bring together current thinking and perspectives on NRM impact assessment and define directions for future research, covering such important areas as economic valuation methods, measurable performance indicators and applicable impact evaluation approaches together with other special features in evaluating the impacts of NRM interventions.

The book brings together a number of peer-reviewed papers many of which were originally presented and discussed at the international workshop on 'Methods for Assessing the Impact of Natural Resource Management Research' held at the International Crops Research Institute for the Semi-Arid Tropics (ICRISAT), 6–7 December 2002. The workshop aimed: a. to deliberate the special features and methodological difficulties of NRM impact assessment, b. to examine the strengths and weaknesses of alternative impact assessment methodologies and suggest options for pilot testing, and c. to identify data requirements for developing impact indicators. The book was conceived and inspired by the issues discussed during the workshop.

Given the multi-faceted and complex nature of NRM impacts, contributions come from a multidisciplinary team that included economists, agro-ecologists, and soil and water management scientists, from a range of institutions covering the CGIAR, universities and research institutions. The main theme of the book is rooted in agricultural and resource economics as applied to the evaluation of multi-dimensional outcomes from NRM interventions. The book contains 16 chapters organised into five parts. The methodological sections are rigorous but well exposed and should be readable to impact assessment practitioners and graduates in applied economics. The applied sections are treated carefully to make them available to general practitioners, development agents, and advisors interested in evaluating the impacts of interventions that affect NRM.

The volume can serve as a valuable reference for economists, impact assessment practitioners, agronomists, resource management specialists, rural development advisors, and researchers and academics interested in the impacts of NRM interventions. The key recommendations and policy findings may also be of interest to policy advisors, planners, development agencies, and research managers, both in national and international agricultural research systems. We hope that the book will add usefully to the scant literature on evaluation of NRM impacts for all those interested in understanding the social benefits of such investments and developing suitable evaluation skills.

The editors would like to thank ICRISAT for providing the necessary funds for the background workshop and for publication of the book. We wish to express our sincere gratitude to the external reviewers for their comments and suggestions that were instrumental in excluding some of the initially suggested chapters and in improving the quality of those that finally appear.

A list of all reviewers is given after the list of contributors. We are very grateful to the authors and co-authors of all the chapters for their contributions and for their efforts in responding to reviewer and editorial comments.

The editors also express thanks for the support and efforts of ICRISAT and CAB International staff in the production of this book. Sincere thanks are due to Tim Hardwick and Rachel Robinson for their patience and support. Special thanks are due to Sue Hainsworth, our technical editor, and her assistants T.N.G. Sharma and Deanna Hash, for their tireless and remarkable efforts to enhance the presentation and readability of the manuscripts, and to P.N. Jayakumar for his editorial and scientific support from the book's inception.

Bekele Shiferaw, H. Ade Freeman and Scott M. Swinton

Foreword

In many poor regions of the world, lack of technological progress and increasing population pressure are taking a heavy toll on the productive resource base. Water scarcity, soil degradation and productivity loss are becoming global challenges to the eradication of poverty, especially in many less-favoured areas where there is a strong nexus between poverty and environmental degradation. Depletion of the resource base diminishes the capabilities of poor people and increases their vulnerabilities to drought and other natural disasters. Sustainable productivity growth and natural resource management are indeed inextricably linked, and strategies aiming to enhance livelihood security should identify ways to enhance the productivity of the natural resource assets of the poor. Semiarid tropical agriculture is characterised by high risks from drought, pest and disease incidence and pervasive poverty. It is here that knowledge-based agro-ecosystem management holds the key to sustainability and livelihood security.

Attainment of the Millennium Development Goals will simply not be possible without sufficient technological progress and improved policies to address the global challenges that face the resource-poor regions of the world. Coupled with efforts to increase agricultural productivity in such regions, natural resource management (NRM) has become one of the cornerstones of research and development efforts within the national, sub-regional and international agricultural research systems. The Consultative Group on International Agricultural Research (CGIAR) has devoted significant resources into this area of research. Development investors, policy makers and researchers alike are keen to assess and evaluate investments in NRM. In the past, progress has been limited by the lack of scientifically valid ways to evaluate the complex economic and environmental outcomes associated with these interventions that need new methods and techniques to enhance their effectiveness.

This book focuses on these felt needs and synthesises recent methodological advances in the evaluation of the impacts of integrated genetic and natural resource management interventions. The overlapping problems of poverty, resource degradation, and the threats of climate change and desertification are real concerns for the future of semiarid tropical agriculture, on which the livelihoods for millions of poor families depend. Methods that enhance the effectiveness of interventions to address such challenges worldwide are urgently needed. The diverse topics covered in this book, contributed by leading researchers, will make a significant contribution to enhancing the impact of resource management interventions in many regions. The authors and editors are to be commended for making positive progress in this difficult area. This work will provide a sound basis for further refinement and future research.

William D. Dar
Director General
International Crops Research Institute for the Semi-Arid Tropics

Contributors

K.V. Anupama, *Scientific Officer (Economics), International Crops Research Institute for the Semi-Arid Tropics, Patancheru 502 324, Andhra Pradesh, India, sepp@cgiar.org*

M.C.S. Bantilan, *Principal Scientist (Economics) and Theme Leader (SAT Futures), International Crops Research Institute for the Semi-Arid Tropics, Patancheru 502 324, Andhra Pradesh, India, c.bantilan@cgiar.org*

Kimberly R. Chung, *Assistant Professor, Department of Community, Agriculture, Recreation and Resource Studies, 323 Natural Resources Building, Michigan State University, East Lansing, Michigan 48823, USA, kchung@msu.edu*

Boru Douthwaite, *Technology Policy Analyst, Centro Internacional de Agricultura Tropical, AA 6713, Cali, Colombia, b.douthwaite@cgiar.org*

Pay Drechsel, *Senior Researcher, Head, West Africa Office, International Water Management Institute, PMB CT 112, Cantonments, Accra, Ghana, p.drechsel@cgiar.org*

R.S. Dwivedi, *Head, Sustainable Agriculture Division, National Remote Sensing Agency, Department of Space, Government of India, Balanagar, Hyderabad 500 037, Andhra Pradesh, India, dwivedi_rs@nrsa.gov.in*

Javier M. Ekboir, *Consultant Economist – at time of meeting – Centro Internacional de Mejoramiento de Maíz y Trigo, AP 6-641 06600 Mexico, D.F. Mexico, jekboir@prodigy.net.mx*

Thomas Enters, *National Forest Programme Facilitator, Food and Agriculture Organization of the United Nations, Regional Office for Asia and the Pacific, Maliwan Mansion, 39 Phra Atit Road, Bangkok 10200, Thailand, thomas.enters@fao.org*

H. Ade Freeman, *Director, Targeting Opportunities Theme, International Livestock Research Institute, PO Box 30709, Nairobi, Kenya, h.a.freeman@cgiar.org*

Mark Giordano, *Senior Researcher (Resource Policy), International Water Management Institute, PO Box 2075, Colombo, Sri Lanka, mark.giordano@cgiar.org*

H.M. Gregersen, *Chair, Standing Panel on Impact Assessment, Science Council, CGIAR, PO Box 498, Solvang, California 93464, USA, h.gregersen@cgiar.org*

Stein T. Holden, *Professor, Department of Economics and Resource Management, Agricultural University of Norway, PO Box 5033, N-1432 Ås, Norway, stein.holden@ior.nlh.no*

P.K. Joshi, *Research Fellow and South Asia Coordinator, Markets, Trade and Institutions Division, International Food Policy Research Institute, Pusa Campus, CG Center Block, NASC Complex, Deva Prakash Sastri Marg, New Delhi 110 012, India, p.joshi@cgiar.org*

J.D.H. Keatinge, *Deputy Director General (Research), International Crops Research Institute for the Semi-Arid Tropics, Patancheru 502 324, Andhra Pradesh, India, d.keatinge@cgiar.org*

T.G. Kelley, *Senior Officer, CGIAR Science Council Secretariat, Sustainable Development Division, Food and Agriculture Organization of the United Nations (FAO), vialle Terme di Caracalla, Rome, Italy 00100, timothy.kelley@fao.org*

John M. Kerr, *Assistant Professor, Department of Community, Agriculture, Recreation and Resource Studies, 323 Natural Resources Building, Michigan State University, East Lansing, Michigan 48823, USA, jkerr@msu.edu*

H. Lofgren, *Senior Research Fellow, Development Strategy and Governance Division, International Food Policy Research Institute, 2033 K Street, NW, Washington, DC, 20006-1002, USA, h.lofgren@cgiar.org*

R.R. Navalgund, *Director, National Remote Sensing Agency, Department of Space, Government of India, Balanagar, Hyderabad 500 037, Andhra Pradesh, India, rangnath@nrsa.gov.in*

S. Navrud, *Associate Professor, Department of Economics and Resource Management, Agricultural University of Norway, PO Box 5033, N-1432 Ås, Norway, stale.navrud@ior.nlh.no*

K.V. Padmaja, *Visiting Scientist (Environmental Sciences), International Crops Research Institute for the Semi-Arid Tropics, Patancheru 502 324, Andhra Pradesh, India, k.padmaja@cgiar.org*

P. Pathak, *Principal Scientist (Soil and Water Management), International Crops Research Institute for the Semi-Arid Tropics, Patancheru 502 324, Andhra Pradesh, India, p.pathak@cgiar.org*

John Pender, *Senior Research Fellow, Environment and Production Technology Division, International Food Policy Research Institute, 2033 K Street, NW, Washington, DC, 20006-1002, USA, j.pender@cgiar.org*

A. Ramakrishna, *Senior Scientist (Agronomy), International Crops Research Institute for the Semi-Arid Tropics, Patancheru 502 324, Andhra Pradesh, India, a.ramakrishna@cgiar.org*

T.J. Rego, *Principal Scientist (Soil Science), International Crops Research Institute for the Semi-Arid Tropics, Patancheru 502 324, Andhra Pradesh, India, t.rego@cgiar.org*

K.L. Sahrawat, *Visiting Scientist (Soil Chemistry), International Crops Research Institute for the Semi-Arid Tropics, Patancheru 502 324, Andhra Pradesh, India, k.sahrawat@cgiar.org*

Bekele Shiferaw, *Senior Scientist (Resource and Development Economics), International Crops Research Institute for the Semi-Arid Tropics, PO Box 39063, Nairobi, Kenya (formerly at ICRISAT, Patancheru 502 324, Andhra Pradesh, India), b.shiferaw@cgiar.org*

Piara Singh, *Principal Scientist (Soil Science), International Crops Research Institute for the Semi-Arid Tropics, Patancheru 502 324, Andhra Pradesh, India, p.singh@cgiar.org*

Scott M. Swinton, *Professor, Department of Agricultural Economics, 304 Agriculture Hall, Michigan State University, East Lansing, Michigan 48824-1039, USA, swintons@msu.edu*

Steve Twomlow, *Principal Scientist (Soil Science) and Theme Leader (Agro-ecosystem Development), International Crops Research Institute for the Semi-Arid Tropics, PO Box 776, Bulawayo, Zimbabwe, s.twomlow@cgiar.org*

S.P. Wani, *Principal Scientist (Watersheds) and Regional Theme Coordinator (Agro-ecosystem Development), International Crops Research Institute for the Semi-Arid Tropics, Patancheru 502 324, Andhra Pradesh, India, s.wani@cgiar.org*

Reviewers

Tony Addison, *World Institute for Development Economics Research, Finland*
Jeffrey Apland, *University of Minnesota, USA*
Samuel Benin, *International Food Policy Research Institute (IFPRI), Uganda*
Marrit van den Berg, *Wageningen University and Research Centre, The Netherlands*
Bruce Campbell, *Centre for International Forestry Research (CIFOR), Bogor, Indonesia*
Peter Cooper, *International Development Research Centre (IDRC), Canada*
Robert Delve, *Centro Internacional de Agricultura Tropical (CIAT), Uganda*
John Dixon, *Food and Agriculture Organization of the United Nations (FAO), Italy*
Simeon Ehui, *The World Bank, USA*
Thomas Enters, *Food and Agriculture Organization of the United Nations (FAO), Thailand*
Robert Evenson, *Department of Economics, Yale University, USA*
Berhanu Gebremedhin, *International Livestock Research Institute (ILRI), Ethiopia*
Francis Gichuki, *International Water Management Institute (IWMI), Sri Lanka*
Meredith Giordano, *International Water Management Institute (IWMI), Sri Lanka*
Solveig Glomsråd, *Central Bureau of Statistics, Norway*
Douglas Gollin, *Williams College, USA*
Fitsum Hagos, *Mekelle University, Ethiopia*
Andy Hall, *United Nations University – Institute for New Technologies (UNU–INTECH), The Netherlands*
Stein Holden, *Agricultural University of Norway, Norway*
Nancy Johnson, *Centro Internacional de Agricultura Tropical (CIAT), Colombia*
John M. Kerr, *Michigan State University, USA*
Gideon Kruseman, *Wageningen University and Research Centre, The Netherlands*
Mywish Maredia, *Michigan State University, USA*
Kelly F. Millenbaugh, *Michigan State University, USA*
Michael Morris, *Centro Internacional de Mejoramiento de Maíz y Trigo (CIMMYT), Mexico*

George Norton, *Virginia Polytechnic Institute and State University, USA*

Suresh Pal, *National Centre for Agricultural Economics and Policy Research (NCAP), India*

Julian Park, *The University of Reading, UK*

John Pender, *International Food Policy Research Institute (IFPRI), USA*

Frank Place, *World Agroforestry Centre (ICRAF), Kenya*

Roberto Quiroz, *Centro Internacional de la Papa (CIP), Peru*

David Rohrbach, *International Crops Research Institute for the Semi-Arid Tropics (ICRISAT), Zimbabwe*

Eirik Romstad, *Agricultural University of Norway, Norway*

Ruerd Ruben, *Wageningen University and Research Centre, The Netherlands*

John Sanders, *Department of Agricultural Economics, Purdue University, USA*

Nteranya Sanginga, *Tropical Soil Biology and Fertility Institute (TSBF)–Centro Internacional de Agricultura Tropical (CIAT), Kenya*

Christopher Scott, *International Water Management Institute (IWMI), India*

Keith Shepherd, *World Agroforestry Centre (ICRAF), Kenya*

Katar Singh, *India Natural Resource Economics and Management Foundation, India*

S. Subrahmanyam, *Centre for Economic and Social Studies, India*

Aysen Tanyeri-Abur, *Food and Agriculture Organization of the United Nations (FAO), Italy*

Hermann Waibel, *Universität Hannover, Germany*

Roger White, *International Centre for Integrated Mountain Development (ICIMOD), Nepal*

Ada Wossink, *North Carolina State University, USA*

T.J. Wyatt, *United States Environmental Protection Agency, USA*

Part I.

Introduction

1 Assessing the Impacts of Natural Resource Management Interventions in Agriculture: Concepts, Issues and Challenges

H.A. Freeman[1], B. Shiferaw[2] and S.M. Swinton[3]

[1] *International Livestock Research Institute (ILRI), Nairobi, Kenya*
[2] *International Crops Research Institute for the Semi-Arid Tropics (ICRISAT), Nairobi, Kenya*
[3] *Michigan State University, East Lansing, Michigan, USA*

Introduction

One of the greatest development challenges facing the world in the 21st century is meeting the rising demand for food while maintaining the sustainability of the natural resource base. Increases in per capita income, population growth and urbanisation are expected to double global food demand in the next 40–50 years. The demand for cereals is estimated to increase from 1.9 billion tonnes (t) in 1997 to 2.5 billion t by 2020 and for meat from 209 million t to 327 million t (Rosegrant *et al.*, 2001). These trends in food demand have important implications for natural resources that provide essential support to life and economic processes.

Natural resource management (NRM) aims for the efficient and sustainable utilisation of renewable and non-renewable natural resources. In the context of this book, NRM in agriculture refers to human administration and sustainable utilisation of biophysical resources for the production of food, feed, fibre and fuel. Production in this sense entails direct husbandry, including such activities as aquaculture and planted forests, but does not include hunting, fishing and gathering of uncultivated species. Natural resources of interest include all those affected by the production process (e.g. soil, water, biodiversity, fish and forests). Accordingly, depending on the resource and environmental service flows affected, impact assessment of NRM in agriculture includes the associated changes in the environmental impacts of agricultural production.

Well-managed natural resources generate flows of benefits that provide the basis for maintaining and improving livelihoods, improve the quality of life, and contribute to sustainable growth. Agricultural production worldwide mostly depends on soil, providing the most important source of livelihoods for the majority of rural people in the developing world. Water is essential for sustaining human populations and, indeed, all species. It is also a key input in agricultural and industrial production and processing as well as an important sink for discharging waste. Fish are an important biological resource that account for 20% of the animal-derived protein consumption in low-income countries and about 13% in the developed countries (Delgado *et al.*, 2003). With increasing intensification of food production, aquaculture is becoming an important source of income and livelihoods in many parts of the world. Forests and forest resources, including agroforestry and tree crops, provide a source of livelihoods for over 1.6 billion people worldwide. Forests also contain at least 80% of the remaining global biodiversity, they help to protect water resources, and they are a significant carbon sink mitigating climate change (World Bank, 2001). Biodiversity enables animal and crop improvement programmes that maintain and increase productivity. Properly managed natural resources provide an essential foundation for reducing poverty and promoting sustainable growth.

However, the combined effects of population growth, higher levels of economic activity per capita, and mismanagement are putting increasing pressure on the natural resource base. There is abundant evidence of natural resources degradation worldwide. Over the past 45 years an estimated 1.2 billion ha has been degraded as a result of human activity. This affects more than 900 million people in 100 countries. Erosion, salinisation, compaction, and other forms of degradation afflict 30% of the world's irrigated lands, 40% of rainfed agricultural lands, and 70% of rangelands. Every year an additional 12–15 million ha of forests are lost to deforestation. The world is facing a systemic water crisis resulting from the unsustainable use and management of water resources. New threats and challenges to water supplies arise from urbanisation, over-extraction of surface and ground water, pollution, and loss of aquatic biodiversity (World Bank, 2001).

Degradation of natural resources has real economic, social, and human costs with substantial impacts on national economies. It also directly threatens the long-term growth of agricultural productivity, food security, and the quality of life, particularly in developing countries. Investments in agricultural research have resulted in dramatic increases in food production generated from higher-yielding crop varieties with improved resistance to pests and diseases, mostly in areas of high agricultural potential in developing countries. The dramatic increase in production of rice, maize and wheat, referred to as the Green Revolution – was credited with averting widespread per capita food shortages and starvation in the later half of the 20th century, particularly in Asia and Latin America. The short-term crop productivity gains of the Green Revolution are however associated with long-term degradation of soils, water, biodiversity, and marginal lands. Pingali and Rosegrant (1998) provided empirical evidence linking the intensification of rice–wheat systems in the Indo-Gangetic plains of South Asia to the build up of salinity and waterlogging, depletion of groundwater resources, formation of hard pans, soil nutrient deficiencies, and increased incidence of soil toxicity. Thus, while improving agricultural productivity is an essential component in many poverty-reduction and growth strategies, degradation of natural resources can threaten the achievement of this objective.

Natural resource degradation is particularly costly for the poor. Poor people often depend directly on natural resources for their livelihoods, making them especially vulnerable when natural resources lose their productive potential. There is growing awareness that sustainable use of natural resources can contribute to poverty alleviation and improvements in human welfare. Project, programme, or policy interventions that improve the management of natural resources can lead to significant economic gains that directly benefit poor people, resulting in substantial improvements in their welfare.

The linkages between sustainable management of natural resources and improvements in the well being of the poor have contributed to a resurgence in development lending and research investments on environment and NRM over the past two decades. The World Bank, for example, is increasing lending for environment and NRM issues after a period of decline over the last few years. In 2003 US$1.1 billion was allocated for environmental and NRM issues, representing 6% of overall lending, an increase from 4.7% in 2002 (World Bank, 2003). Similarly, international organisations focusing on sustainable increase in agricultural productivity and improvement in rural livelihoods such as the Consultative Group on International Agricultural Research (CGIAR), have increased the share of NRM research in their overall research portfolio (Kelley and Gregersen, Chapter 15, this volume). Between 1994 and 2001, CGIAR research investments in protecting the environment rose from 15 to 19% of total resource allocation, while investment on biodiversity almost doubled from 6 to 11% (Barrett, 2003). These trends in resource allocation generally reflect the growing consensus that the objectives of poverty alleviation, food security, and sustainable management of natural resources are highly interdependent.

This chapter identifies key issues involved in assessing the impacts of NRM interventions. Such interventions include adoption of changed NRM practices arising from investments in research and outreach that are implemented through NRM projects, programmes, and policies. The focus is on impact analysis of NRM interventions, not on conducting NRM projects *per se*. The next sections discuss the purposes of impact assessment, followed by the underlying concepts and techniques for conducting impact assessment. This is followed by a discussion of the special challenges that complicate impact assessment of NRM interventions. The chapter ends by providing an overview of the conceptual and empirical approaches for NRM impact assessment.

Why Assess NRM Impacts?

Impact assessment should enhance the understanding of the extent to which project, programme, and policy interventions affect the target population and the magnitude of these intervention effects on the welfare of the intended beneficiaries. Resources are limited and managers in research and development institutions are under pressure to allocate available resources efficiently and effectively.

Impact assessment, whether it is backward-looking, evaluating the impact of past research and development (R&D) investments (*ex post* impact assessment) or forward-looking, evaluating the impact of current and future R&D investments (*ex ante* impact assessment) should help in setting priorities over competing interventions and inform policy decisions on efficient allocation of scarce resources.

Impact assessment can be used to measure the outcomes and impact of development interventions, aiming to discern intervention effects from the influence of other external factors. As noted above, this is particularly challenging with NRM interventions.

Donors, policy makers, and development managers need information to monitor progress in achieving outputs and outcomes, providing a basis to demonstrate results, and strengthening accountability for results that may justify continued funding. Often, broad indicators of impact such as aggregate rates of returns to investments and benefit–cost ratios are used as indicators to provide evidence of the effectiveness of past and future interventions. These indicators are used to make decisions on whether to expand, adjust, or drop project, programme, or policy interventions. *Ex post* evaluation also provides lessons that could be used to improve the design and management of service delivery and other future interventions. Comprehensive impact assessment that includes both productivity and environmental and sustainability impacts provides an objective basis for comparing the effectiveness of alternative interventions in achieving the stated welfare and sustainability objectives. Such information is useful for planning, setting priorities, and allocating resources to alternative interventions. However, evaluating the actual livelihood and poverty impacts of agricultural and NRM interventions would require analysis of distributional and equity impacts in addition to computation of such simple efficiency indicators as net present values, benefit–cost ratios, and internal rates of return. New methods and approaches are needed to extend traditional impact assessments to address such policy-relevant concerns.

R&D organisations are increasingly interested in assessing a broad range of impacts from NRM interventions. This, however, requires examining a range of multi-dimensional impacts that may include impacts on the quality of the resource base as well as the flow of ecosystem services that provide basic life support functions in agro-ecosystems. These non-market benefit objectives imply that conventional economic impact analyses are fundamentally incompatible with measuring the benefits that NRM projects seek to obtain. Methodological development in the approaches and techniques for valuation of ecosystem and environmental goods and services is enabling assessment of environmental impacts associated with NRM interventions that have been largely neglected in past impact assessment studies.

Nevertheless, methods for assessing the multi-faceted impacts from NRM interventions are far less developed than methods for assessing impact for crop improvement research (Izac, 1998; Shiferaw and Freeman, 2003). This explains, in part, the dearth of credible quantitative evidence, *ex ante* or *ex post*, that assesses the impact of NRM research compared to the evidence on the effects of crop improvement research. For example, of the 1886 rates of return on research investment reviewed by Alston *et al.* (2000) over 50% were for crops research, while NRM research accounted for less than 5%. The limited number of studies on NRM impact assessment, despite the increased interest on sustainability issues, suggests that tracing the practical linkages between NRM interventions with changes in the resource base, the environment, and human welfare is fraught with complexities (Nelson and Maredia, 1999). The specific challenges and empirical difficulties that impact evaluators face in undertaking valid and plausible assessment of NRM impacts are discussed below.

Impact Assessment: Concepts and Processes

In the literature, the term 'impact assessment' is used interchangeably with 'impact evaluation'. Impact assessment determines the welfare changes from a given intervention on individuals, households and institutions and whether those changes are attributable to the project, programme, or policy intervention (Baker, 2000; World Bank, 2002).

Impact assessments are often undertaken *ex ante*, evaluating the impact of current and future interventions, or *ex post*, evaluating the impact of past intervention. Impact assessment can also be made concurrently within the project cycle. *Ex ante* assessment intends to inform policy decisions as to whether a proposed project or programme intervention should be carried out at all. Such evaluations gather information on the likely economic and environmental impacts and how the flow of costs and benefits is distributed across the affected populations. The distributional impacts and identification of winners and losers are critical elements in evaluating the social impacts of proposed interventions. The *ex ante* assessment compares the expected benefits and costs over time along with the anticipated social impacts. Such information is often used to prioritise interventions and inform policy choice as to whether the expected social benefits would outweigh the costs – to justify implementation of proposed interventions. *Ex post* impact assessments gener-ally intend to measure realised benefits and costs of programme interventions to see whether stated objectives have been met and whether the realised benefits indeed outweigh the direct and indirect costs incurred. *Ex post* assessment also attempts to understand the pathway through which observed impacts have occurred and why interventions fail or succeed in attaining stated objectives. Hence, *ex post* assessments can inform policy choices as to whether related planned programme interventions should be discontinued, modified, improved or sustained in the future.

An important aspect of impact assessment is to understand how interventions affect the beneficiaries or affected populations and whether any outcomes and improvements are a direct result of the intervention. An intervention will not enhance economic efficiency unless the realised or anticipated benefits exceed the overall costs. In cases where the desired impact is not being achieved, the evaluation can also provide useful information on how the programme design could be improved.

Measuring project outcomes alone is not sufficient to assess impacts. In many cases, there may be other factors or events that affect outcomes other than the project itself. For example, if an agroforestry outreach project is initiated and shortly thereafter the national government ceases to subsidise imported fertiliser, farmers may begin to rely upon agroforestry methods to meet crop nutritional needs. In order to measure the real impact of the agroforestry outreach intervention, it is important to control for other confounding factors such as the subsidy termination, and to net out those outcomes that can be attributed only to the intervention itself. This means that impact assessment must estimate the counterfactual, i.e. what would have happened had the intervention never taken place.

Determining the counterfactual is at the core of evaluation design (Baker, 2000). Three broad quantitative methods can be used to identify an appropriate counterfactual (Heckman and Robb, 1985; Heckman and Smith, 1995), including estimation methods used with randomised experimental design, non-randomised quasi-experimental methods, and non-experimental designs.

In the experimental design approach, groups are selected randomly from the same population as the programme participants, while the control group is randomly assigned among those who do not receive the programme. The control group should resemble the treatment group in every sense, with the only difference between the two being the presence of the programme intervention in the treatment group. The main benefit of this technique is the simplicity in interpreting the results – intervention impact can be estimated by the mean difference between the treatment and control groups. While the experimental design is considered the ideal and most robust approach to estimating intervention impacts, it has several disadvantages. Firstly, randomisation, which involves denial of benefits for a certain group of people, may not be ethically acceptable for many interventions. Secondly, randomisation may not be politically acceptable. Thirdly, the proposed project, programme or policy may have economy-wide effects that make randomisation unfeasible. Fourthly, experimental designs may be technically impossible (e.g. due to mobile populations) or expensive and tedious to implement.[1] These difficulties often limit the practical usefulness of the experimental design approach for establishing a valid counterfactual.

Quasi-experimental designs such as matching, reflexive comparison, and double difference methods, and non-experimental designs, such as instrumental variables methods, can be used when it is not possible to construct treatment and comparison groups through experimental design. Matching involves identifying non-programme participants comparable in essential characteristics to programme participants to be matched on the basis of common characteristics that are believed to influence programme outcomes. The propensity score matching approach that is based on the predicted probability of participation given observed characteristics is the most commonly used approach for matching. The reflexive comparison method compares programme participants before and after the programme. The double difference method compares both programme participants and non-participants before and after the programme. Instrumental variables consist of using 'instruments' that matter to participation but not to outcomes given participation, allowing identification of exogenous variation in outcomes attributable to the programme, while recognising that its placement may not be random but purposive. Instrumental variables are first used to predict programme participation; then the programme impact is estimated using predicted values from the first equation (Baker, 2000).

Selection bias is a major challenge to measuring programme impacts in non-experimental settings. Selection bias occurs when pre-existing conditions skew outcomes in a way that is not truly attributable to the programme intervention. For example, if farmers with the best land adopt a practice of soil conservation faster than farmers with poor land, the yield gain they achieve may exceed what other farmers could expect, due to their higher land quality. When bias exists, the assessment may provide inaccurate results that could lead to erroneous inferences and conclusion about the impacts of the intervention (Friedlander and Robins, 1995). Randomised experiments avoid selection bias through random selection. The quasi-experimental and non-experimental designs must rely upon statistical methods to minimise bias due to non-random data. Certain statistical methods allow comparison of programme participants and non-participants while controlling for the process of selection (Pender, Chapter 6, this volume; Greene, 1997; Baker, 2000). However, these methods tend to be less robust statistically than ones that use experimental data. Moreover, the statistical methods for correcting selection bias can be quite complex (e.g. Kerr, 2001), and it is often difficult to fully correct for it in practice (Baker, 2000).

Qualitative methods are also used for impact assessment. Such methods seek to determine impacts by relying on methods other than the counterfactual (Mohr, 1995). Qualitative approaches involve understanding the processes, behaviours and conditions surrounding NRM interventions. Often qualitative methods are participatory, relying upon the perceptions of the individuals or groups being studied (Valadez and Bamberger, 1994). Qualitative approaches tend to use open-ended designs for data collection, including focus group discussions, key informant surveys, and participatory appraisals. Examples can be found in Chapters 11 (Bantilan *et al.*) and 14 (Douthwaite *et al.*) in this volume. Commonly used analytical tools include stakeholder analysis and beneficiary assessment. Qualitative approaches provide insights into the way in which households and communities perceive a project and how they feel affected by it. Qualitative methods can be simple, quick, flexible, and tailored to specific socio-economic conditions. However the subjectivity involved in data collection, the lack of a counterfactual and limited statistical rigour make the results less conclusive and more difficult to generalise than quantitative assessments.

Qualitative approaches are increasingly used in conjunction with quantitative approaches (Baker, 2000), and such combinations can enhance the validity and reliability of impact evaluations (Bamberger, 2000). While quantitative approaches allow statistical tests for causality and isolation of programme effects from other confounding influences, qualitative methods allow in-depth study of selected issues and help the evaluator find explanations for the results obtained in the quantitative analysis. In short, quantitative methods excel at answering impact assessment questions about 'what' and 'how much', whereas qualitative methods are preferred for exploring questions of 'how' and 'why'. A mix of quantitative and qualitative approaches is ideal because it provides the quantifiable impacts of the intervention as well as an explanation of the processes and relationships that yielded such outcomes.

The evaluation design chosen for NRM impact assessment needs to capture the special features, complexities and multiple outcomes associated with such interventions. For example, assessing the impacts of NRM technology and policy interventions requires accounting for both the tangible and the less-tangible and diffuse productivity and environmental impacts. The process of tracking these relationships and impact pathways may involve several steps. Nelson and Maredia (1999) discussed five steps in assessing environmental costs and benefits in NRM projects. These steps involve:

- Understanding the causes and impact of changes in the use of natural resources such as declining soil fertility, land degradation, water pollution, deforestation, loss of biodiversity, etc.
- Identifying the main types of economic costs and benefits. Economic costs could include depletion of the stock of natural resources and species losses. An important consideration is to identify the distribution of the burden of these costs over time and space and across affected communities
- Determining whether or not there is a means to measure costs and benefits in monetary terms
- Assessing the extent of changes in the use of natural resources and the environmental consequences resulting from these changes. This includes collecting data to estimate the impact of environmental effects on such indicators as productivity, income, and human health
- Using economic techniques to place values on environmental changes.

Key biophysical processes and related indicators of NRM status are explored in this volume with foci on the soil (Pathak *et al.*, Chapter 3), water resources (Sahrawat *et al.*, Chapter 4), and ecosystem services (Wani *et al.*, Chapter 5). Shiferaw *et al.* (Chapter 2), discuss several methods for placing economic values on non-market ecosystems services, while Drechsel *et al.* (Chapter 9) provide examples of applying some of the commonly used valuation methods to valuing changes in soil fertility.

Challenges in NRM Impact Assessment

Apart from the general challenges of attribution and selection bias in impact analysis, there are special conceptual and methodological challenges that arise from several unique features of natural resource management. NRM impact assessment needs to address important challenges of attribution, measurement, spatial and temporal scales, multidimensional outcomes, and valuation. The cross-commodity and integrated nature of NRM interventions makes it very challenging to attribute impact to any particular one among them. In crop genetic improvement where the research outputs are embodied in an improved seed, it is less difficult to attribute yield improvements to the investment in research. Changes in NRM frequently involve observable research products adopted by farmers as well as qualitative information about recommended management practices. Knowledge about such improved management practices may be transmitted through formal and informal outreach activities and by the self-experimentation and indigenous knowledge of the farmers themselves. In many cases, for such knowledge and information-based changes in NRM practices, it is difficult to identify the impacts attributable to the intervention. Also, it is not uncommon for different agencies to be involved in the development and promotion of new NRM technologies, making it hard to separate the impacts attributable to specific programmes. For example, in the evaluation of watershed programmes in India, it was difficult to attribute improvements in resource conditions and farm incomes to specific interventions, since increased participation and collaboration among a range of R&D partners was identified as a significant determinant of success (Kerr, 2001). The fact that most agricultural NRM interventions are information-based but not embodied in an easily measured package vastly complicates the attribution of observed impacts.

Identifying an appropriate counterfactual in NRM interventions is particularly challenging because quantifying the biophysical impacts of interventions on natural resources can be costly, imprecise, and slow. For NRM interventions that aim to halt resource degradation, the counterfactual may be a significant productivity decline. Hence, a properly measured counterfactual may reveal that achieving non-declining productivity represents a major gain over what would otherwise have occurred.

Identifying appropriate spatial boundaries for assessing NRM impact is often fraught with difficulty (Campbell *et al.*, 2001; Sayer and Campbell, 2001). Agricultural NRM typically involves different spatial scales, from farmers' fields to entire watershed catchments, implying that many levels of interaction may need to be considered in assessing the impacts of research interventions. Multiple scales of interaction create upstream and downstream effects that complicate impact assessment. For example, assessing the impact of land use interventions in a watershed may need to take into account multiple interactions on different scales because erosion and runoffs in the upper watershed may not have the same impact on water quality downstream. It is also likely that interventions could have different effects, which in some cases can generate opposite impacts on different spatial scales. For example, soil and water conservation interventions can have a positive impact on crop yields upstream but negative impacts by reducing water availability downstream when water is a limiting factor for production, or positive impacts by reducing sedimentation, runoff and flooding when water is not a limiting factor.

In the temporal dimension, methodological challenges for NRM impact assessment arise from slow-changing variables and substantial lags in the distribution of costs and the benefits. For example, soil loss, exhaustion of soil fertility, and depletion of groundwater resources take place gradually and over a long period of time. In some cases it may be difficult to perceive the costs or the benefits of interventions to reverse these problems. In other cases, assessing the full range of the impacts of investments related to these slow-changing variables in a holistic manner may involve intensive monitoring of multiple biophysical indicators on different spatial scales over long periods of time. These factors make impact monitoring and assessment of NRM interventions a relatively slow and expensive process. Differences in time scale for the flow of costs and benefits are translated into lags in the distribution of costs and benefits that complicate impact assessment. Typically, costs are incurred up-front while delayed benefits accrue in incremental quantities over a long period of time (Pagiola, 1996; Shiferaw and Holden, 2001). For example, the benefits from the biodiversity that is used in genetic improvement of crop and animal varieties accrue in the long term but costs of *in situ* and *ex situ* conservation are incurred in the short term. The timing of an intervention can also affect its impact. This is, for example, the case for improved crop management practices that require optimising sowing date, fertiliser application, weeding and harvesting.

When outcomes are delayed and tend to vary according to local biophysical conditions, simulation models can facilitate the *ex ante* evaluation of NRM technology options that fit micro-climatic and agro-ecological niches. Biophysical process models are mainly used to explore the biophysical and productivity impacts of changes in agricultural and NRM practices (Wani *et al.*, Chapter 5, this volume). Bioeconomic models, on the other hand, interlink economic and biophysical information to simulate optimal resource use and investment behaviour (Holden, Chapter 8, this volume; Shiferaw and Holden, Chapter 12, this volume). Both kinds of models require biophysical and experimental agronomic data to calibrate and validate them to local conditions.

NRM interventions may generate multidimensional biophysical outcomes across resource, environmental and ecosystem services. These might include changes in the quality and movement of soil, quantity and quality of water, sustainability of natural resources, and conservation of biodiversity. Appropriate indicators are needed to monitor the impacts of NRM interventions on the biophysical conditions of the soil (Pathak *et al.*, Chapter 3, this volume), water resources (Sahrawat *et al.*, Chapter 4, this volume), and the flow of ecosystem services that support agro-ecosystems (Wani *et al.*, Chapter 5, this volume). The multidimensionality of outcomes from NRM interventions means that impact assessment often faces difficult measurement challenges, including very different measurement units and potentially the integration of very different natural resource outputs into some kind of uniform aggregate yardstick (Byerlee and Murgai, 2001).

The multidimensionality of NRM outcomes extends to those directly or indirectly affecting human beings. NRM interventions can generate environmental and health benefits whose values might not be reflected in current markets, but on which society places a value for multiple reasons. For example, water and water-based ecosystems provide not only direct values in consumptive uses (e.g. fishing, irrigation) and non-consumptive uses (e.g. aesthetic value), but also indirect use values such as ecosystem functions and services, option values for possible future uses and applications and non-use values for intrinsic significance (existence and heritage value). Empirical valuation of non-market benefits is explored by Shiferaw *et al.* (Chapter 2, this volume). But depending on how NRM ideas are conveyed, the human outcomes may extend even further. Integrated NRM projects engage in participatory activities that may empower individuals and communities in ways that extend far beyond the realm of agricultural NRM, as discussed by Douthwaite *et al.* (Chapter 14, this volume).

Approaches for Assessing NRM Impacts

Impact assessment for NRM interventions ultimately needs to show the social costs and benefits associated with the research, promotion, and adaptation of these interventions. Given the complexities and challenges associated with measuring, monitoring and valuing such changes, more innovative assessment methods are required. An important factor that needs to be considered in the selection of appropriate methods is the capacity to account for non-monetary impacts that NRM interventions generate in terms of changes in the flow of resource and environmental services that affect sustainability and ecosystem health. As discussed earlier, a mix of quantitative and qualitative methods may be the optimal approach for capturing on-site and off-site monetary and non-monetary impacts. The economic surplus approach is the commonly used method for evaluating the impacts of agricultural research investments, particularly for crop improvement technologies. This approach estimates benefits as changes in 'economic surplus' (the aggregate value that consumers are willing to pay above and beyond what it costs producers to supply the good or service in question). The cumulative benefits are then compared to cumulative R&D costs over time. Specifics and the challenges of incorporating non-marketed on-site effects and off-site externalities are discussed by Swinton (Chapter 7, this volume), with Bantilan *et al.* (Chapter 11, this volume) providing an empirical application.

Promising analytical methods that can be used to quantify economic changes due to NRM interventions include econometrics (Alston *et al.*, 1995) and bioeconomic optimisation modelling. For example, econometric methods can be used in empirically estimating the demand for marketed or certain non-marketed goods and services, providing elasticities for calculations of economic surplus. Econometric methods can also be used to link a time-series of measures of output, costs and profits directly to past R&D investments (Alston *et al.*, 1995). Likewise, they can be used to establish statistical relationships between changes in NRM practices and measured performance indicators, such as land productivity, total factor productivity, production costs, net farm income, or income volatility. Pender (Chapter 6, this volume) discusses the conceptual and empirical issues while Kerr and Chung (Chapter 10, this volume) provide an empirical application of this method.

Bioeconomic modelling nests essential biophysical processes within economic behavioural models. Their constrained optimisation perspective allows evaluating how technological and/or policy changes would affect economic welfare, sustainability, and environmental conditions over time. The integrated framework captures biophysical process evolution along with rational human management responses. Holden offers a conceptual treatment of bioeconomic modelling (Chapter 8, this volume), while Shiferaw and Holden provide an empirical application for a farm household (Chapter 12, this volume) and Holden and Lofgren demonstrate the use of an economy-wide computable general equilibrium model for evaluating NRM technology and policy impacts (Chapter 13, this volume).

As a response to the complexities that impact assessment practitioners face in evaluating the multi-faceted impacts of NRM, there is an increasing interest in developing more holistic and 'softer' assessment methods. Integrated natural resource management (INRM) calls for participatory NRM interventions at multiple scales with frequent adaptive feedback and multiple stakeholders (who often hold contrasting objectives) (Campbell *et al.*, 2001; Sayer and Campbell, 2001). Douthwaite *et al.* (Chapter 14, this volume) explore the conceptual underpinning of the INRM framework and its implications for evaluating NRM impacts.

Organisation of the Book

The chapters in this book address the conceptual framework, methodological challenges and selected empirical experiences of NRM impact assessment. In so doing, they explore many of the complexities identified in this introductory overview. The book's 16 chapters are organised into five parts. Following this initial part that introduces the challenges and approaches to NRM impact assessment, Part II includes four chapters that deal with the valuation of ecosystem services and the measurement of biophysical indicators of NRM impacts. Part III introduces advances in methods used to evaluate the economic and environmental impacts of NRM technology and policy interventions. Part IV deals with NRM impact assessment in practice. Five case studies illustrate the methodological advances discussed in Part III. The final part of the book (Part V) highlights some of the existing controversies and outlines best practices, research issues, and recommendations for NRM impact assessment into the future.

Endnote

[1] One way to enhance ethical and political acceptability of randomisation is to phase the intervention such that some groups gain access to programme benefits at a later stage. In this way the random selection determines when a given group gains access to the benefit, not if they receive it.

References

Alston, M.J., Norton, W.G. and Pardey, P.G. (1995) *Science Under Scarcity. Principles and Practice for Agricultural Research Evaluation and Priority Setting.* Cornell University Press, Ithaca, New York, USA, and London, UK, 585 pp.

Alston, M.J., Chan-Kang, C., Marra, M.C., Pardey, P.G. and Wyatt, T.J. (2000) A meta-analysis of rates of return to agricultural R&D: exe pede herculem? *Research Report 113.* International Food Policy Research Institute (IFPRI), Washington, DC, 159 pp.

Baker, J.L. (2000) *Evaluating the Impacts of Development Projects on Poverty. A Handbook for Practitioners.* World Bank, Washington, DC, 230 pp.

Bamberger, M. (2000) *Integrating Quantitative and Qualitative Methods in Development Research.* World Bank, Washington, DC, 189 pp.

Barrett, C.B. (2003) *Natural Resources Management Research in the CGIAR: A Meta-Evaluation in The CGIAR at 31: An Independent Meta-Evaluation of the Consultative Group on International Agricultural Research.* World Bank, Washington, DC, 217 pp.

Byerlee, D. and Murgai, R. (2001) Sense and sustainability revisited: the limits of total factor productivity measures of sustainable agricultural systems. *Agricultural Economics* 26(3), 227–236.

Campbell, B., Sayer, J.A., Frost, P., Vermeulen, S., Ruiz-Perez, M., Cunningham, A. and Ravi, P. (2001) Assessing the performance of natural resource systems. *Conservation Ecology* 5(2), 32. [online] http://www.ecologyandsociety.org/vol5/iss2/art22/index.html

Delgado, C.L., Wada, N., Rosegrant, M.W., Meijer, S. and Ahmed, M. (2003) *Outlook for Fish to 2020. Meeting the Global Demand.* International Food Policy Research Institute, Washington, DC, 28 pp.

Friedlander, D. and Robins, P.K. (1995) Evaluating program evaluations: New evidence on commonly used non-experimental methods. *American Economic Review* 85, 923–937.

Greene, W.H. (1997) *Econometric Analysis,* 3rd edn. Prentice Hall, Upper Saddle River, NJ, 1075 pp.

Heckman, J. and Robb, R. (1985) Alternative methods of evaluating the impact of interventions: An overview. *Journal of Econometrics* 30, 239–267.

Heckman, J. and Smith, J.A. (1995) Assessing the case for social experiments. *Journal of Economic Perspectives* 9(2), 85–110.

Izac, A.N. (1998) *Assessing the Impact of Research in Natural Resources Management.* The World Agroforestry Centre (ICRAF), Nairobi, Kenya, 38 pp.

Kerr, J. (2001) Watershed project performance in India: conservation, productivity and equity. *American Journal of Agricultural Economics* 83, 1223–1230.

Mohr, L.B. (1995) *Impact Analysis for Program Evaluation,* 2nd edn. Sage Publications, Thousand Oaks, California, 336 pp.

Nelson, M. and Maredia, M. (1999) Environmental Impacts of the CGIAR: An Initial Assessment. Paper presented at the International Centers Week, October 25–29, 1999, Washington, DC. Standing Panel for Impact Assessment, Food and Agriculture Organization of the United Nations, Rome, Italy, 71 pp. [online] http://www.sciencecouncil.cgiar.org/publications/pdf/envimp.pdf

Pagiola, S. (1996) Price policy and returns to soil conservation in semi-arid Kenya. *Environmental and Resource Economics* 8, 251–271.

Pingali, P. and Rosegrant, M. (1998) Supplying wheat for Asia's increasingly westernized diets. *American Journal of Agricultural Economics* 80(5), 954–959.

Rosegrant, M., Paisner, M.S., Meijer, S. and Witcover, J. (2001) *Global Food Projections to 2020: Emerging Trends and Alternative Futures. IFPRI 2020 Vision.* International Food Policy Research Institute (IFPRI), Washington, DC, 206 pp.

Sayer, J.A. and Campbell, B. (2001) Research to integrate productivity enhancement, environmental protection and human development. *Conservation Ecology* 5(2), 32. [online] http://www.ecologyandsociety.org/vol5/iss2/art32/index.html.

Shiferaw, B. and Holden, S.T. (2001) Farm-level benefits to investments for mitigating land degradation: empirical evidence from Ethiopia. *Environment and Development Economics* 6, 335–358.

Shiferaw, B. and Freeman, H.A. (eds) (2003) *Methods for Assessing the Impacts of Natural Resource Management Research. A Summary of the Proceedings of an International Workshop, 6–7 December 2002, International Crops Research Institute for the Semi-Arid Tropics (ICRISAT), Patancheru, India,* 136 pp.

Valadez, J. and Bamberger, M. (eds) (1994) *Monitoring and Evaluating Social Programmes in Developing Countries.* The World Bank, Washington, DC, 536 pp.

World Bank (2001) *Making Sustainable Commitments: An Environment Strategy for the World Bank.* World Bank, Washington, DC, 280 pp.

World Bank (2002) *Understanding Impact Evaluation.* World Bank, Washington, DC. [online] http://www.worldbank.org/poverty/impact/index.htm

World Bank (2003) *Environment Matters: Annual Review, July 2002 – June 2003.* World Bank, Washington DC. [online] http://lnweb18.worldbank.org/ESSD/envext.nsf

Part II.

Valuation of Ecosystem Services and Biophysical Indicators of NRM Impacts

2 Valuation Methods and Approaches for Assessing Natural Resource Management Impacts

B. Shiferaw[1], H.A. Freeman[2] and S. Navrud[3]

[1] *International Crops Research Institute for the Semi-Arid Tropics (ICRISAT), Nairobi, Kenya*

[2] *International Livestock Research Institute (ILRI), Nairobi, Kenya*

[3] *Agricultural University of Norway, Ås, Norway*

Introduction

Impact assessment in natural resource management (NRM) is nascent but developing rapidly in response to felt needs. Determining how to value changes in NRM is a major challenge. The value of some NRM investments can be estimated via the value of increases in yield of marketable products or the value of savings in production costs, but many NRM investments generate benefits that are hard to measure because they are not bought and sold in markets. Environmental and resource service flows that offer indirect use and non-use benefits to society certainly have value, but their measurement is a challenge. This chapter focuses on valuation methods and associated issues for measuring the social benefits that result from NRM investments.

Several thorny issues are associated with valuation of the productivity and environmental impacts of NRM investments. These include incomplete understanding of ecosystem functions and difficulties in predicting the effect of interventions on major ecosystem functions and services; lack of measurable performance indicators when effects are relatively well understood; and problems in relating changes in the flow of ecosystem services to human welfare. The non-tangibility of the benefits, time lags, and spatial (scale) effects further complicate the measurement of social, economic, and environmental impacts from NRM interventions. Farmer investments in NRM often provide non-excludable and non-consumptive public goods to the local community and beyond. For example, vegetative barriers and trees planted on the upper reaches of a watershed by a private land-user provide watershed protection, biodiversity conservation and carbon sequestration services to the local community, some of the benefits of which may even

 Natural Resource Management in Agriculture: Methods for Assessing Economic and Environmental Impacts (eds B. Shiferaw, H.A. Freeman and S.M. Swinton)

extend beyond the micro-watershed to regional and global levels. Although precise estimation of the full economic value of such investments is costly and difficult, the application of appropriate valuation methods may provide useful estimates for the direct and indirect, marketed and non-marketed ecosystem services generated by NRM investments.

Despite the extensive work on environmental valuation and benefit–cost analysis, there is a dearth of literature on methods for valuation of ecosystem services from NRM technology adoption and a serious lack of empirical examples in the developing countries that estimate the social impacts of NRM research and development efforts. This chapter provides an overview of the valuation methods and methodological approaches used to evaluate the economic and environmental impacts of NRM interventions. How NRM investments affect the flow of ecosystem services, the issues involved in translating changes in service flows to welfare gains, and some promising approaches for valuation of welfare changes are discussed. The suitability of the methods described depends on such specific circumstances as: resource types and interventions, anticipated economic and environmental effects, and interaction of biophysical changes with socio-economic conditions. The second part of the chapter summarises the multiple ecosystem services associated with NRM. In the third part, the core issues involved in the valuation of ecosystem services are discussed. This is followed by a presentation of the theoretical foundations and overview of valuation methods along with some examples on applications in the area of natural resources. The next part summarises how economic and environmental impacts can be integrated to provide an assessment of the social net benefits from NRM interventions, and the conclusion highlights the major issues and most promising valuation methods.

Agro-ecosystem Services and Functions

Agro-ecosystems are communities of plants and animals interacting with their physical and chemical environments that have been modified by people to produce food, fibre, fuel and other products for human consumption and processing (Altieri, 2002). Watersheds and agro-ecosystems offer a number of ecosystem services of value to society (for simplicity ecosystem goods and services are referred to as ecosystem services). Ecosystem services consist of flows of materials, energy, and information from the natural capital of ecosystems that provide direct and indirect human welfare benefits. In many cases, such services are public goods that cannot be privatised at low cost (high costs of exclusion) and whose consumption by one consumer does not reduce the amount available for others (non-rival). Hence, self-interested private individuals may lack the economic incentive to provide such services in socially optimal quantities.

Ecosystem services that embody public goods include: biodiversity conservation, flood and erosion control, carbon sequestration, nutrient recycling, and water retention and storage (Bingham *et al.*, 1995). In other cases, the ecosystem service may involve high costs to privatise and may also be congestible (consumption by one reduces consumption by others). Costs of exclusion are often high for ecosystem services that arise from such common property resources as groundwater, community pastures and woodlots.

Technological interventions for NRM may have diverse effects on an ecosystem at various levels. The first step towards evaluating the economic and environmental impacts of NRM interventions requires an understanding of how such investments change ecosystem functions. Each ecosystem function can be conceived as a subset of complex ecological processes that provide specific goods and services that directly or indirectly satisfy human needs. The diverse ecosystem services generated through ecosystem functions therefore provide various economic, environmental, and socio-cultural benefits and values to people. De Groot *et al.* (2002) developed a typology for the classification of ecosystem functions and services. Although their general typology is meant for natural ecosystems, it can be adapted for use in agro-ecosystems to understand the likely impacts of NRM interventions. Changes in the scale and intensity of managing natural resources in agriculture will change the flow of agro-ecosystem services, which will in turn influence the quantity and/or quality of goods and services produced. Depending on the type of NRM technology used, the typology developed by de Groot *et al.* (2002) suggests that valuable ecosystem services may be generated through any of the following ecosystem functions:

- Production
- Regulatory services
- Habitat
- Socio-cultural (information) services.

Table 2.1 summarises the major ecosystem functions and services together with selected indicators of change due to NRM interventions in the context of agro-ecosystems.

Production of food and raw materials is a major ecological function of agro-ecosystems that includes food, feed, fuel, raw materials and medicines. This function is transmitted through the conversion of solar energy into edible plants by autotrophs for human and animal consumption. Farm animals convert fodder and herbaceous material into economic goods and services for use by humans. Natural resource investments may also influence the ability of the agro-ecosystems to produce products for ornamental and medicinal use, and the conservation of biological diversity. As shown later in this chapter, when data are available, simulation models and statistical methods can be used to establish the relationships between NRM investments and changes in the flow of goods and services (see also Chapter 5, this volume). These effects are typically realised on-site and create economic incentives for resource users to adopt new technologies. When the productivity effects are limited, farmers' direct economic benefits and the incentives for adoption and adaptation of NRM technologies will be low.

The regulation function relates to the role of agro-ecosystems in the maintenance of essential ecological processes and life-support systems. Such ecosystem services may be transmitted through changes in land cover that

Table 2.1. Ecosystem functions and potential indicators of change in agro-ecosystem services associated with natural resource management (NRM) interventions.

Ecosystem services	Ecosystem functions (processes and components)	Indicators for changes in agro-ecosystem services
A. *Production functions* – Provision of natural resources as factor inputs in production activities		
Food	Conversion of solar energy into edible plants and animals for humans	Changes in land productivity (crop and livestock)
Raw materials	Conversion of solar energy into biomass for feed, construction and other uses	Changes in fodder, fuelwood, timber, etc., output
Genetic resources	Conservation of genetic materials	Changes in agro-biodiversity
Medicinal resources	Bio-chemical substances, medicinal uses	Changes in availability of medicinal plants or changes in use benefits from medicinal plants
Ornamental resources	Ornamental use	Changes in economic benefits from ornamental plants and animals
B. *Regulation functions* – Maintenance of essential ecological processes and life support systems		
Climate regulation	Influence of land cover and carbon sequestration on climate	Changes in land cover and carbon sequestration
Water regulation	Role of land cover in regulating runoff and river discharge	Changes in runoff and sediment loss
Water supply	Filtering, retention and storage of fresh water	Changes in water availability and quality
Soil retention	Role of vegetation root matrix and soil biota in soil retention	Changes in rates of soil erosion and sediment loss
Soil formation	Weathering of rock, accumulation of organic matter	Changes in soil depth
Nutrient regulation	Role of biota in storage and recycling of nutrients	Changes in nutrient balances, soil fertility and organic matter
Pollination	Role of biota in movement of floral gametes	Changes in pollinating insects
C. *Habitat functions* – Providing habitat for wild plant and animal species		
Refugium function	Suitable living space for certain desirable species	Changes in the stock of wildlife, soil flora and fauna
Nursery function	Suitable reproduction habitat for certain desirable species	Changes in rates of reproduction
D. *Sociocultural functions* – Providing opportunities for cognitive development		
Aesthetic information	Attractive landscape features	Changes in landscape and scenery
Recreation	Variety in landscapes with (potential) recreational uses	Changes in recreational benefits (agro-ecotourism, outdoor sports, etc.)

Table 2.1 Continued.

Ecosystem services	Ecosystem functions (processes and components)	Indicators for changes in agro-ecosystem services
Cultural and artistic	Features with cultural and artistic value	Changes in cultural and artistic use (e.g. motivation for books, films, advertising, etc.)
Spiritual and historic	Agro-ecosystem types with spiritual and historic value	Changes in use for religious and historical use (e.g. heritage, spiritual symbol)
Science and education	Agro-ecosystem types with scientific and educational value	Recognition for scientific or educational purposes

Source: Updated based on Costanza *et al.* (1998) and de Groot *et al.* (2002)

influence and regulate: climate change (e.g. through carbon sequestration), water flows (runoff and river discharges), and protect soils from erosion, water supply through filtering, retention and storage of fresh water (e.g. wetlands, check dams, etc.); soil formation through decomposition of organic matter and weathering of rocks; nutrient regulation through storage and recycling of nutrients; biological control of pests; pollination through the role of fauna in the movement of floral gametes. A number of useful ecological (biophysical) indicators can be developed to monitor the NRM technology impacts on these kinds of ecosystem services (Chapters 3–5, this volume).

The habitat function indicates the useful services provided by agro-ecosystems in the provision of habitat (suitable living space) and nursery (reproductive space) services for uncultivated and cultivated plant and animal species. People derive non-material well-being from the flow of these services. It is difficult to develop simple indicators to monitor NRM impacts on these ecosystem services. The number of species in a given habitat (species richness) and the species diversity can be measured using different biological indices (Chapter 5, this volume).

Natural resource investments also provide such socio-cultural services as aesthetic information (e.g. attractive landscape), recreational services (e.g. ecotourism), and scientific and spiritual services. These are mainly public goods that provide useful services to society or the community as a whole.

When markets exist, changes in some of these agro-ecosystem services resulting from NRM investments can be quantified and valued in monetary terms. For public goods (e.g. changes in biodiversity, water and air quality) markets are either missing or often imperfect. The quantification of benefits and valuation therefore presents special difficulties. Before valuation methods are considered, the major issues and challenges surrounding valuation of ecosystem services are briefly described.

Issues in Valuation of Agro-ecosystem Services

There are two fundamental steps in the valuation of impacts from NRM investments: firstly, understanding and predicting the changes in the flow of ecosystem services attributable to the technological or policy intervention, and secondly, devising acceptable methods for valuing these changes. The first helps identify and quantify *what is to be valued* while the second one helps develop *suitable methods* for valuing the changes. In this section the issues involved in uncovering what is to be valued and how it is to be valued are discussed. As described above, ecosystems are very dynamic and complex, and human knowledge about them is very incomplete. This limits the ability to understand and quantify the changes in the ecosystems service flows associated with human interventions. The effects of NRM interventions can be physical, chemical or biological, and may take different forms over temporal and spatial scales. However, understanding and predicting the impacts of interventions on ecosystem functions is the prerequisite to economic valuation. Good valuation depends on sound agroecological information on the effects of policy and management interventions. Functional inter-linkages and feedback effects make it difficult to determine the causal relationship between human interventions and changes in ecosystem functions and processes (Bingham *et al.*, 1995). Any sensible effort to assess the impacts of NRM interventions requires a reasonable understanding of how and to what extent the different ecosystem service flows will change as a result of human interventions. This implies an interdisciplinary effort involving agroecologists, agronomists, biophysical scientists and economists. Bingham *et al.* (1995) argue that if there is no agreement on the effect of changes on the flow of ecosystem services, there can be no agreement on valuation of the impacts.

If changes can be predicted or quantified reasonably, the next question will be – which of these changes can be valued in monetary terms? The choice of which changes to value is an important challenge for the economist. Before values can be placed on the impacts, it is necessary to know what is to be quantified and how it can be measured. Indicators of changes in the service flows (immediate impacts) are critical for valuation. Indicators can be developed through experimentation and appropriate monitoring of changes over a sufficient period of time, or through the application of exploratory and predictive simulation models. The latter approach is most useful when changes are slow to evolve or when complexity of anticipated interactions makes actual experimentation very difficult. NRM combines both features and involves multiple interventions that make it problematic to isolate partial effects. In the absence of good counterfactuals, experimental data might not provide useful insights about the anticipated impacts. Oriade and Dillon (1997) provide a good review of applied simulation models used in agricultural systems.

There are various efforts to develop measurable indicators for changes in the flow of agro-ecosystem services (Dumanski and Pieri, 2000; Arshad and Martin, 2002). The next three chapters in this volume provide a detailed account of the measurable indicators for soil, water, and other agro-ecosystem services. The challenge is to develop indicators that could be easily monitored on-farm as part of the project cycle. Good indicators are those that capture major elements in a complex interactive system while simultaneously showing how the value obtained relates to some ideal or desired level. Smyth and Dumanski (1993) reported the use of participatory rural appraisal techniques for developing land-quality indicators for sustainable land management for sloping lands in Indonesia, Thailand, and Vietnam. The framework for evaluating sustainable land management was used to develop threshold levels for the sustainability of land-management indicators. Campbell *et al.* (2001) proposed linking indicators to changes in five livelihood assets (natural, physical, financial, social, and human capital). They suggest a collective measure for each of the capital assets that could be used to develop an aggregate index. Although the aggregate index gives little guidance as to what needs to be included under each asset category, it might serve as an organising framework to develop a few indicators under each asset category for those projects expected to have wide-ranging impacts.

Once the relevant changes are identified and quantified through appropriate indicators, the next question becomes – how to value these changes? There are many vexing issues on how ecosystem service flows are valued. Even if effects can be predicted and monetary valuation is possible, many still argue if money values could adequately inform decision-making, especially when irreversible changes, trade-offs, and distributional effects are involved. The term 'value' may also have different concepts and meanings for different disciplines (Bingham *et al.*, 1995; Bockstael *et al.*, 2000; Farber *et al.*, 2002). In common usage it means 'importance' or 'desirability'. An economic value measures the change in well-being associated with the change in the quantity or quality of the service flow. Changes in resource and environmental service flows can affect human welfare in complex ways and through marketed or non-marketed activities. The most common approach to translating these changes into monetary units is to express the welfare change as the amount a person would pay or be paid (in compensation) to be as well-off with the change as without it. The amount that individuals are willing to pay, or might accept as compensation, is not an absolute value; it will vary across individuals depending on property rights, perceived welfare gains/losses, the context, and the availability of substitutes.

There are two key questions that need to be answered in the process of economic valuation of ecosystem services. The first is how to construct a measure of how much better or worse-off an individual is because of the change in the quantity or quality of the service flow. The second is how to add up the individual welfare changes (gains and losses) to assess the value of this change for society as a whole. Recent advances in economic theory provide answers to these two fundamental questions and offer useful methods for the valuation of many ecosystem service flows regardless of the functioning of markets.

Valuation Techniques

Unlike agricultural products harvested in fixed time periods, environmental and ecosystem services associated with NRM interventions flow in real time on a continuous basis. Understanding the changes in ecosystem service flows, measuring and monitoring outcomes across time and space is very important for quantifying environmental impacts. The basic principles that guide valuation exercises and the different valuation methods relevant for NRM, including their strengths and weaknesses are discussed.

Theoretical foundation

The economic approach to the valuation of resources is based on the contribution of the resource to human welfare. Whether the good or service is marketed or non-marketed, its unit economic value is determined by the welfare contributions that it makes to humans. Changes in welfare are measured in terms of each individual's personal assessment of changes in well-being (Bockstael *et al.*, 2000). For traded commodities, the demand curve depicts the marginal willingness to pay (WTP) (or marginal benefit) for the good or service. The height of the demand curve at each point of the quantity demanded shows the maximum WTP for the commodity. The household will consume all units of the commodity where the marginal WTP exceeds the market price. The consumer enjoys a consumer surplus for all points where the marginal WTP is higher than the market price. The welfare change associated with a change in the price of a marketed commodity is often measured using the change in consumer surplus, derived from the Marshallian demand curve with a constant level of income. For a non-marketed ecosystem service, the maximum WTP for an improvement in quantity or quality is the area between the initial and new levels of the resource under the demand (marginal benefit) curve. Value estimation then involves determining directly or indirectly the shapes of these marginal WTP curves for the ecosystem services (Freeman, 1993).

Environmental and resource service flows typically exhibit public-good characteristics of high costs of exclusion and non-rivalry. This makes it very difficult for markets to value these ecosystem goods and services accurately, and leads to a market failure and non-tradability. In order to illustrate how the values for such non-marketed resources could be estimated, let us assume that a given household maximises its welfare (U) from consumption of a vector of marketed goods (c), ecosystem goods and services (q) and has a fixed budget y, such that:

$$Max\ U = U(c, q) + \lambda(y - p'c) \qquad (1)$$

The standard utility-maximising solution to this problem will give the Marshallian demand function for the tradable commodity:

$$c^* = c(p, q, y) \qquad (2)$$

which is a function of a vector of market prices (p), the disposable income (y) and the ecosystem services (q) considered to be a public good. If this is substituted into the utility function, the indirect utility function could be derived:

$$v(p, q, y) = U(c(p, q, y), q) \tag{3}$$

The marginal effect of the change in the level of the public good q_i on household welfare can be derived as:

$$\frac{\partial v(p, q, y)}{\partial q} = \frac{\partial U\big(c(p, q, y), q\big)}{\partial q} \tag{4}$$

This is equal to the marginal valuation of the environmental good in question. It is a measure of the marginal welfare benefit (demand curve) for the public good q (Johansson, 1987). For a given change in q from q^0 to q^1, the welfare effect on household h can be estimated as:

$$\Delta U^h = v^h(p, q^1, y) - v^h(p, q^0, y) = \int_{q^0}^{q^1} \left(\frac{\partial v^h(p, q, y)}{\partial q} \right) \tag{5}$$

The total welfare effect (WTP) summed over all the affected households (h) can be calculated as:

$$\sum_h \Delta U^h = \sum_h \int_{q^0}^{q^1} \left(\frac{\partial v^h(p, q, y)}{\partial q} \right) \tag{6}$$

In general utility functions are unobserved and it would be useful to convert Equation 6 into a monetary measure of welfare change. This is done by assuming constant marginal utility of income (λ_h) for each household, and dividing the marginal valuations in Equation 6 by (λ_h). This is the same as vertical summation of the demand curves and will provide the aggregate uncompensated WTP for all the affected households given the change in q from q^0 to q^1. The compensating surplus (*CS*) and equivalent surplus (*ES*) measures (analogous to the *CV* and *EV* measures for price changes) can also be directly derived from the indirect utility function. For an improvement in q from q^0 to q^1 the *CS* and *ES* measures can be computed as:

$$v(p, q^1, y - CS) = v(p, q^0, y) \tag{7.1}$$

$$v(p, q^1, y) = v(p, q^0, y + ES) \tag{7.2}$$

In terms of empirical applications, the expenditure function that can be derived from the indirect utility function using the envelope theorem by solving for the expenditure level (y) that will provide a given level of utility, can be very useful in directly estimating the monetary measure of the welfare change associated with provision of the public good (q). The expenditure function for household h is given by $e^h(p, q, \bar{u})$. The aggregate welfare change measure for a change in q from q^0 to q^1 for *CS* can be given as:

$$CS = \Sigma_h \left(e^h(p, q^0, u^0) - e^h(p, q^1, u^0)\right) = \Sigma_h \int_{q^0}^{q^1} \left[\frac{\partial v^h(p, q, u^0)}{dq} \right] dq \tag{8}$$

The aggregate ES measure for a change in q from q^0 to q^1 can be given as:

$$ES = \Sigma_h \left(e^h(p, q^0, u^1) - e^h(p, q^1, u^1)\right) = \Sigma_h \int_{q^0}^{q^1} \left[\frac{\partial v^h(p, q, u^1)}{dq}\right] dq. \tag{9}$$

The CS is the maximum amount of money that the individual is willing to pay to secure an increased provision of the public good q. The ES measures the minimum sum of money that must be given to individuals before the change to make them as well-off as they would have been following an increase in q. This forms the basis for valuation of non-marketed ecosystem services. When the environmental and resource flows serve as inputs in production of market goods by producers, the equivalent welfare measure of the change in productivity is the change in producer and consumer surplus (Ellis and Fisher, 1987). Freeman (1993) demonstrates the other indirect benefit estimation approaches where q enters the production function as a factor input or as an input in the household production of utility-yielding commodities. Before the various methods and approaches used for eliciting values for ecosystem services are discussed, the components of the total economic value and the effect of markets and externalities on the choice of valuation methods are briefly demonstrated.

Valuation of impacts

The valuation of changes in ecosystem services needs to take into account intended and unintended outcomes. Individuals may attach values for such changes because of the use benefits derived, or any anticipated or conceived non-use welfare benefits. Agricultural activities often impose external costs on society mainly because individual resource-use decisions occur at points that equate marginal private benefits and costs. Soil erosion and sedimentation, and use of fertiliser, pesticides and other chemicals are some examples that impose costs on other agents and ecosystems. Unintended economic effects that spill over to other agents are often called externalities. More formally, externalities are unintended effects on the production or consumption activities of an economic agent resulting from the activities of another economic agent that are not mediated through markets. Adoption of 'best practice' NRM technologies like integrated pest management (IPM) or upland watershed management reduces external costs imposed on ecosystems and on other farmers in the lower reaches of the watershed. Hence, NRM investments may provide multiple ecosystem services to different economic agents across spatial scales as illustrated in Table 2.2 (Pagiola *et al.*, 2002). The use value (UV) of a given NRM investment includes the sum of direct and indirect use benefits (marketed and non-marketed) that accrue to all beneficiaries on-site and off-site. The challenge is how these dispersed benefits could possibly be valued. This requires good knowledge about the nature of the effect, how long the effect will last, its spatial dispersion, and the affected parties. The use value of the resource to different groups of economic agents cannot exceed the perceived benefits accruing to the group. Therefore, local forest managers,

Table 2.2. Perceived (on-site and off-site) benefits from integrated watershed management investments (soil, water and vegetation).

Ecosystem goods and services	Local communities	Downstream water and land users	Distant stakeholders and global community
Supplemental irrigation			
Improved agricultural productivity			
Fuelwood, pasture and construction materials			
Reduction in flooding and siltation			
Water purification			
Carbon sequestration			
Biodiversity preservation			

for example, will not consider the water quality captured by downstream water users, and biodiversity benefits to the global community. In this case the total use value of the resource will be given as:

$$UV_i = \sum_{i=1}^{n}\sum_{j=1}^{m} \gamma_{ij} Y_i \qquad (10)$$

where γ_{ij} is the distribution parameter reflecting the ecosystem service i captured at location j, and Y_i is the total use value of the ecosystem service i. Thus, the UV is the sum of all ecosystem services captured by all the beneficiaries across locations. Obviously, this poses practical difficulties in mapping out the benefit dispersion and in elicitation of values from different agents.

The total economic value (TEV) of a given resource may however include non-use values. The non-use values (NUV) include what are called option value (OV), bequest value (BV) and existence value (ExV). Figure 2.1 illustrates the components of the total economic value. OV is a measure of how much individuals are willing to pay for the option of preserving the asset for future personal use. BV is the value that individuals are willing to pay to ensure that the resource will be preserved for future generations. ExV is the value that individuals attach to the mere existence of a given natural resource or environmental asset unrelated either to current or optional use. Thus:

$$\begin{aligned} TEV &= UV + NUV \\ &= (\textit{Direct use value} + \textit{Indirect use value}) + (OV + ExV + BV) \qquad (11) \end{aligned}$$

The nature of the externality and the structure of markets will have substantial implications on the choice of effects to be valued and the valuation methods to be used. This can be seen by relating the anticipated benefits across spatial scales and the existence of markets to value these benefits (Dixon *et al.*, 1994). As can be seen from Table 2.3, the benefits from goods and services in Group I are both tradable within the local economy and are captured on-

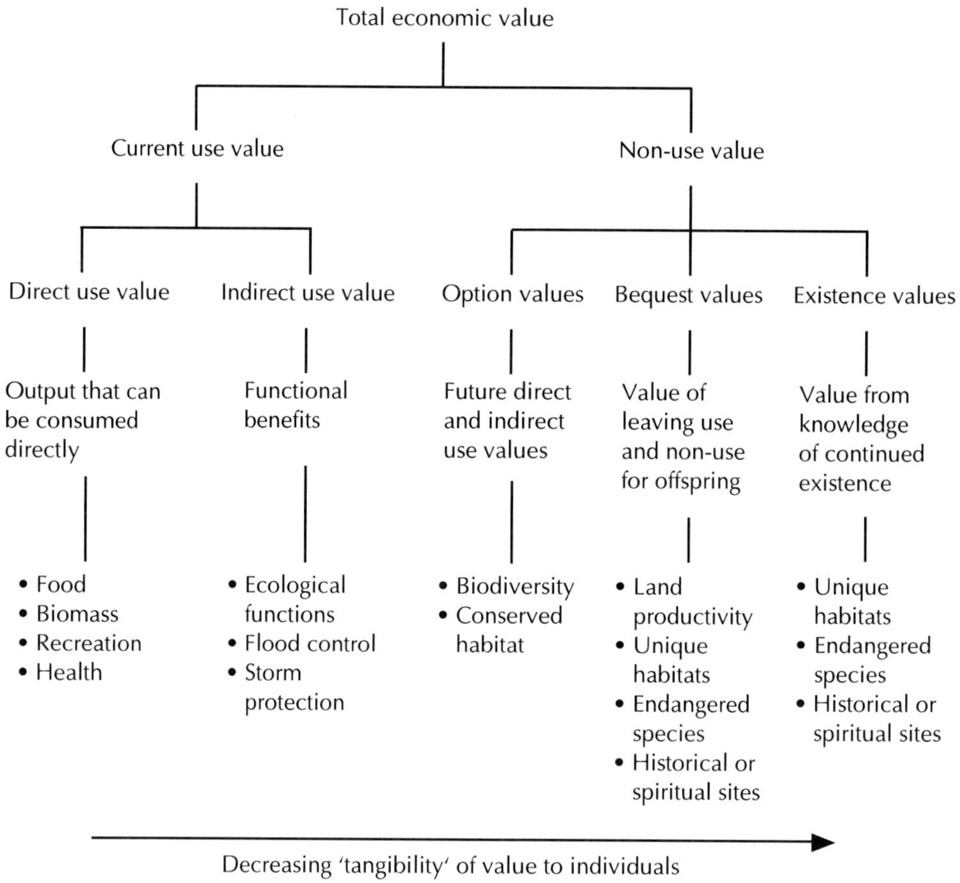

Fig. 2.1. Decomposition of total economic value of ecosystem goods and services (Munasinghe and Lutz, 1993).

site. These goods and services could be valued using market prices, with adjustments for any distortions that may exist (e.g. monopoly, externalities, and existing taxes and subsidies not reflecting external costs and benefits). For goods and services in Group II, market prices may exist, but local producers do not capture benefits, i.e. the lion's share of such benefits is 'externalised'. To the extent that these external benefits can be quantified, they could be incorporated into the social economic analyses of NRM impacts.

For those in Group III, benefits accrue within the local economy (of the household or village) but many of the goods and services are non-tradable. Missing markets mean that such benefits, however large, are seldom included in empirical impact assessments. Even the most difficult for valuation are goods and services generated from NRM investments, which are neither captured by the producers nor traded through markets (Group IV). Examples in this category include benefits of climate regulation (carbon sequestration) and biodiversity conservation resulting from upland

Table 2.3. Valuation of ecosystem goods and services from tree planting: The role of markets and externalities.

Tradability	Location of goods and services: On-site	Location of goods and services: Off-site
Marketed	**Group I** Benefits accrue on-site (e.g. fuelwood, fodder, timber, etc.) and are tradable – Usually included in impact assessment (IA)	**Group II** Off-site tradable benefits (e.g. higher crop yields or more hydropower resulting from reduced siltation in dams) – Sometimes included in IA
Non-marketed	**Group III** Benefits accrue on-site but are highly non-tradable (e.g. soil and water conservation, recreation, regulation of micro-climate, etc.) – Seldom included in IA	**Group IV** Off-site non-tradable benefits (e.g. Carbon sequestration, reduced flooding, biodiversity conservation) – Usually ignored in IA

Source: Modified based on Dixon *et al.* (1994) and others

tree planting. In the absence of markets, such non-tradable and external benefits need to be estimated by other methods using surrogate markets or constructed (hypothetical) markets in areas where the benefits are captured. Because of non-excludability, the level of investment by local producers to generate such services may be socially sub-optimal, necessitating many governments to intervene in their production and management.

Valuation methods

Advances in resource and environmental economics in the last few decades have provided many useful methods that can be employed to value use and non-use values of ecosystem goods and services, both marketed and non-marketed (Mitchell and Carson, 1989; Braden and Kolstad, 1991; Freeman, 1993; Bateman and Willis, 1999). The intention here is not to review the extensive literature on environmental valuation but to highlight briefly the methods that can be applied for valuation of NRM impacts in the context of developing countries. Table 2.4 presents an overview of some of these methods that can be potentially applied to value ecosystem services resulting from NRM investments. The methods can be distinguished by the type of market used, as well as the implied behaviour of the economic agent in the valuation of goods and services. Changes in productivity, replacement cost, avoided cost and opportunity cost methods use actual markets, but

Table 2.4. Valuation methods for ecosystem goods and services.

Implied behaviour	Conventional market	Surrogate market	Constructed market
Based on potential or stated behaviour	Productivity change approach (PCA) Replacement cost approach (RCA) Avoided costs Opportunity costs		Contingent valuation (CV) method Choice modelling (CM)
Actual or revealed behaviour	Provision costs Defensive expenditures Relocation costs	Hedonic pricing (HP) (land value) Wage differentials Travel cost	Experimental markets

behaviour is *assumed*, i.e. no actual outlays or market transactions have occurred. Defensive expenditures, provision costs and relocation costs also use actual markets but behaviour is *expressed* or *observed* through conventional markets. Methods like hedonic pricing and wage differentials use surrogate markets to value non-tradable goods and services indirectly through marketed goods and services that embody their values. In the extreme case, it may be possible to construct experimental markets (where behaviour is *revealed* with actual WTP or accept compensation for a change) or hypothetical markets (where behaviour is *stated* without actual transactions as in the case of the contingent valuation method). Methods that use information from conventional markets are presented below followed by those that use surrogate markets and those requiring constructed markets. The treatment gives more emphasis to methods that offer high potential for application in NRM impact valuation.

Productivity change approach (PCA)

Some agricultural resource-improving investments lead to changes in productivity and/or production costs. This means that physical changes in production or overall farm profits derived from adoption of such technologies can be established and valued using market prices.[1] This approach is quite attractive and suitable for evaluating NRM impacts, because physical changes in productivity can be observed and measured. Production functions, erosion damage functions, fertiliser response functions, crop–water responses to supplementary irrigation, and simulation models that relate changed resource conditions to productivity are good examples of PCA. A number of studies on the economic costs of land degradation, soil erosion, etc. have used this method to value the benefits from resource management investments. Magrath and Arens (1989) used detailed erosion–yield relationships to measure the on-site costs of soil erosion in Java, Indonesia. The capitalised cost of a 1% productivity decline is estimated to amount to 4% of the total value of dryland crops in Java. They also estimated off-site costs of sedimentation in reservoirs, irrigation systems and harbours, and found that these costs are about a quarter of the total erosion damage costs. Bojo (1991) used this approach to value the economic cost of soil degradation in Lesotho. However, Barbier (1998) noted these case studies probably overestimated the scarcity value of soil, because the economic value of conservation was not assessed as a net gain in profitability over the erosive (conventional) system. When there are no economically viable options for mitigation, soil degradation might not have on-site opportunity costs (i.e. on-site costs of soil degradation exist only when conservation is profitable on-farm). Shiferaw and Holden (2001) estimated various erosion–yield functions to evaluate conservation benefits and the net gain to farmers from adoption of conservation methods. Gebremedhin *et al.* (1999) also estimated returns to investments in terracing in the Ethiopian highlands, using experimental data on crop yields under different conservation methods. This is a good approach for valuing the economic cost of soil degradation or for evaluating conservation benefits.

Although production functions with resource conditions as factor inputs (along with other usual input factors) can be used to estimate the economic value of the resource, lack of such data often limits the application of this approach. One major difficulty is that it takes a long time for NRM investments to have an observable effect on the flow of ecosystem services. The first step in applying this method is to quantify the effect of changes in NRM on the quantity or quality of the resource base that affects resource productivity. For example, changes in NRM may affect rooting depth, water-holding capacity or organic matter levels in the soil. In this way, the condition of the resource stock (S) will be a function of the conservation effort and other exogenous characteristics such that:

$$S = g(K, Z) \quad (12)$$

where K is the level of NRM investment per ha, and Z is a set of exogenous factors (e.g. soil type, agroecological zone, rainfall, etc.). When the experimental data needed to estimate this statistically are lacking, simulation models may be used to estimate the effect of the change in K on the condition of the resource or the flow of ecosystem services (assuming that suitable parameters are available for the simulation models).

Moreover, the changes in productivity associated with changes in K may take a long time to be visible to farmers. Use of such other inputs as fertilisers and high-yielding varieties also often mask NRM investment benefits to farmers. When data that relate crop productivity with input use and biophysical conditions (e.g. soil depth, soil moisture, soil types) are available, econometric methods can be used to establish useful relationships such that:

$$Q = f(X, S, Z) \quad (13)$$

where Q is the productivity of land, X is a vector of inputs used, S is a vector of resource quality indicators, and Z is a vector of other exogenous factors that influence crop productivity. Controlling for variable inputs and fixed exogenous factors, the marginal effect of the anticipated change in the quality of the natural resource will be given as:

$$\partial Q(\cdot)/\partial S = f_S(X,S,Z)\Big|_{\{X=\overline{X},\, Z=Z_i,\, \forall i\}} \tag{14}$$

In some cases, the level of input use may not remain unchanged, e.g. improvement in soil fertility may prompt credit-constrained farmers to reduce the demand for fertilisers to produce the same level of output. In such cases, the productivity change associated with changes in NRM can be measured using the savings in input costs.

As shown in Equation 14, the marginal effect of NRM investments will depend on the quality of the resource stock and other exogenous factors. The total effect on productivity can be estimated by integrating over the level of change in *S* resulting from the change in *K*.

$$\Delta Q = \int_{S_0}^{S_1} \left(\frac{\partial Q(\cdot)}{\partial S} \right) dS \tag{15}$$

where S_0 is the old and S_1 the new levels of the resource condition associated with changes in *K* (NRM investments). The economic value of the change in NRM needs to be calculated as a producer surplus by including the opportunity cost of the variable inputs used in production (including the cost of *K*). If the change in output does not induce price changes, the producer surplus will be the value of the change in output minus the cost of production. The productivity changes may flow at different rates as the resource condition changes from S_0 to S_1 over a period of time. If the productivity changes are long-lasting, the present value of net productivity benefits (producer surplus) can be computed using the social rate of discount. While the approach is attractive and widely used, it has some disadvantages. These include high data requirements (when the econometric approach is used), lagged effects of NRM that hinder reliable assessment of productivity changes on-site, and difficulties in accounting for any off-site (externality) effects of the change in NRM. This last limitation is perhaps the greatest, because the PCA approach does not measure the value of non-marketed environmental goods and services.

Replacement costs approach (RCA)

Under this approach, potential expenses that may be needed to replace or restore the damaged natural resource asset are estimated using the prices of marketable products. The resulting estimate is not a measure of the benefits of avoiding the damage in the first place, since the damage cost may be higher or lower than the replacement cost. The implied expenditure to restore a given resource to a pre-damaged state or baseline condition may however be different from the costs of replacing its functions. Because of this, the RCA is mainly used in the latter context where the estimated resource values reflect the potential expenses needed to replace the services of the damaged resource through some substitutes (e.g. use of fertilisers or other fertility management practices to replace lost soil nutrients). Replacement costs can be a valid measure of economic value when the following conditions are met (Dixon *et al.*, 1994; Bockstael *et al.*, 2000):

- The magnitude of the damage is measurable and there are no secondary benefits associated with the replacement expenditure
- The substitute provides functions that are similar to the lost ecosystem service
- The substitute is the least-cost option for replacing the lost service
- Affected individuals in aggregate would, in fact, be willing to incur these costs if the natural functions were no longer available
- When the replacement costs are greater than the aggregate WTP or the social value of the productive resource destroyed, it will be economically inefficient to replace the damaged ecosystem service.

As Barbier (1998) noted, when these conditions are not fulfilled and least-cost replacement options are not known, simplistic application of the approach could lead to overestimated and misleading values. For the case of soil erosion, he noted that some of the eroded soil may be deposited on-farm and cannot be considered lost completely. Moreover, all the eroded soil might not have economic value if its marginal productivity effect is negligible. In these situations, the RCA can lead to overestimated resource values. By definition, the RCA includes only the costs of replacing damaged ecosystem services on-site, but the concept is equally applicable for valuation of any associated external effects. While the full restoration costs may include non-use values, the replacement costs reflect the use value of the resource or ecosystem service.

A number of studies have used this method. One example is the case study by Kim and Dixon (1986), which assessed the viability of alternative soil conservation techniques in upland agriculture in Korea. The difference in the estimated cost of physically replacing lost soil and nutrients (estimated based on differences in soil erosion) was taken as a measure of the potential benefits of preventing soil erosion. With the assumption that the value of retaining productive soil is higher than the replacement cost, the study found that preventive measures were more economical than physically replacing lost soil and nutrients.

Provision costs

Economic values for non-market ecosystem services that contribute to human welfare can sometimes be derived from people's decisions to use related resources or to substitute other resources where the quality of the service flow is impaired. The provision costs approach (PCA) refers to the actual expenditures that farmers or communities may incur to provide vital environmental goods and services. Unlike the mitigating expenditures, these expenses are directly targeted in the provision and production of the required good or service. While it can be considered as a variant of the RCA, the PCA does not refer to restoration of the ecosystem service, but to costs of providing the damaged service through alternative means. Some examples include farmers' expenses on drilling wells for irrigation and drinking water when water regulation services of watersheds are damaged, and the costs of alternative sources of household energy after deforestation. The strength of the method is in trying to value the resource in question using the actual cost outlays in producing the required good or service. However, the costs may also serve other purposes, and external benefits are excluded when private provision costs are considered. The method also relies on existence of markets for major inputs used in the production of the environmental good or service.

Defensive expenditures

Farmers, communities, and governments often incur actual expenditures to mitigate or prevent productivity loss or reduce degradation problems. When the extent and potential effect of resource degradation or improvement is difficult to assess, actual preventive or defensive expenditures may be used to assess a rough value of the change in the resource quality. Kim and Dixon (1986) use lowland farmers' defensive expenditures to prevent deposition of silt on rice fields to evaluate alternative soil management techniques designed to stabilise upland soils. There are several problems in the use of this method. Firstly, defensive expenditures, like all WTP, are limited by income and the value so obtained may not reflect the social scarcity value of the resource. It may at best be a lower-bound estimate. Secondly, the value tends to be quite arbitrary as actual expenditures may be targeted to meet several objectives (e.g. conservation of multiple resources).

For use in NRM impact assessment, it is important to determine the anticipated change in resource conditions attributable to the intervention, and how much farmers often spend to prevent an equivalent deterioration in the resource. If defensive expenditures on-site and off-site can be estimated, they may provide a rough indication of the value of the improvement in the resource. In some cases, relocation costs associated with environmental change can be considered part of defensive expenditures. Hence, the relocation costs approach is not discussed separately.

Hedonic pricing (HP)

The theory of hedonic prices is based on the premise that market prices reflect a bundle of observable characteristics and attributes of differentiated products (Rosen, 1974). Different attributes of the same product reflecting differences in its inherent worth will have an associated price, and consumers can easily identify what they are paying for in selecting various options. When goods and services contain non-priced environmental attributes embedded in them, consumers may also place implicit values on each of the attributes so that market prices are composed of environmental and non-environmental attributes. Therefore, when the good or service provided by NRM investments cannot be directly valued using conventional markets, behaviour revealed through surrogate markets can be used for valuation.[2] For example, the value of access to clean water and air can be estimated indirectly through the differences in market prices for houses in polluted and clean localities, after controlling for their structural and other attributes (Harrison and Rubinfeld, 1978). Wage differentials for occupations with different levels of health or environmental risk have also been used to estimate certain environmental values. The HP method is designed to control for certain non-environmental attributes so that the remaining property value differentials can be 'surrogate' values of the non-priced environmental goods and services.

To the extent that surrogate markets are competitive, the HP approach can therefore be very useful for valuing NRM impacts. For example, land values in competitive markets can be used to value differences in land quality. If prices for agricultural land reflect quality changes, the hedonic function for a given parcel with a vector of biophysical (environmental) characteristics $L = (l_1, l_2, \ldots, l_n)$ and socio-economic characteristics of the location and the buyer $Y = (y_1, y_2, \ldots, y_n)$ can be estimated econometrically as:

$$P = g\,(L, Y) \tag{16}$$

where P is the market price of a unit of land. The socio-economic characteristics include such variables as buyer characteristics, population density, distance to urban areas, distance to markets, and type of crops grown. The coefficients of this model can be used to determine the implicit price associated with land characteristic, holding all other factors constant. For example, for soil characteristic l_i (e.g. soil depth) the implicit price is the partial derivative with respect to soil depth such that:

$$\frac{\partial P}{\partial l_i} = g_l(L, Y) \tag{17}$$

If the impact of NRM investments on the biophysical conditions of the resource is known, market prices can be used to value indirectly the changes in resource attributes. One disadvantage of this method is that it requires extensive information on selling or rental prices and associated socio-economic and biophysical characteristics of the property. Even when such data are available, market prices may not be competitive or may not fully reflect such non-observable quality differentials as changes in nutrient balances or biophysical attributes of the soil. The method works quite well if markets reflect quality differentials. Even when they do, market prices may reflect only the capitalised value of future on-site productivity gains from using the land. Changes in non-productivity benefits (e.g. biodiversity, carbon sequestration) and reductions in off-site effects from NRM investments might not be reflected in market prices. In a recent study Shiferaw *et al.* (2003) found that farmers' perceived value of land parcels in semi-arid Indian villages were able to clearly reflect soil and farm characteristics that affect land productivity. Factors such as irrigation, soil depth, soil fertility levels, and soil type had significant effects on perceived land values. For example, irrigated plots, *ceteris paribus*, were perceived to have values 45% higher than non-irrigated plots, whereas a one-level rise in ordinal soil depth increased land values by 5% and in soil fertility by 18%. Such other factors as conservation investments and erosion risk were found to have no significant effects on land values. This shows that the land value method can be used as an alternative to PCA for valuing the effect of NRM investments on land quality aspects that influence productivity. Due to market failures and imperfections, including incomplete land tenure rights, changes in other attributes like public goods aspects and non-use values cannot be easily valued using the land value approach. The contingent valuation (CV) method is useful for valuing such changes.

Contingent valuation (CV) method

In cases where people's preferences are not revealed directly or indirectly through conventional markets, the CV method is used to assess their WTP for marginal changes in quantity or quality of goods and services by posing hypothetical questions. The CV method is a direct stated preference method that involves asking a sample of a relevant population questions about their WTP or willingness to accept (WTA). The monetary value of the change in NRM is acquired by asking respondents about their WTP for a benefit, or what they are WTA by way of compensation to tolerate a cost or forgo a benefit. The name *contingent valuation* originates from the fact that the valuation is contingent on the hypothetical scenario put forward to the respondent. CV is mainly used for valuation of non-marketed ecosystem services and the non-use values associated with non-excludable and non-divisible resource and environmental flows. Unlike the indirect methods that use observed or revealed behaviour, the CV method relies on stated or potential behaviour as expressed in hypothetical markets. An important advantage of the CV method is that responses to WTP and WTA questions provide theoretically correct measures of welfare change as defined in Equations 5–9.

As discussed earlier, the appropriate welfare measures for changes in environmental quantity or quality are compensating surplus (CS) and equivalent surplus (ES) measures. Theoretically, an individual can be asked about WTP or WTA for either an improvement or a deterioration (Table 2.5). Which question is appropriate depends on the implied property right for the specific situation. The CS measure relates to the initial welfare level and implies entitlements to the *status quo*. Thus, asking about WTP to secure an improvement, or WTA compensation to tolerate a loss, implies that the individual is entitled to the existing level. The ES measure relates to the welfare level after the change and suggests the implied property rights in the change. Asking about WTA compensation to forgo an improvement implies an entitlement to the higher level, while WTP to avoid deterioration implies an entitlement to the lower level. WTP is also constrained by income whereas WTA is not. As a result, estimates of WTA tend to be higher than WTP. Some authors suggest using WTP for situations where individuals are expected to gain from an improvement and WTA in situations where people are forced to give up or suffer some damage to their welfare (Carson, 1991). Mitchell and Carson (1989) discuss ways to frame the payment questions to elicit WTP. Arrow *et al.* (1993) provide a guide for best-practice CV studies.

Table 2.5. Welfare measures for environmental quality and quantity changes.

	Compensating surplus (CS)	Equivalent surplus (ES)
Improvement	WTP[a] for the change to occur (to secure a benefit)	WTA[b] compensation for the change not occurring (to forgo a benefit)
Deterioration	WTA compensation for the change occurring (tolerate a loss)	WTP for the change not to occur (to prevent a loss)

[a]WTP = willingness to pay.
[b]WTA = willingness to accept.

In a nutshell, application of the CV method requires the following steps:

- Create a survey instrument to elicit WTP/WTA and the means of payment or compensation
- Administer the survey instrument with a sample population
- Analyse the responses and estimate the average and marginal WTP/WTA
- Estimate the total WTP/WTA for the population of interest.

In developed countries, various survey methods including mail and telephone surveys have been used. In developing countries, in-person interviews remain the most feasible and reliable option. Such surveys often start with discussions with key informants and focus groups followed by pilot testing of the survey format. The actual data collection should introduce the changes in the resource or environmental conditions being valued and the expected benefits or trade-offs to society resulting from this change. Pictures and maps can be used to illustrate these points. The survey should also include standard data on the socio-economic condition of the respondent (e.g. age, education, assets, income, etc.). Various approaches to eliciting WTP or WTA are suggested. Open-ended questions like 'What is the most you are willing to pay for…' or 'What is the minimum that you are willing to accept as compensation for …' have been commonly used in the past. This approach has been criticised for inviting strategic bias, by which respondents may use their replies to influence a more favourable research outcome (e.g. to reduce a payment they might expect to have to pay). In actual markets, buyers are offered a price and may bargain from there to arrive at the selling price. Many respondents find the open-ended approach difficult and fail to provide any bids. The iterative bidding approach that starts with an initial amount to be revised up or down until a no-change point is reached, has been used as an alternative to open-ended questions. This approach is now being abandoned because of a starting-point bias, i.e. the WTP/WTA amount tends to be systematically related to the initial bid value. An alternative approach that is gaining popularity is the binary choice or referendum format, where respondents are asked whether they would vote in support of a proposed change in policy or environmental condition that would cost a US$*x* increase in tax payments. The offered amount can be varied and randomly assigned to the sample. Follow-up questions to the binary choice payment questions have also been used to identify the upper and lower bounds for the bids. It seems that depending on the design, a discrete-choice format with follow-up questions can mimic a bargaining process, commonly used in transactions in developing countries (FAO, 2000).

Once the data from a representative sample are collected, statistical analyses will be needed to estimate the average WTP/WTA and the aggregate value of the ecosystem service. The type of analysis of CV responses depends on the elicitation format used. If the payment question is open-ended, the stated WTP/WTA bids can simply be averaged.[3] The sample average is an unbiased estimator of the population mean. In cases where outliers influence the average bid, the median is a best estimate of a representative central value. As defined earlier, the WTP can be given as:

$$WTP_h = e(p, q^0, u^0, x) - e(p, q^1, u^0, x) \quad (18)$$

where $e(.)$ is the expenditure required to attain a given level of utility, WTP_h is the WTP for household h and x represents the socio-economic characteristics of the respondent and other exogenous variables that affect the WTP. In order to check the internal validity of the CV method, a regression model can be fitted as:

$$WTP_h = X_h \beta + \eta_h \quad \eta_h \sim (0, \sigma^2) \quad (19)$$

where X_h is a vector of explanatory variables and η_h is the error term distributed normally with means 0 and standard deviation σ. This function is often called the valuation function. As shown below, the valuation function is especially relevant for use in benefit transfer studies. It allows the new user to plug in mean values of explanatory values to predict the benefit value for a new setting. If the binary choice payment format is used, alternative methods for estimating the mean bid are discussed in the literature (Hanemann, 1984). Additional analyses will be needed to estimate the average bid and aggregate values for the change. The binary response is an indicator for the WTP/WTA that is observed only when the respondent's WTP/WTA is less than the offered bid value. A maximum likelihood probit model can be estimated using these binary responses to identify the factors that determine the probability of a positive response to a given bid. The mean WTP/WTA can then be obtained by calculating the predicted value of the valuation function at the mean values of the covariates.

Once the average WTP or WTA values for a representative group of people have been determined, they are aggregated to a total value directly dependent on the number of individuals affected. For ecosystem services that provide international public goods, the number of people with a positive WTP is likely to be large, and a modest estimate of the population size needs to be made. In principle, scaling up the average WTP/WTA across the affected population is analogous to the vertical summation of individual compensated demand curves for public goods.

As the examples in Box 2.1 demonstrate, carefully designed and administered CV surveys can provide useful estimates of the value of the changes in non-marketed ecosystem services resulting from NRM investments. The reliability of estimates and validity of results depend on the design and implementation. Of course, they also share the weaknesses of all stated-preference methods.

Box 2.1. NRM-related CV studies in the developing countries.

Today, there are several examples and good reviews of CV applications in developing countries (Munasinghe and Lutz, 1993; FAO, 2000; Pearce *et al.*, 2002). Whittington (1998) examines issues and lessons learned in administering CV surveys in developing countries. Two studies relate to NRM impact assessment in agriculture. Holden and Shiferaw (2002) applied the CV approach to estimate farmers' WTP to mitigate soil degradation in Ethiopia. In the light of increasing land degradation in the highlands, the intention was to elicit the farmers' WTP for NRM technologies that might not provide immediate benefits to farmers. The survey questions were framed to reflect the attributes of available and proposed NRM technologies with three alternative scenarios. Farmers were asked about their WTP for new NRM technologies that: a. sustain land productivity at current levels, b. enhance productivity by a fixed amount from the second year onwards, and c. enhance productivity by a fixed amount from the sixth year onwards. Teff (*Eragrostis tef*), the locally grown cash and staple cereal, was used as numeraire. The WTP surveys were administered as part of a larger survey where broader socio-economic data were collected that allowed estimating regression equations to identify the WTP covariates and check for internal validity. Farmers' expressed WTP for land management options was significantly lower than those implied by experimental and econometric estimates of soil erosion and productivity decline. Shyamsundar and Kramer (1996) applied the CV method to value forest ecosystem services in Madagascar using a binary choice payment format to elicit the local people's WTA compensation for welfare losses associated with land-use restrictions and loss of access to forests. Due to the extreme poverty of farmer respondents, Shyamsundar and Kramer used WTA questions specified in terms of bags of rice, the local staple food. They estimated a probit function and a valuation function to infer the WTA for specific households and the mean for the sample. This was used to estimate the aggregate use value of the forest service flows to the local people.

Although the approach has been widely applied for benefit–cost analysis of projects with environmental impacts, its use in assessment of technology and policy impacts in agriculture and natural resources has been scanty. CV surveys can be very useful for generating information that will inform policy choices in developing countries where market failures are more pervasive. The method is a relatively simple and cost-effective means (especially when literacy is widespread) to estimate the value of public goods and non-market ecosystem services associated with NRM investments.

Choice modelling (CM)

Choice modelling (also called choice experimentation) is an indirect stated preference method that arose from conjoint analysis and has been employed in marketing, transportation and psychology. Bennett and Blamey (2001) provided a collection of papers on the theory and application of CM in environmental valuation. Alpizar *et al.* (2003) provided a good review of using CM for non-market valuation. It differs from typical conjoint methods in that individuals are asked to choose from alternative bundles of attributes (alternatives) instead of ranking or rating them. Under the CM approach, respondents are asked to choose their most-favoured choice out of a set of three or more alternatives, presenting variations in the attributes of the item being valued (Adamowicz *et al.*, 1998). The *status quo* is given as one of the alternatives in the choice set. Each respondent gets the same number of choice sets, but the composition of the choice sets varies across respondents. This allows the researcher to value changes in attributes and the trade-offs compared to the *status quo* and different alternatives. Furthermore, in the case of damage to a particular attribute, compensating amounts of other goods (rather than compensation based on money) can be calculated. While several statistical methods can be used, multinomial and conditional logit models are commonly used to analyse the choices that people make. This approach can provide substantially more information about a range of possible alternative policies and can reduce the sample size needed compared to the CV method. However, survey design issues with the CM approach are often much more complex due to the number of goods that must be described and the statistical methods that must be employed. This may limit its application for valuation of NRM impact in the context of developing countries.

Comparison of alternative valuation methods

This section has reviewed the promising methods that can be used in valuation of NRM impacts. The choice of valuation methods depends on the existence of markets, the spatial and temporal diffusion of the impact, and whether the values relate to use or non-use values. Direct-use values such as productivity changes can be measured indirectly using data from observed or stated market behaviour of producers and consumers in conventional or surrogate markets. For non-use values, like benefits captured by future generations (sustainability) and indirect-use values such as ecosystem regulation functions, there is no observable market behaviour that contains relevant information, hence hypothetical behaviour in constructed markets must be used. The PCA, RCA, HP and CV methods are the most commonly applied in relation to environmental resources, and they offer promising opportunities for valuation of NRM impacts. Each of these methods measures different aspects of the total economic value (see Fig. 2.1) and has its strengths and weaknesses. The PCA and RCA use observed market information to measure use values indirectly. HP is also an indirect method that uses surrogate markets to measure use values. The CV is the direct stated preference method mainly used in respect to non-use values, but it could also be applied for use values.

Perhaps because of their relative ease and cost-effectiveness, the PCA and RCA are most commonly used in NRM valuation exercises. These two approaches measure different aspects of resource degradation focusing on productivity change and the costs of replacing damaged ecosystem services; hence, they often provide divergent estimates. The relative size of the two estimates may also be useful for NRM technology choice and farmer-investment decisions. Farmers are unlikely to adopt resource management practices unless the productivity benefits are higher than the investment costs. Drechsel *et al.* (Chapter 9, this volume) discuss this and the related pros and cons of these methods in more detail. The PCA needs to be computed as a net gain over the less-conserving alternative. When data on changes in resource conditions and productivity are available, the PCA is a recommended method for measuring the values of marketed productivity impacts. A major weakness is its inability to value external effects and non-market benefits.

The RCA imposes strict assumptions – that the substitute be the least-cost alternative and that the cost be less than the aggregate WTP. The RCA cannot be used to value ecosystem services that do not have marketed substitutes, and it cannot measure non-use values. When production data are limiting, RCA can be a useful alternative to value changes in certain resources like soil quality.

If markets reflect changes in environmental quality and resource conditions, the HP method is another promising technique for estimating benefits. The disadvantages of HP applied to land markets include lack of transaction data and failure of land markets to fully reflect non-productivity related changes in ecosystem services. When sales transactions are limited, land rental markets may provide an alternative source of relevant information.

When existing markets cannot be used to acquire the necessary information, the CV and CM methods can be the most useful approaches for NRM valuation. The strength of these approaches is their flexibility to generate information from constructed markets to measure both use and non-use values relevant to a given situation. When properly applied, the WTP/WTA estimates provide theoretically correct measures of welfare change. Although the survey design is more complex, the CM requires less data and provides more policy-relevant information than the CV method. However, these methods are criticised for their reliance on hypothetical markets where true behaviour is unobservable and also for survey implementation problems that may bias results. Several approaches can be used to reduce bias. If non-market ecosystem services and non-use values are a significant part of NRM impacts (as is often the case), these are the only conceptually justified approaches, and should be carefully applied depending on the availability of technical and financial resources.

When new valuation studies cannot be made due to time or financial constraints, the *benefit transfer approach* can be used to apply valuation estimates from other studies of similar changes in environmental quality at a new site. Although termed '*benefit transfer*', damage estimates can also be transferred. Four benefit transfer approaches exist: unit value transfer (e.g. direct transfer of mean WTP per household), adjusted unit value transfer (e.g. corrected for differences in per capita income levels), value function transfer, and meta-analysis. Value function transfer uses regression equations estimated for one location to predict resource values in another location, while meta-analysis uses independent case studies to synthesise and provide a summary estimate of resource value for specific conditions. Value transfer generally increases the uncertainty in the estimated environmental value. The early examples of benefit transfer were conducted in an uncritical manner, often lacking sound theoretical, statistical and empirical basis, and did not question the validity and reliability of the transferred values.

Recently, there has been growing interest in the development of benefit transfer methods and statistical techniques (Navrud, 2004; Navrud and Ready, 2004). Results from validity tests have shown that the uncertainty in spatial and temporal benefit transfer can be quite large, especially when economic and ecological conditions are quite different. Thus, care should be taken in using benefit transfer in policy uses where the demand for accuracy is high.

At present, there is a dearth of both benefit transfer applications in developing countries and sufficient valuation studies for meta-analyses. There is also a lack of validity tests of benefit transfer between developing and developed countries. One such study underlines the considerable uncertainty in using benefit transfer estimates (Barton and Mourato, 2003). Correcting for differences in gross domestic product (GDP) per capita seems to improve benefit transfer, but the actual difference in income levels in the two samples does not typically correspond to the differences in GDP, so correcting for income levels in unit-value transfers often makes things worse (Barton and Mourato, 2003; Navrud and Ready, 2004). Since the explanatory power of WTP functions is often poor, value function transfers may not do a better job in transferring benefits than simple value transfers.

Impact Evaluation

Since NRM interventions are expected to provide multiple economic and environmental benefits to various stakeholders including smallholder farmers, NRM impact evaluation should include non-marketed ecosystem goods and services along with marketed economic benefits. The market and non-market values of changes in goods and environmental services estimated using the valuation methods discussed in this chapter are vital in estimating costs and benefits that are used to evaluate the overall impact of the intervention. This requires a more holistic approach that would expand conventional impact assessments (Baker, 2000) to include non-tradable goods and environmental services. The welfare gains from NRM investments associated with direct economic benefits (e.g. yield gains or cost savings) can be assessed using a conventional approach. Unfortunately, as shown earlier, NRM investments generate other sustainability benefits and ecosystem services that have use and non-use values to people. Indirect welfare gains from such environmental improvements are legitimate parts of the welfare changes associated with NRM interventions and need to be considered in impact evaluation. The total welfare gain to people can then be decomposed into direct economic benefits derived from productivity changes and indirect environmental economic components. When NRM technologies generate productivity (including cost-saving) benefits in addition to changes in resource quality and sustainability, both sources of welfare gain are likely to be significant. In cases where the impact is expressed mainly in terms of non-tangible ecosystem service flows, the environmental and sustainability benefits could become a major part of the total welfare gain. Although precise estimation of non-market ecosystem service flows is always difficult, the valuation methods discussed above can be used to estimate the multiple welfare benefits associated with NRM interventions.

The conventional economic surplus approach (Alston *et al.*, 1995; Swinton, Chapter 7, this volume) includes changes in consumer surplus and producer surplus associated with supply shifts and price changes from changes in agricultural technology. As shown above, welfare gains associated with changes in environmental conditions are measured using the ES and CS measures of welfare gains to consumers and the producer surplus benefits to producers. When these extended economic and environmental welfare benefits are known, the social impact of research and development (R&D) investments in NRM can be evaluated using the benefit–cost analysis approach. The economic welfare gains from NRM can be given as:

$$\pi_t^P = \pi_t^{PN} - \pi_t^{PT} \tag{20}$$

where π_t^P is the period t productivity-related economic gain from change in NRM that can be calculated as the difference in net benefits between the new (π_t^{PN}) and the traditional (π_t^{PT}) NRM practices. π_t^P is essentially the flow of consumer and producer surpluses associated with productivity changes generated by NRM interventions. The environmental welfare gains from NRM can similarly be given as:

$$\pi_t^E = \pi_t^{EN} - \pi_t^{ET} \tag{21}$$

where π_t^E is the period t environmental welfare gains that can be calculated as the difference between environmental benefits from the new (π_t^{EN}) and the traditional (π_t^{ET}) NRM practices. This is the total WTP/WTA measure of welfare change resulting from changes in the flow of non-productivity related ecosystem services valued by people. π_t^E is essentially the social WTP for better NRM to enhance agricultural sustainability and the flow of ecosystem services (environmental quality). These values can also include the changes in external or off-site effects of NRM interventions. In order to assess the social impact of NRM interventions, additional information on the research, development and extension costs will be needed. If it is assumed that the flow of these costs is given by RE_t, such costs incurred up front could be quite significant, especially when the benefit flow is delayed because of the long time required for technology development and adaptation and when a positive discount rate is used in the calculation of net present values from the investment. The net welfare gain from NRM interventions will then be estimated as:

$$NPV = \sum_{t=1}^{n} \left(\pi_t^P + \pi_t^E - RE_t\right)\left(1 + r\right)^{-t} \tag{22}$$

where NPV is the social net present value of the NRM intervention, r is the real social rate of discount. Some of the changes in ecosystem services (e.g. soil fertility) may be reflected in productivity changes. The additive framework given in Equation 22 is valid when the productivity benefits and non-productivity related environmental or sustainability outcomes are clearly separable. When such separation is not possible, the approach can lead to double counting and overestimation of NRM impacts. An important area for further research is on the mechanisms used to separate productivity and sustainability effects, and under which conditions the estimated productivity and environmental values can be additive. If the social benefits of interventions are higher than the costs of the interventions, i.e. the total benefit is higher than the costs and gainers from the intervention can overcompensate losers, then NRM is considered to be socially beneficial. This may not be the case when externalities are ignored in the analysis. When environmental net benefits (π_t^E) cannot be estimated, the impact could be assessed in terms of the required tradeoffs and implications for sustainability of productivity gains. This could also include situations where the impact of NRM interventions is reflected in terms of reductions in production risk, improved stability of production, and reduced vulnerabilities of rural households to droughts, floods and other environmental shocks.

Parameters estimated for linking NRM changes with ecosystem goods and services (e.g. to estimate the effect of soil and water conservation on productivity) can also be integrated into bioeconomic models. The integration of important biophysical information and ecological processes with economic decision behaviour through bioeconomic modelling allows simultaneous assessment of welfare effects and environmental and distributional outcomes. Holden (Chapter 8, this volume) and Shiferaw and Holden (Chapter 12, this volume) further discuss these issues. One innovative approach for future research to evaluate the social impacts of NRM interventions is to compare the stream of aggregated net benefits (estimated based on optimised values derived from the model) with R&D investment costs.

Conclusions

The changes in environmental and resource service flows associated with NRM investments accrue over different temporal and spatial scales. Many of these ecosystem services generate valuable direct and indirect welfare benefits to people. When NRM investments generate private and public goods benefits, valuation of such changes is a crucial first step in the evaluation of overall social impacts. A prerequisite to effective valuation of NRM impacts is the ability to predict the changes in ecosystem service flows that can be attributed to the intervention itself. This requires a strong partnership between agroecologists and economists. The scientific understanding of ecosystem functions and services and how they are affected by human interventions is still incomplete. More work is needed to understand and quantify the effect of NRM interventions on ecosystem functions and services. Appropriate indicators are needed to measure selected changes in ecosystem services. Without reliable data, valuation efforts will not provide any useful economic values. With advances in agroecology and biophysical simulation modelling, the ability to predict the likely effects of certain interventions has improved. The economic approach to valuation of ecosystem services is based on the trade-offs that people are prepared to make in exchange for these services. The changes in the flow of ecosystem services can affect human welfare in complex ways and through marketed or non-marketed activities.

This chapter has offered an overview of ecosystem services from NRM investments, the need for indicators of ecosystem condition, key challenges to valuation of environmental services, and recent advances in the methods available for valuation of economic and environmental benefits. There is a dearth of examples in valuation of NRM impacts, especially in the context of developing countries. However, the recent progress in developing valuation methods has created new opportunities. For NRM impact assessment, the estimated values of changes in ecosystem services need to be social scarcity prices that account for non-marketed outcomes and external effects. Impact assessment of agricultural technologies has often ignored external effects and environmental impacts. However, resource management interventions typically generate non-marketed sustainability and environmental benefits. The greatest challenge in valuation of NRM impacts is in quantification and measurement of these non-productivity related outcomes and non-market benefits. Such standard techniques as the productivity change approach or revealed preference methods like defensive expenditures, provision costs or hedonic pricing can be used to measure productivity-related outcomes. However, markets and observed behaviour cannot be used for valuation of impacts on non-use and indirect use values related to regulation and habitat provision functions. When the impacts can be quantified using measurable indicators, stated preference methods could be used for the valuation of such effects.

The CV method is the most appropriate option when the indirect and surrogate market options cannot be used to value the change in ecosystem services. It is most appropriate for valuing non-use values and non-tradable use values of ecosystem services. However, the CV method has only rarely been applied to NRM impact assessment in the developing countries. Since poverty limits the ability to pay, the WTA compensation is a preferred approach for valuation of ecosystem services in poor communities. Choice modelling is an alternative and promising sated preference method. It is important to test and enhance these methods for valuation of non-market ecosystem services associated with NRM. Case studies are required to gain experience and develop improved protocols in the application of CV and/or CM methods for NRM impact assessment. In some situations benefit transfer approaches can be used to inform urgent policy decisions. However, more research is needed to enhance the transferability of benefits between countries or eco-regions.

Once the values of changes in ecosystem services are estimated, impact evaluation needs to compute the overall social gains from NRM interventions. Many NRM interventions imply supply shifts for both market and non-market goods and environmental services. This implies the need to estimate the size of the supply shift and the resulting effect on estimated unit resource values. However, more work is needed to understand how such values can be effectively integrated into impact assessment studies. In some cases, high uncertainties about the nature and magnitude of changes and the temporal and spatial impacts of NRM interventions may limit the policy relevance of monetary values. More research is needed to improve the validity and reliability of these estimates for use in policy analysis and impact assessments.

Acknowledgement

We have substantially benefited from valuable comments and suggestions from two anonymous reviewers and from Scott Swinton, the editor for this chapter. The usual disclaimer applies.

Endnotes

[1]In *ex ante* assessments, prevailing market prices may not be an appropriate way to value productivity changes unless a 'small project' assumption is imposed so that prices remain largely unaffected. If productivity changes are expected to affect market prices, appropriate adjustments can be made using the general equilibrium framework.

[2]The travel cost method is another surrogate market approach for valuation of recreational use values of ecosystem services. It has been applied widely for the valuation of wildlife in protected areas. Since typical agricultural NRM investments do not provide marketable recreational benefits, the method is not discussed here.

[3]Apart from average WTP, marginal WTP is also of interest. This can be determined by estimating an inverse demand curve with price as a function of quantity. An inverse demand curve is also essential for estimating economic surplus rather than assuming constant average WTP.

References

Adamowicz, V., Boxall, P., Williams, M. and Louviere, J. (1998) Stated preference approaches for measuring passive use values: Choice experiments and contingent valuation. *American Journal of Agricultural Economics* 80, 64–75.

Alpizar, F., Carlsson, F. and Martinsson, P. (2003) Using choice experiments for non-market valuation. *Economic Issues* 2003, 83–110.

Alston, M.J., Norton, W.G. and Pardey, P.G. (1995) Science under scarcity. *Principles and Practice for Agricultural Research Evaluation and Priority Setting.* Cornell University Press, Ithaca, New York, and London, UK, 585 pp.

Altieri, M. (2002) Agroecology: the science of natural resource management for poor farmers in marginal environments. *Agriculture, Ecosystems and Environment* 93, 1–24.

Arrow, K.J., Solow, R., Portney, P.R., Leamer, E.E., Radner, R. and Shuman, H. (1993) Report of the National Oceanographic and Atmospheric Administration panel on contingent valuation. *Federal Register* 58, 4601–4614.

Arshad, M.A. and Martin, S. (2002) Identifying critical limits for soil quality indicators in agro-ecosystems. *Agriculture, Ecosystems and Environment* 88, 153–160.

Baker, J. (2000) *Evaluating the Impact of Development Projects on Poverty: A Handbook for Practitioners.* The World Bank, Washington, DC, 217 pp.

Barbier, E.B. (1998) The economics of soil erosion: theory methodology and examples. In: Barbier, E.B. (ed.) *The Economics of Environment and Development. Selected Essays.* Edward Elgar, Cheltenham, UK, pp. 281–307.

Barton, D. and Mourato, S. (2003) Transferring the benefits of avoided health effects from water pollution between Portugal and Costa Rica. *Environment and Development Economics* 8(2), 351–372.

Bateman, I.J. and Willis, K.G. (eds) (1999) *Valuing Environmental Preferences. Theory and Practice of the Contingent Valuation Method in the US, EU, and Developing Countries.* Oxford University Press, Oxford, UK, 645 pp.

Bennett, J. and Blamey, R. (eds) (2001) *The Choice Modelling Approach to Environmental Valuation.* Edward Elgar, Cheltenham, UK, 69 pp.

Bingham, G., Bishop, R., Brody, M., Bromley, D., Clark, E.T., Cooper, W., Costanza, R., Hale, T., Hayden, G., Kellert, S., Norgaard, R., Norton, B., Payne, J., Russell, C. and Suter, G. (1995) Issues in ecosystem valuation: improving information for decision making. *Ecological Economics* 14, 73–93.

Bockstael, N.E., Freeman, A.M., Kopp, R.J., Portney, P.R. and Smith, V.K. (2000) On measuring economic values for nature. *Environmental Science and Technology* 34, 1384–1389.

Bojo, J. (1991) *The Economics of Land Degradation: Theory and Applications to Lesotho.* Stockholm School of Economics, Stockholm, Sweden.

Braden, J.B. and Kolstad, C.D. (1991) *Measuring the Demand for Environmental Quality.* Elsevier Science Publisher B.V., Amsterdam, The Netherlands, 370 pp.

Campbell, B., Sayer, J.A., Frost, P., Vermeulen, S., Ruiz-Perez, M., Cunningham, A. and Ravi, P. (2001) Assessing the performance of natural resource systems. *Conservation Ecology* 5(2), 22. [online] http://www.consecol.org/vol5/iss2/art22/index.html

Carson, R.T. (1991) Constructed markets. In: Braden, J. and Kolstad, C. (eds) *Measuring the Demand for Environmental Commodities.* North-Holland, Amsterdam, The Netherlands, pp. 121–162.

Costanza, R., d'Arge, R., de Groot, R., Farber, S., Grasso, M., Hannon, B., Limburg, K., Naeem, S., O'Neill, R.V., Parueol, J., Raskin, R.G., Sutton, P. and van den Belt, M. (1998) The value of the world's ecosystem services and natural capital. *Ecological Economics* 25, 3–15.

Dixon, J.A., Scura, L.F., Carpenter, R.A. and Sherman, P.B. (1994) *Economic Analysis of Environmental Impacts.* Earthscan Publications, London, UK, 210 pp.

Dumanski, J. and Pieri, C. (2000) Land quality indicators: research plan. *Agriculture, Ecosystems and Environment* 81, 93–102.

Ellis, G.M. and Fisher, A.C. (1987) Valuing the environment as input. *Journal of Environmental Management* 25, 149–156.

FAO (Food and Agriculture Organization of the United Nations) (2000) Applications of the contingent valuation method in developing countries: a survey. *FAO Economic and Social Development Paper 146.* FAO, Rome, Italy, 70 pp.

Farber, S.C., Costanza, R. and Wilson, M.A. (2002) Economic and ecological concepts for valuing ecosystem services. *Ecological Economics* 41, 375–392.

Freeman, A.M. (1993) *The Measurement of Environmental and Resource Values: Theory and Methods.* Resources for the Future, Washington, DC, 516 pp.

Gebremedhin, B., Swinton, S.M. and Tilahun, Y. (1999) Effects of stone terraces on crop yields and farm profitability: results of on-farm research in Tigray, Northern Ethiopia. *Journal of Soil and Water Conservation* 54, 568–573.

de Groot, S.R., Wilson, M.A. and Boumans, R.M.J. (2002) A typology for classification, description and valuation of ecosystem functions, goods and services. *Ecological Economics* 41, 393–408.

Hanemann, W.M. (1984) Welfare evaluations in contingent valuation experiments with discrete responses. *American Journal of Agricultural Economics* 66, 322–341.

Harrison, D. and Rubinfeld, D.O. (1978) Hedonic housing price and demand for clean air. *Journal of Environmental Economics and Management* 5, 81–102.

Holden, S.T. and Shiferaw, B. (2002) Poverty and land degradation: peasants' willingness to pay to sustain land productivity. In: Barrett, C.B. and Place, F. (eds) *Natural Resources Management in African Agriculture: Understanding and Improving Current Practices.* CAB International, Wallingford, UK, pp. 91–102.

Johansson, P.-O. (1987) *The Economic Theory and Measurement of Environmental Benefits.* Cambridge University Press, Cambridge, UK, 236 pp.

Kim, S.H. and Dixon, J.A. (1986) Economic valuation of environmental quality aspects of upland agricultural projects in Korea. In: Dixon, J.A. and Hufschmidt, M.M. (eds) *Economic Valuation Techniques for the Environment: A Case Study Workbook.* Johns Hopkins University Press, Baltimore, Maryland.

Magrath, W. and Arens, P. (1989) The case of soil erosion on Java: A natural resource accounting approach. *Environment Department Working Paper 18.* The World Bank, Washington, DC.

Mitchell, R.C. and Carson, R.T. (1989) *Using Surveys to Value Public Goods: The Contingent Valuation Method.* Resources for the Future, Washington, DC, 67 pp.

Munasinghe, M. and Lutz, E. (1993) Environmental economics and valuation in development decision making. In: Munasinghe, M. (ed.) *Environmental Economics and Natural Resource Management in Developing Countries.* The World Bank, Washington, DC, 240 pp.

Navrud, S. (2004) Value transfer and environmental policy. In: Tietenberg, T. and Folmer, H. (eds) *The International Yearbook of Environmental and Resource Economics 2004/2005. A Survey of Current Issues.* Edward Elgar, Cheltenham, UK, and Northampton, Massachusetts, pp. 189–217.

Navrud, S. and Ready, R. (eds) (2004) *Environmental Value Transfer: Issues and Methods.* Kluwer Academic Publishers, Dordrect, The Netherlands (in press).

Oriade, C. and Dillon, C.A. (1997) Developments in biophysical and bioeconomic simulation of agricultural systems: A review. *Agricultural Economics* 17, 45–58.

Pagiola, S., Bishop, J. and Landell-Mills, N. (eds) (2002) *Selling Forest Environmental Services: Market Based Mechanisms for Conservation and Development.* Earthscan Publications, London, UK, 299 pp.

Pearce, D., Pearce, C. and Palmer, C. (eds) (2002) *Valuing the Environment in Developing Countries: Case Studies.* Edward Elgar, Cheltenham, UK.

Rosen, S. (1974) Hedonic prices and implicit markets: product differentiation in perfect competition. *Journal of Political Economy* 82(1), 34–55.

Shiferaw, B. and Holden, S.T. (2001) Farm level benefits to investments for mitigating land degradation: empirical evidence from Ethiopia. *Environment and Development Economics* 6(3), 335–358.

Shiferaw, B., Reddy, R.V., Wani, S.P. and Rao, G.D.N. (2003) Watershed management and farmer conservation investments in the semi-arid tropics of India: analysis of determinants of resource use decisions and land productivity benefits. *Socioeconomics and Policy Working Paper Series 16.* ICRISAT, Patancheru, Andhra Pradesh, India, 28 pp.

Shyamsundar, P. and Kramer, R.A. (1996) Tropical forest protection: an empirical analysis of the costs borne by local people. *Journal of Environmental Economics and Management* 31, 129–144.

Smyth, A.J. and Dumanski, J. (1993) FESLM: An international framework for evaluating sustainable land management. *World Soil Resources Report 73*. Food and Agriculture Organization of the United Nations, Rome, Italy, 77 pp.

Whittington, D. (1998) Administering contingent valuation surveys in developing countries. *World Development* 26(1), 21–30.

3 Measurable Biophysical Indicators for Impact Assessment: Changes in Soil Quality

P. Pathak, K.L. Sahrawat, T.J. Rego and S.P. Wani

International Crops Research Institute for the Semi-Arid Tropics (ICRISAT), Patancheru, India

Introduction

Soil plays a key role as the interface between terrestrial and aquatic ecosystems on the one hand and the atmosphere on the other. The importance of soil in meeting food, feed and fibre needs and maintaining environmental sustainability cannot be overemphasised. A healthy or good quality soil acts as an environmental filter in cleaning air and water. Soil is a major sink for global gases and its appropriate management favourably affects the carbon dioxide (CO_2) balance that is important in combating global warming. If mismanaged, soil can work against us; it can pollute the air and water and lead to a fall in agriculture production.

Decline in soil quality has occurred worldwide, particularly in the semi-arid tropical (SAT) regions and is manifested as adverse changes in physical, chemical and biological soil properties and its contamination by inorganic and organic chemicals (Arshad and Martin, 2002; Lal, 2004). In many parts of world production of major cereals is declining mainly due to soil degradation coupled with inadequate soil and water management (Steer, 1998).

Natural resource management (NRM) interventions in terms of fertility, soil and water management practices in various farming systems have become necessary to address the problem of soil degradation and hence increasing investments in NRM research and development are being made worldwide. To diagnose and quantify the impacts of various NRM interventions, reliable soil quality indicators are necessary. Impact assessment is essential for the development of suitable management strategies for soil quality and to maximise productivity and sustainability for the benefits of society.

Appropriate and measurable soil quality indicators are needed to assess the impact of various NRM interventions on soil quality in agricultural lands. Measurable and simple soil quality indicators are important because many of the conventional soil attributes used to characterise soils become useful only

 Natural Resource Management in Agriculture: Methods for Assessing Economic and Environmental Impacts (eds B. Shiferaw, H.A. Freeman and S.M. Swinton)

after soil degradation has already taken place. To have soil quality indicators together with the soil quality thresholds needed to monitor and assess the impact of NRM technologies seems rather a tall order. Modern agricultural practices used to intensify agriculture have complicated the selection of such indicators, but several measurable indicators are available and can be used to assess the biophysical impact of NRM practices. Unfortunately, there is no universal set of indicators that is equally applicable in all cases, so the selection of those relevant to specific conditions is extremely important.

The objective of this chapter is to identify and discuss with examples from recent literature the use of biophysical indicators in monitoring the impact of NRM interventions on soil quality attributes. The use of simulation modelling to assess the long-term effects of NRM interventions on soil quality and future research needs are also covered.

Soil Quality Indicators

Soil quality indicators are measurable soil attributes that influence the inherent capacity of the soil to perform its production and environment-related functions. Attributes that are management-responsive are most desirable as indicators. During the past 10 years many definitions of soil quality with similar elements have been proposed (Arshad and Coen, 1992; Doran and Parkin, 1994; Karlen *et al.*, 1997). A recent definition was proposed by Karlen *et al.* (2003) and a committee of the Soil Science Society of America: "the fitness of a specific kind of soil, to function within its capacity and within natural or managed ecosystem boundaries, to sustain plant and animal productivity, maintain or enhance water and air quality, and support human health and habitation"; this seems to be inclusive and appropriate for the objectives of this chapter.

It should, however, be mentioned that the soil quality paradigm has received several criticisms because of its general lack of sufficient quantification and scientific rigour (Letey *et al.*, 2003; Sanchez *et al.*, 2003; Sojka *et al.*, 2003). These authors believe that in assessing soil quality emphasis should be directed towards using available technical information to motivate and educate farmers on 'quality soil management' involving management practices that optimise the combined goals of high crop production, low environmental degradation, and sustained resource use (Sojka *et al.*, 2003). However, several scientists believe that with further refinement, soil quality indicators could provide a more useful tool for assessing soil quality. It may be useful to note that indicators for monitoring soil quality could also help towards developing quality soil management.

Scientists aim to develop a set of basic soil characteristics to serve as key soil quality indicators (Stott *et al.*, 1999) that are sensitive to climatic and management interventions. Ideally, the best soil quality indicators are those attributes or characteristics that show observable and significant changes between 1 to 3 years, with 5 years being an upper limit to usefulness.

Given the complex nature of soil and the exceptionally large number of soil properties that may have to be determined, it is important to be able to select properties that are appropriate and practical. Stephen (2002) grouped attributes that can be used as indicators of soil quality into four broad groups: physical, chemical, biological and visible indicators. Karlen and Stott (1994) proposed a framework for evaluating physical and chemical indicators of soil quality. Turco *et al.* (1994) discussed the various microbial indicators of soil quality. Arshad and Martin (2002) proposed selected physical, chemical and biological soil quality indicators (Table 3.1). In the light of diverse soil functions for which indicators are used, the quality indicators listed may not be sufficient to evaluate the changes in soil quality resulting from various agricultural and NRM interventions. Depending upon the local conditions, some may have to be added or excluded. These are discussed in turn below.

Table 3.1. Selected physical, chemical and biological soil quality indicators used to assess soil quality.

Soil quality indicator	Rationale for selection
Physical	
Top soil-depth	Estimates moisture availability, rooting volume for crop production and erosion
Aggregation	Indicates status of soil structure, erosion resistance, crop emergence can be an early indicator of soil management effect
Texture	Indicates retention and transport of water and chemicals
Bulk density	Shows plant root penetration and air-filled porosity
Infiltration	Indicates runoff, leaching and erosion potential
Biochemical	
pH	Indicates nutrient availability, sorption and desorption of molecules
Organic matter	Affects fertility, structure, water retention and sorption and desorption of molecules
Electrical conductivity	Defines salt content, crop growth, soil structure and water infiltration
Suspended pollutants	Affects food quality, water quality and human and animal health
Soil respiration	Indicates biological activity, biomass activity and organic matter quantity and quality
Form of soil N	Defines availability to crops, leaching potential, mineralisation/ immobilisation rates
Extractable N, P and K	Indicates capacity to support plant growth and serve as an environmental quality indicator

Source: Adapted from Arshad and Martin, 2002

Physical quality indicators

Physical indicators are principally concerned with the physical arrangement of solid particles and pores. They include soil texture, moisture-holding capacity, bulk density, porosity, aggregate strength and stability, crusting, surface sealing, compaction and depth.

Chemical quality indicators

The list of potential soil chemical indicators attributes is very large and the final selection will depend upon the soil function and process under consideration. These attributes include: pH, salinity (electrical conductivity), organic matter content, cation-exchange capacity (CEC), plant nutrient status, concentrations of potentially toxic elements, and – possibly the most important attribute – the capacity of the soil to buffer against change.

Biological quality indicators

Biological parameters are relatively dynamic and sensitive to changes in both soil management and climate. This gives biological indicators a comparative advantage over physical or chemical parameters, so they can be used as indicators of soil quality at an early stage. Some of the parameters that could serve as such indicators are: populations of micro-, meso-, and macro-organisms, soil respiration rate, enzyme activities, rate of nutrient mineralisation, microbial biomass, and more detailed characterisation of soil organic matter fractions.

Visible quality indicators

It is often the observation of visible attributes that brings to attention the changes in soil quality and causes public awareness and, at times, alarm. But in many cases, when there is visible evidence of decline in soil quality, the process of decline may have proceeded too far, and the chance to restore quality may have already been lost. The visible attributes include evidence of erosion in the form of rills and gullies, exposure of subsoil, surface ponding of water, surface runoff, and poor plant growth (Stephen, 2002).

NRM and Soil Quality Indicators

Changes in physical quality indicators

The recent developments in soil quality research emphasise the importance of identifying key soil indicators and their threshold values in relation to specific soil functions. The potential of a soil to support crop growth is largely determined by the environment that the soil provides for root growth. Roots need air, water, nutrients and adequate space to develop and make water and nutrients accessible to plants.

Such physical attributes as bulk density, porosity, air-filled porosity, crusting, sealing, water-holding capacity and depth all determine how well roots develop. Changes in these physical soil attributes directly affect the health and productivity of crops. The influence of agricultural practices, specifically NRM interventions, on changes in some soil physical attributes/indicators are discussed in the following examples.

Bulk density

Compact soil layers with high bulk density in the soil profile impede root growth by reducing the effective soil rooting volume. Measurements of soil bulk density along with penetration resistance (interpreted with respect to water contents) are used to identify root-impeding layers in the soil profile. When the bulk density of a soil increases to a critical level, it impedes root penetration and restricts root growth and the soil volume explored by roots. For example, in many Alfisols in SAT regions, soil compaction is one of the major constraints to crop establishment and productivity (El-Swaify *et al.*, 1985). Pierce *et al.* (1983) reported critical values of bulk density for soils varying in texture. Compaction by wheeled traffic has direct and at times irreversible effects on soil structure.

A long-term experiment conducted at the International Crops Research Institute for the Semi-Arid Tropics (ICRISAT), Patancheru, India examined the impact of improved management options on soil physical attributes (Table 3.2). It was found that management practices in Vertisol

Table 3.2. The effect of management practices on physical properties of Vertisols at ICRISAT, Patancheru, India (1975–99).

	Improved land management technology		Traditional technology
Soil properties	Broadbed	Furrow	Flat
1. Texture (0–10 cm soil layer)			
Clay (%)	50.8		46.3
Silt (%)	21.5		21.4
Fine sand (%)	15.5		15.4
Coarse sand (%)	12.2		16.9
Gravel (%)	4.8		14.5
2. Bulk density (g/cm^3)	1.2	1.5	1.5
3. Total porosity (%)	52.1	39.5	41.5
4. Air-filled porosity (%)	41.0	33.0	32.0
5. Penetration resistance (MPa)	1.1	9.8	8.5
6. Sorptivity (mm/h$^{1/2}$)	121.2	100.6	88.5
7. Cumulative infiltration in 1 h (mm)	347.2	205.7	264.7

Source: ICRISAT, experimental results

watersheds caused significant differences in soil bulk densities. Throughout the soil profile, the bulk density was found to be significantly lower in the watershed with improved technology than in the watershed where a traditional system was used (Fig. 3.1).

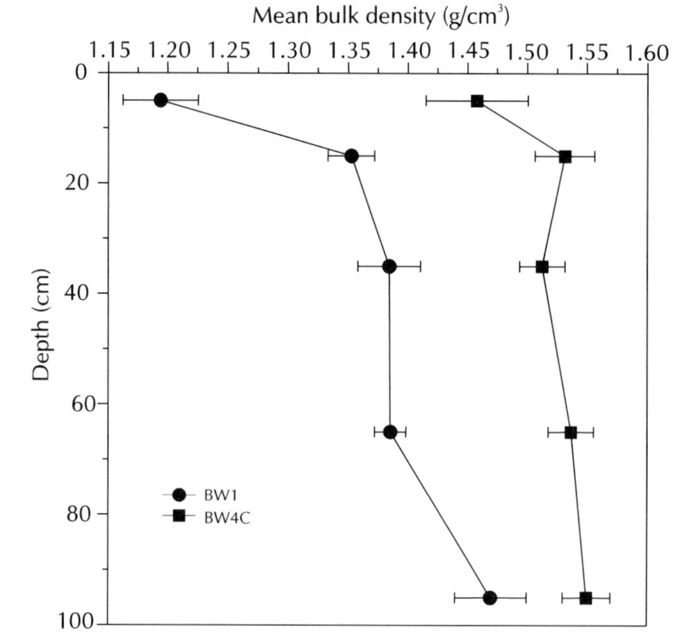

Fig. 3.1. Long-term effects of improved (BW1) and traditional (BW4C) management practices on bulk density of Vertisols, ICRISAT, Patancheru, India, 1975–98.

However, the differences in bulk densities were relatively greater in the top 15 cm layer. The maximum difference in bulk density was recorded in the 0–5 cm layer. The data clearly show the advantage of the improved technology where the soil is kept loose. This has major implications for root growth and tillage operations especially when tillage operations are done using animal power.

Penetration resistance

Penetration resistance measurement can be measured to identify root-impeding layers. When the penetration resistance of a soil increases to a critical level, it becomes more difficult for roots to penetrate and their growth is impeded. The long-term ICRISAT experiment showed that the soil in a watershed where improved land and water management was practised had a lower penetration resistance in the cropping zone than the corresponding zone in one traditionally managed (Fig. 3.2). The penetration resistance however increased with depth, but it was consistently lower with improved management. In the long term the broadbed-and-furrow (BBF) land configuration in the improved system led to progressive improvement in soil tilth in the bed zone. The BBF land configuration and reduced penetration resistance allows timely tillage operations that are crucial for Vertisols because they are difficult work, both in dry and very wet conditions. Klaij (1983) reported similar results on the positive effects of land surface treatments for Alfisols where lower penetration resistance is crucial for crop emergence and root growth.

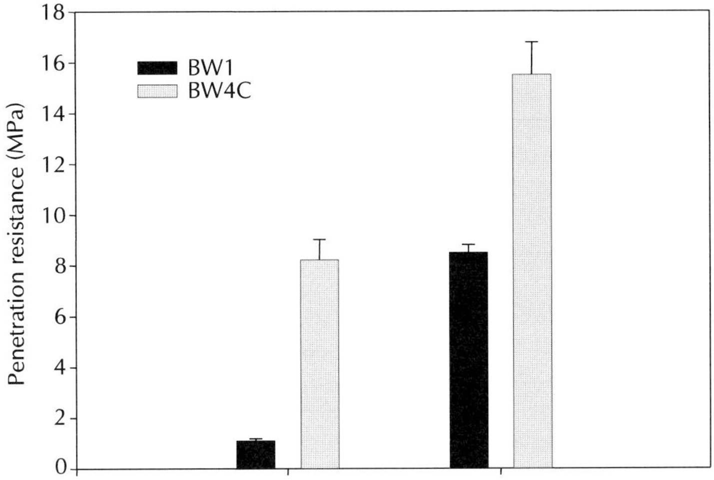

Fig. 3.2. Long-term effects of improved (BW1) and traditional (BW4C) management practices on soil penetration resistance of Vertisols, ICRISAT, Patancheru, India, 1975–98.

Porosity/air-filled porosity

The problem of temporary waterlogging and the resulting lack of adequate aeration is quite common in many soils. In medium to high rainfall areas, crops on Vertisols often suffer extensively from temporary waterlogging and poor soil aeration (El-Swaify *et al.*, 1985). In such situations, maintaining high air-filled porosity is crucial to increasing crop productivity. A long-term experiment on Vertisols at ICRISAT showed the improved system had higher air-filled porosity than the traditional system (Fig. 3.3). In the improved system, the air-filled porosity in the top 10 cm layer improved by 28% during 1975–98. This improvement contributed to better crop growth and higher yields.

Rooting depth

Rooting depth is the depth in the soil profile to which roots penetrate and access water and plant nutrients. Rooting depth is especially important in dryland agriculture where the shortage of both water and nutrients limit plant growth and productivity. Exploration of large volumes of soil by roots can increase the accessibility of water and nutrients to growing plants. Deep-rooted crops are considered better at extracting water and nutrients from

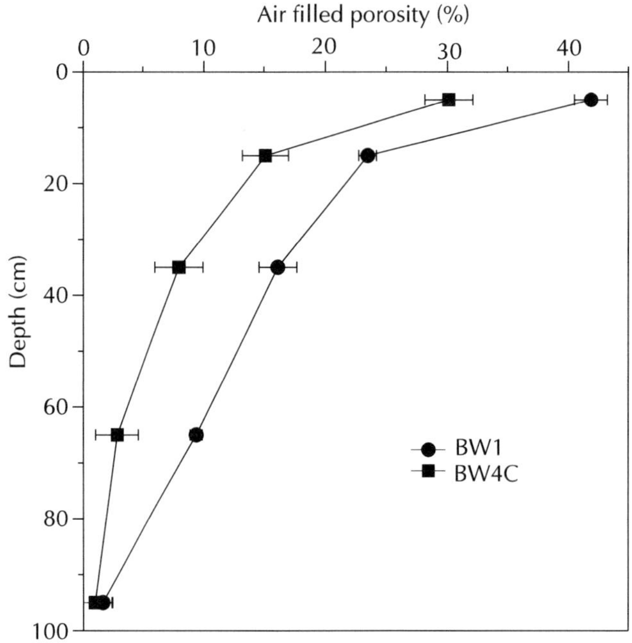

Fig. 3.3. Long-term effects of improved (BW1) and traditional (BW4C) management practices on air-filled porosity of Vertisols, ICRISAT, Patancheru, India, 1975–98.

deeper layers in the soil profile. Dryland crops such as sorghum may send their roots over 1 m deep into the soil in search of water (El-Swaify *et al.*, 1985). Irrigated crops such as rice have relatively shallow rooting depths (up to 60 cm). The threshold values for rooting depth vary with crop and the irrigated or non-irrigated conditions under which the crop is grown. The deeper rooting depths of dryland crops need to be considered while using indicators.

It is not surprising that rooting depth has been related to crop productivity. Crops grown in soils in which the rooting depth is limited by the presence of a physical or chemical constraint are generally less productive. As limiting layers move closer to the soil surface where erosion removes the topsoil, crop productivity generally declines. The effect of rooting depth on crop productivity varies with crop type (Taylor and Terrell, 1982). Soil management practices can have important effects on rooting depth. For example, erosion reduces rooting depth by removing the top soil layer while compaction reduces it by bringing to the surface layers in the soil that are impenetrable by crop roots (National Research Council, 1993). Changes in soil management practices influence root mass and length in the soil that are indicative of changes in rooting depth and can be monitored by sampling the roots at various depths in the profile.

Water-holding capacity

An important attribute of a soil is its ability to store and release water to growing plants. The water-holding capacity of soils is measured as the total amount of water stored in the different soil layers of a given profile. Plant-available water capacities of soils are required as inputs for nearly all crop simulation models. Water-holding capacity is directly related to soil structure and texture. The rate and direction of water flow through the soil is an important factor determining the effect of farming practices on soil quality (Sahrawat *et al.*, Chapter 4, this volume).

Management of the soil can have significant effects on its water-holding capacity by changing the depth and texture of surface layers (through soil erosion), the structure and compactness of surface and subsurface layers, and by affecting the rate of infiltration of rainfall. Ritchie (1981) discussed the importance of water available to plants and the techniques for measuring plant-available water in soils. Plant-available water capacities are determined at the depth of rooting considering temporal changes in plant-available water capacities during the growing season. The water-holding capacity of soil is estimated by the difference in water content at field capacity and wilting point of soil. Both these parameters can be measured in the laboratory or field using methods described by Singh and Vittal (1997).

Soil loss

Soil erosion has an overriding influence on soil characteristics that determine soil quality for productivity and environment-related functions. Eroded sediments usually contain higher amounts of plant nutrients than do bulk soils, thus soil erosion depletes the soil of nitrogen (N), phosphorus (P), potassium (K), and total organic carbon (C) reserves (Barrows and Kilmer, 1963; Young *et al.*, 1985; Lal, 2004). Erosion can also bring subsoil horizons closer to the surface of the soil profile. These horizons might have different pH, low available water-holding capacities, and high bulk densities and can thus influence soil quality. For example using the productivity index model, Pierce *et al.* (1983) and Larson *et al.* (1985) determined which of the four soil attributes in the subsoil – available water-holding capacity, bulk density, pH, or rooting depth – would cause the greatest decline in soil productivity on 75 major soils of the Corn Belt of the USA, assuming that erosion removed 50 cm (20 inches) of soil from the surface. Of the 75 soils tested, the productivity index decreased significantly in 37. This was associated with a significant degradation in the available water-holding capacity in the subsoil (13 soils), increased bulk density (4 soils), decreased rooting depth (7 soils), and increased bulk density combined with decreased rooting depth (13 soils) (Larson *et al.*, 1985).

Soil erosion removes organic carbon along with sediments. Since organic carbon content is an important indicator of soil quality, it is suggested that current rates of erosion may have significant effects on long-term soil quality (National Research Council, 1993).

Soil loss can be measured using suitable sediment samplers (Pathak *et al.*, 2002). Soil loss is commonly estimated using equations such as the Universal Soil Loss Equation (USLE) or the Water Erosion Prediction Project (WEPP) model. These models require data on soil properties, slope, erosion control practices in use, vegetative cover, rainfall and other climatic parameters.

The effects of management practices on soil loss can also be measured using field experiments. For example, El-Swaify *et al.* (1985) reported a long-term watershed experiment on Vertisols at ICRISAT, Patancheru, soil loss from erosion from watersheds under improved and traditional systems (Table 3.3).

The annual soil loss gives a good indication of the long-term effect of soil erosion on the productive capacity of soils. It is also useful in determining off-site sediment damages and the effectiveness of conservation technologies. The changes in soil physical quality indicators reported in Table 3.2 took place only in the long term. They might have been partly due to a differential loss of soil under improved and traditional management practices (Table 3.3).

Table 3.3. The effect of management practices on runoff and soil loss in watersheds at ICRISAT, Patancheru, India 1974–82.

	Improved			Traditional		
Year	Rainfall (mm)	Runoff (% of seasonal rainfall)	Soil loss (t/ha)	Rainfall (mm)	Runoff (% of seasonal rainfall)	Soil loss (t/ha)
1974	811	14.3	1.30	811	27.5	6.60
1975	1041	15.6	1.39	1055	24.0	5.21
1976	687	10.6	0.98	710	33.3	9.20
1977	585	0.2	0.07	586	9.0	1.68
1978	1125	24.3	2.93	1117	36.7	9.69
1979	690	10.6	0.70	682	29.6	9.47
1980	730	15.9	0.97	688	24.1	4.58
1981	1126	29.5	5.04	1126	38.6	11.01
1982	615	1.6	0.20	615	3.3	0.70
Mean	823	13.6	**1.51**	821	25.1	6.46

Source: Adapted from El-Swaify *et al.*, 1985

Changes in chemical quality indicators

The objective of using appropriate chemical indicators is to sustain agricultural productivity without adversely affecting soil quality. Chemical quality indicators used to monitor soil quality include organic matter, cation exchange capacity (CEC), soil acidity and exchangeable bases, soil salinity and sodicity, total and available P, total exchangeable and non-exchangeable K, total and available sulphur (S) and soil reserves of total and available micronutrients (Table 3.1).

In the following section, examples are given of the use of some chemical indicators for monitoring soil quality in crop production systems.

Organic carbon

Organic matter is an important component of soil and consists of organic C and total N. Generally, organic C constitutes 58% of soil organic matter, and is used as a measure of when to convert organic C to organic matter in soils.

Organic matter plays a critical role in maintaining physical, chemical and biological integrity of soils. Total organic C is measured using wet digestion or combustion methods in the laboratory. The dynamics of soil organic matter are controlled by management practices and agroclimatic factors, especially rainfall, temperature and soil-water regime. The maintenance of organic matter status in soils, especially in arable production systems in tropical regions, is a challenge. In contrast, it is relatively easy to maintain organic matter levels in wetland rice soils because compared to arable systems, organic matter preferentially accumulates in soils under wetland paddy culture (Jenny and Raychaudhari, 1960; Sahrawat, 2004). Although the decomposition of organic matter is fast in tropical conditions, the primary productivity of wetlands is much higher than that of arable soils; this, combined with several other factors, results in a preferential accumulation of organic matter in wetlands (see Sahrawat, 2004).

Soil organic C influences the physical, chemical and biological characteristics of soil which directly or indirectly influence crop productivity. Compared to soils in the temperate or humid tropics, soils in SAT regions have relatively low contents of soil organic matter. The traditional farming practices followed by farmers in the dryland areas do not maintain sufficient soil organic matter content (El-Swaify *et al.*, 1985). The changes in soil organic C that can be measured accurately take a long time to occur, and depend on the determination methods used and their precision. However, the changes in soil organic C can be an important soil quality indicator for evaluating the impact of management practices in both agricultural and forest lands.

The long-term effects of improved and traditional management on soil chemical and biological properties of Vertisols are shown in Table 3.4. Soil organic C, total N and available N, P, and K, microbial biomass C and N were higher in the improved than in the traditional system (Wani *et al.*, 2003).

Total nitrogen

Like organic matter, total N is an important indicator for soil chemical quality. Total N with organic C constitutes soil organic matter. Total N consists of organic and inorganic N; organic N is the source of N supply to growing plants. Total soil N is commonly measured in the laboratory using digestion but combustion methods can also be used for its determination.

A long-term (1985–97) experiment on a Vertisol at ICRISAT studied the effects of introducing different legumes into cropping systems and their rotation to improve system productivity through the supply of N by legumes. The total soil N concentration in the 0–15 cm layer increased by 125 μg N/g of soil in 12 years in pigeonpea-based systems that had no input of N (Rego and Rao, 2000). In the traditional (rainy-season fallow, postrainy-season sorghum) and non-legume based system, the total soil N declined compared to the baseline.

Cereal-N requirements are large and an increase in N use efficiency is highly desirable, not only for economic considerations, but improved N-use efficiency also reduces chances of surface and groundwater resources

Table 3.4. Biological and chemical properties of semi-arid tropical Vertisols in 1998 after 24 years of cropping under improved and traditional systems in catchments at ICRISAT, Patancheru, India.

Properties	System	Soil depth (cm) 0–60	Soil depth (cm) 60–120	Standard error
Organic carbon (t C/ha)	Improved	27.4	19.4	0.89
	Traditional	21.4	18.1	
Soil respiration (kg C/ha)	Improved	723	342	7.8
	Traditional	260	98	
Microbial biomass C (kg C/ha)	Improved	2676	2137	48.0
	Traditional	1462	1088	
Microbial biomass N (kg N/ha)	Improved	86.4	39.2	2.3
	Traditional	42.1	25.8	
Non-microbial organic N (kg N/ha)	Improved	2569	1879	156.9
	Traditional	2218	1832	
Total N (kg N/ha)	Improved	2684	1928	156.6
	Traditional	2276	1884	
Mineral N (kg N/ha)	Improved	28.2	10.3	2.88
	Traditional	15.4	26.0	
Net N mineralisation (kg N/ha)	Improved	-3.3	-6.3	4.22
	Traditional	32.6	15.4	
Olsen P (kg P/ha)	Improved	6.1	1.6	0.36
	Traditional	1.5	1.0	

Source: Wani *et al.*, 2003

pollution with N (Sahrawat *et al.*, Chapter 4, this volume). Hence, N-use efficiency should be considered an important soil and water quality indicator for monitoring the biophysical impacts of NRM.

Changes in available soil nutrient reserves

In addition to the use of organic C and N as chemical quality indicators, several other soil attributes are used for soil quality for agricultural and environment-related functions. These include changes in CEC and total and extractable nutrient status with regard to major (N, P, and K), secondary (calcium (Ca), magnesium (Mg), and sulphur (S)) and micronutrients (iron (Fe), zinc (Zn), manganese (Mn), boron (B), and molybdenum (Mo)).

Nutrient balances in production-systems can also be effectively used to ascertain the sustainability of the systems. Soils have a nutrient reserve controlled by their inherent fertility and management. A negative balance of such nutrients as N, P and K indicates nutrient mining and non-sustainability of the production systems.

Diffuse reflectance spectroscopy

Assessments of soil attributes normally rely on laboratory data resulting from the analysis of large numbers of samples required to adequately characterise spatial variability beyond the plot scale. Methods for rapid estimation of soil properties are needed for quantitative assessment of soil quality parameters. Shepherd and Walsh (2002) developed a promising approach that estimates several soil properties simultaneously, directly from diffuse reflectance spectra in rapid non-destructive ways. The method is based on scanning air-dried samples using a portable spectrometer (0.35–2.5 μm wavelength) with an artificial light source. Soil properties are calibrated to reflectance using multivariate adaptive regression splines and screening tests are developed for various soil fertility constraints using classification trees. At random, one-third of the soil samples are used for validation purposes (using standard and the proposed methods). Using this technique from about 3000 African soils belonging to nine orders, Shepherd and Walsh found that the soil attributes could be calibrated directly to soil reflectance spectra with validation R^2 values ranging from 0.70 to 0.88, indicating good agreement between the values obtained by their proposed method and standard laboratory methods. The soil attributes calibrated included: exchangeable Ca; effective cation-exchange capacity (ECEC); exchangeable Mg; organic C concentration; clay content; sand content and soil pH.

The spectral technique provides a tool for generating results of soil assessments that are conducted at a limited number of sites and thereby increase the efficiency of expensive and time-consuming soil-related studies. The rapid nature of the measurement allows soil variability to be more adequately sampled than by the conventional approach.

The spectral library approach of Shepherd and Walsh (2002) provides a coherent framework for linking soil information with remote sensing information for improved spatial prediction of soil functional capacity for agricultural, environmental, and engineering applications. Indeed, as shown below, Sanchez *et al.* (2003) found this approach useful for fertility capability classification (FCC) when assessing soil quality.

Changes in biological quality indicators

The dynamic nature of soil microorganisms makes them sensitive indicators of the soil processes leading to changes in soil quality. Biological indicators based on microbial composition, number and processes provide advanced indication of subtle changes in soil quality. However, changes in soil physical and chemical properties alter the soil environment that supports the growth of the microbial population (Lee and Pankhurst, 1992; Stott *et al.*, 1999).

Total number of microorganisms

Total microbial counts can be used as a good indicator to assess the impact of a particular management treatment on soil biological activity. The microbial population is enumerated by microscopy. Microorganisms are extracted from soil and transferred to an optically suitable background before enumeration. Several studies have recorded increase in microbial numbers in soils soon after adding an available C source (Jenkinson and Ladd, 1981).

Soil respiration

Soil respiration is the oxidation of organic materials by soil microorganisms that generates energy for microbial growth and maintenance, and produces carbon dioxide (CO_2). The soil respiration rate provides a comprehensive picture of total soil biological activity. Soil respiration is measured by determining the amount of CO_2 evolved under well-defined conditions during a given time period. Soil respiration rates were found to be higher in Vertisols under an improved than in a traditionally managed system (Table 3.4).

Microbial biomass carbon (C) and nitrogen (N)

Microbial biomass C and N in soils represent a readily available source of plant nutrients. Because the decay and turnover of microbial biomass in soils is rapid it results in the release of CO_2–C and available N. Thus, measurement of microbial biomass C and N provide a dynamic indicator of soil quality which by accurate standardisation can also be used to measure the extent of soil degradation. Microbial biomass C and N are measured as the net release of C and mineral N (ammonium plus nitrate) that results from fumigation of soil samples (Jenkinson and Ladd, 1981).

Soil and water conservation, tillage, and cropping systems influence microbial biomass C and N (Table 3.4). It has been suggested that soils with a higher proportion of soil organic C as microbial biomass gain C; those with a lower proportion lose C (Anderson and Domsch, 1986).

In a long-term (24-years) experiment on Vertisols, microbial biomass C was about 10.3% of the total soil organic C in the improved system compared to only 6.4% in the traditional system. Improved Vertisols management practices resulted in higher values (10.3 vs. 6.4%) of biomass C as a proportion of soil organic C to 120-cm soil depth, indicating that with improved management these Vertisols would reach a new C-storage equilibrium. The microbial N was 2.6% of the total biomass N in the improved system and 1.6% in the traditional system (Wani *et al.*, 2003).

Potentially mineralisable nitrogen

Along with microbial biomass N, potentially mineralisable N serves as a surrogate for the 'active N fraction' for soil quality impact assessment. The measurement of potentially mineralisable N in soils is based on the net release of mineral N (ammonium plus nitrate) from soil samples incubated for a given period under well-defined moisture and temperature conditions. Cropping systems and inputs of organic matter affect potentially mineralisable N (Wani *et al.*, 1994).

Earthworm activity

Changes in earthworm populations can significantly affect soils by influencing soil structure, nutrient cycling dynamics, and soil microbial populations.

The earthworm population decreases as soil degradation increases, and this can serve as a very sensitive indicator of soil degradation (Tian *et al.*, 2000). The earthworm population can be measured by earthworm counts per soil volume (e.g. number/m^3 of soil) in the cultivated layer.

Integrated Soil Quality Indicators

Whilst there may be doubts about the efficacy of developing integrated indices of soil quality, there is a continuing demand for them, given the complex nature of the soil and the exceptionally large number of soil properties that need to be determined. At the Rodale International Conference on the Assessment and Monitoring of Soil Quality, there was a general consensus that soil quality (Rodale Institute, 1991, cited in Arshad and Martin, 2002) encompasses three broad issues:

1. The ability of the soil to enhance crop production (productivity component)
2. The ability of the soil to function in attenuation of environmental contaminants, pathogens, and off-site damage (environment component)
3. The linkage between soil quality and plant, animal and human health (health component).

It has, therefore, been suggested that any protocol designed to determine soil quality must provide an assessment of the function of soil with regard to these three issues. To do this effectively, soil quality assessment must incorporate specific performance criteria for each of the three elements listed above, and it must be structured in such a way as to allow for quantitative evaluation and unambiguous interpretation using one aggregate soil quality index (that incorporates the above three soil functions). The objective of the proposed approach is in defining a single integrated soil quality index and not to replace past research on specific indicators but to complement it by presenting a more clearly defined framework for the development of mathematical relationships driven by basic soil attributes (Doran and Parkin, 1994).

Soil quality indices

Since soil quality encompasses plant and biological productivity, environmental quality, and human and animal health, it is imperative that the soil quality indicator provides an assessment of these functions. To achieve this objective effectively, the soil quality indicator must incorporate specific performance criteria for each function. This concept gave birth to an index.

Parr *et al.* (1992) proposed a soil quality index (*SQ*) as follows:

$$SQ = f\,(SP, Q, E, H, ER, BD, FQ, MI) \quad (1)$$

where *SP* are the soil properties, *Q* the potential productivity, *E* the environmental factors, *H* the health (human/animal), *ER* the erodibility, *BD* the biological diversity, *FQ* the food quality/safety, and *MI* the management inputs.

There has been some effort to define the exact mathematical form of the generic functional form given in Equation 1. Subsequent to the Rodale Conference, many soil scientists have proposed more detailed procedures for evaluating soil quality functions by combining and integrating specific soil quality elements into soil quality indices (Doran and Parkin, 1994; Karlen and Stott, 1994). These procedures allow for weighting of various soil quality elements, depending upon the user goals, site-specific considerations and socio-economic concerns. For example, Doran and Parkin (1994) proposed the following index of soil quality as a function of six specific soil quality elements:

$$SQ = f(SQ_{E1}, SQ_{E2}, SQ_{E3}, SQ_{E4}, SQ_{E5}, SQ_{E6}) \quad (2)$$

where the specific soil quality elements (SQ_E) are defined as:
SQ_{E1} = food and fibre production
SQ_{E2} = erosivity
SQ_{E3} = groundwater quality
SQ_{E4} = surface water quality
SQ_{E5} = air quality
SQ_{E6} = food quality

The advantage of this approach is that the different functions of soil can be assessed by specific performance criteria established for each element for a given ecosystem: for example, yield goals for crop production (SQ_{E1}); limits for erosion losses (SQ_{E2}); concentration limits for chemical leaching from the rooting zone (SQ_{E3}); nutrient, chemical, and sediment loading limits to adjacent surface water systems (SQ_{E4}); production and uptake rates for trace gases that contribute to ozone (O_3) destruction or the greenhouse effect (SQ_{E5}); and nutritional composition and chemical residue of food (SQ_{E6}).

One suggestion to operationalise this aggregate index is to use a weighted simple multiplicative function:

$$SQ = (K_1SQ_{E1})\,(K_2SQ_{E2})\,(K_3SQ_{E3})\,(K_4SQ_{E4})\,(K_5SQ_{E5})\,(K_6SQ_{E6}) \quad (3)$$

where K_i = weighting coefficients for the different soil quality parameters.

There could be several ways to develop an aggregate index from a set of different soil quality indicators. Campbell *et al.* (2003) propose various approaches including simple additive indices, principal components methods, canonical correlations and simple radar diagrams for evaluating the performance of NRM interventions. For example, to develop a simple additive index, it is necessary to know the maximum and minimum values of each indicator. A standardised value for each indicator is then calculated using the formula: (Indicator value at time x – minimum) / (maximum – minimum). For each indicator the potential values range from 0 (least desirable) to 1 (most desirable). A composite index is then calculated as the average of the indicator values. Weights can also be added if the relative importance of the different performance indicators is known. Details of the advantages and disadvantages of the other approaches can be found in Campbell *et al.* (2003).

Although the proposed indices would seem promising since they integrate several soil attributes in a single index, there are no published reports on their practical application and evaluation in the field. If this approach is going to be useful for NRM impact assessment, further research on the different ways of developing a comprehensive indicator would need to be carried out. When multiple variables are measured to characterise soil quality, it may not be easy to reduce the various indicators into a single and meaningful index.

Fertility capability classification (FCC) approach

Sanchez *et al.* (2003) stated that the soil quality paradigm that was originally developed in the temperate region is not very suitable for the tropics. According to them, soil quality in the tropics should focus on three main concerns: food insecurity, rural poverty and ecosystem degradation. Soil quality in the tropics must be considered a component of the integrated natural resources management (INRM) framework, therefore Sanchez *et al.* (2003) suggested that based on quantitative topsoil attributes and soil taxonomy, the fertility capability soil classification (FCC) system is probably a good starting point for measuring soil quality in the tropics. To overcome certain limitations, they proposed a new FCC version 4 (Sanchez *et al.*, 2003).

The proposed system consists of two categories. The first – type/substrata – describes topsoil and subsoil texture. The second – condition modifier – consists of 17 modifiers defined to delimit specific soil conditions affecting plant growth with quantitative limits. The type/substrata types and condition constitute soil attribute modifiers in terms of their capability for plant growth. The 17 condition modifiers include: soil drought stress (dry); nutrient capital reserves; erosion risk; aluminium toxicity; major chemical limitations; P fixation; waterlogging; leaching potential; calcareous reaction; cracking clays; gravel; shallow depth; salinity; alkalinity; presence of amorphous materials; volcanic soils; high organic content; and sulphidic soils. Like other soil indices, the FCC approach can be used to evaluate and monitor soil quality for soil productivity and sustainability purposes by measuring the FCC index at regular intervals starting with baseline measurement.

There are several important issues not addressed by this new version of FCC. These include nutrient depletion, soil compaction, surface sealing, surface crusting and others related to air and water flow. The FCC-based soil index is still at an initial stage and many more details still need to be worked out before it can be used.

Models to Assess Soil Quality

The development of relationships between soil attributes and the functions of soils is a monumental task. Simulation models can be useful tools in tracking and understanding these relationships. Algorithms in existing simulation models [e.g. the Nitrate Leaching and Economic Analysis Package (NLEAP),

Erosion-Productivity Impact Calculator (EPIC), Chemical, Runoff and Erosion from Agricultural Management Systems (CREAMS), and Water Erosion Prediction Project (WEPP)] may provide a useful starting point (Doran and Parkin, 1994). These models provide a predictive tool for the process such that, given what is known, if one of the parameters that affect the process changes, the associated change in a given indicator can be predicted (Arshad and Martin, 2002). Models are normally constructed using results of detailed long-term data. Because agroclimatic conditions often vary from year to year, reliable long-term data is essential to capture the historical reality and predict future events with some degree of confidence. By using soil process models, the rate of change and the direction of change in selected soil quality indicators can be predicted. Models allow the researcher to simulate various management practices in order to predict their consequences and impacts on biophysical soil conditions and on such economic outputs as grain yield. Wani *et al.* (Chapter 5, this volume) discuss the use of simulation modelling to predict the likely impacts of NRM on various soil quality indicators.

One of the major limitations in using these models is that most of them require calibration and testing before they can be used in a given region. To the extent that the impacts of NRM interventions tend to be location-specific, lack of data from a given location can become a major limiting factor in validating the models to local conditions.

Summary and Conclusions

The intensification of agricultural activities to meet the increasing demands from fast-growing populations, particularly in the developing countries, without sufficient investments to sustain the system has led to rapid soil degradation. There is also increasing conflict among the various agricultural and environmental functions of soils. Various NRM interventions have been designed to counter the process of degradation or to enhance the sustainability of the system. In order to enhance the effectiveness of these interventions, and to attain the desired objectives, suitable indicators are required to monitor the biophysical impacts of management practices on soil conditions. This calls for the development of threshold levels for the various indicators, as these values are likely to vary by ecoregion and soil type.

In practical terms, it is not feasible to recommend the use of a large number or a common set of indicators for all agricultural interventions because of the varying size and complexity of agricultural and watershed development projects. Therefore, the selection of a few relevant indicators based on the purpose and an adequate understanding of various processes at the local level is extremely important. However, there is a general consensus that any assessment of soil quality must include a minimum set of physical, chemical and biological soil parameters. In this context the importance of a baseline characterisation of soils and sites to measure the changes attributable to a given management intervention cannot be overemphasised.

Review of the available literature and empirical examples has indicated that in general, biological indicators, followed by chemical and physical indicators could be successfully used to monitor the impact of soil management options. Changes in such physical indicators as texture, infiltration, moisture holding capacity, bulk density, porosity, aggregate stability, surface crusting and sealing, soil compaction and penetration resistance take considerable time. However, depending on the magnitude of the change, simple physical measurements such as runoff and soil loss can serve as supplementary indicators of changes in soil quality. Such soil chemical indicators as pH, salinity, organic C, organic matter content, CEC, status of plant nutrients, and concentration of potentially toxic elements can also provide good indicators. Changes in soil organic C, CEC, or soil pH or the build up of toxic elements require a long time span and cannot be monitored during a short NRM intervention. Amongst the biological indicators soil respiration, microbial biomass C and N mineralisation are commonly used to monitor changes in soil quality.

Recently, Lal (2004) reviewed the progress made in identifying soil quality indicators, especially those that are relevant to the developing countries. The key soil quality indicators listed in Table 3.5 have been proposed and evaluated for universally monitoring soil quality, although the rate of change in these parameters and threshold values varies between soils in tropical (mostly developing) and temperate (mostly industrialised) countries (Lal, 2004). Obviously, since the purpose of NRM interventions is to enhance or sustain productivity, there is a need to relate the soil quality indicators to agricultural productivity and sustainability indicators.

Table 3.5. Recent developments in identifying physical, chemical and biological indicators of soil quality.

Indicator	Associated soil characteristics/properties
Minimum data set	Aggregate stability, clay content, bulk density, soil organic C content, pH, total N, available P, S, micronutrients, mineralisable N and microbial biomass C and N
Soil N	N use efficiency and INM (integrated nutrient management)
Soil P	Environmental threshold levels of soil P
Soil K	Threshold K values, positive K balance
Cations and acidity	Critical pH and cations (K, Ca, Mg and Na), Al and Mn toxicity
Soil organic matter status	Key indicator of soil quality and environment moderation
Subsoil compaction	Soil strength
Soil structure	Critical values of soil organic C concentration
Erosion	Soil organic C, effective rooting depth, available water capacity
Soil biological quality	Microbial biomass and activity, earthworm and termite biomass

Source: Adapted from Lal (2004)

In the tropics lack of sufficient data often impedes a full assessment of the overall impact of NRM interventions on soil quality. Recently the use of an FCC, based on topsoil quantitative attributes and soil taxonomy, has been proposed. Version 4 of the FCC provides an alternative to qualitative approaches for assessing soil quality. Moreover, the development of such new tools as reflectance spectroscopy to predict soil functional attributes provides techniques for rapid measurement of soil characteristics. Simulation modelling and geographic information systems (GIS) can be useful tools for assessing the impact of NRM interventions on soil quality and the trade-offs between returns and environmental quality, especially when long-term and costly experimentation is not feasible.

References

Anderson, T.H. and Domsch, K.H. (1986) Carbon link between microbial biomass and soil organic matter. In: Megusar, F. and Gantar, M. (eds) Proceedings of Fourth International Symposium on Microbial Ecology, 24–29 August 1986, Ljubljana, Yugoslavia. *Slovene Society for Microbiology*, pp. 467–471.

Arshad, M.A. and Coen, G.M. (1992) Characterization of soil quality: physical and chemical criteria. *American Journal of Alternative Agriculture* 7, 5–12.

Arshad, M.A. and Martin, S. (2002) Identifying critical limits for soil quality indicators in agro-ecosystems. *Agriculture, Ecosystems and Environment* 88, 153–160.

Barrows, H.L. and Kilmer, V.J. (1963) Plant nutrient losses from soils by water erosion. *Advances in Agronomy* 15, 303–316.

Campbell, B.M., Sayer, J.A., Frost, P., Vermeulen, S., Perez, M.R., Cunningham, A. and Prabhu, R. (2003) Assessing the performance of natural resource systems. In: Campbell, B.M. and Sayer, J.A. (eds) *Integrated Natural Resource Management: Linking Productivity, the Environment and Development*. CAB International, Wallingford, UK, pp. 267–292.

Doran, J.W. and Parkin, T.B. (1994) Defining and assessing soil quality. In: Doran, J.W., Coleman, D.C., Bezdicek, D.F. and Stewart, B.A. (eds) *Defining Soil Quality for a Sustainable Environment*. Soil Science Society of America (Special Publication 35), pp. 3–21.

El-Swaify, S.A., Pathak, P., Rego, T.J. and Singh, S. (1985) Soil management for optimized productivity under rainfed conditions in the semi-arid tropics. *Advances in Soil Science* 1, 1–63.

Jenkinson, D.S. and Ladd, J.N. (1981) Microbial biomass in soil: measurement and turnover. In: Paul, E.A. and Ladd, J.N. (eds) *Soil Biochemistry*, Vol. 5. Marcel Dekker, New York, pp. 415–471.

Jenny, H. and Raychaudhari, S.P. (1960) *Effect of Climate and Cultivation on Nitrogen and Organic Matter Reserves in Indian Soils*. Indian Council of Agricultural Research, New Delhi, India, 127 pp.

Karlen, D.L. and Stott, D.E. (1994) A framework for evaluating physical and chemical indicators of soil quality. In: Doran, J.W., Coleman, D.C., Bezdicek, D.F. and Stewart, B.A. (eds) *Defining Soil Quality for a Sustainable Environment*. Soil Science Society of America (Special Publication 35), pp. 53–72.

Karlen, D.L., Mausbach, M.J., Doran, J.W., Cline, R.G., Harris, R.F. and Schuman, G.E. (1997) Soil quality: a concept, definition and framework for evaluation. *Soil Science Society of America Journal* 61, 4–10.

Karlen, D.L., Ditzler, C.A. and Andrews, S.S. (2003) Soil quality: why and how? *Geoderma* 114, 145–156.
Klaij, M.C. (1983) Analysis and evaluation of tillage on Alfisol in a semi-arid tropical region of India. Ph.D. Thesis, University of Wageningen, Wageningen, The Netherlands, 178 pp.
Lal, R. (2004) Soil quality indicators in industrialized and developing countries – similarities and differences. In: Schjønning, P., Elmholt, S. and Christensen, B.T (eds) *Managing Soil Quality: Challenges in Modern Agriculture.* CAB International, Wallingford, UK, pp. 297–313.
Larson, W.E., Pierce, F.J. and Dowdy, R.H. (1985) Loss in long-term productivity from soil erosion in the United States. In: El-Swaify, S.A., Moldenhauer, W.C. and Lo, A. (eds) *Soil Erosion and Conservation.* Soil Conservation Society of America, Ankeny, Iowa, pp. 262–217.
Lee, K.E. and Pankhurst, C.E. (1992) Soil organisms and sustainable productivity. *Australian Journal of Soil Research* 30, 855–892.
Letey, J., Sojka, R.E., Upchurch, D.R., Cassel, D.K., Olsen, K.R., Payne, W.A., Petrie, S.E., Price, G.H., Reginato, R.J., Scott, H.D., Smethurst, P.J. and Triplett, G.B. (2003) Deficiencies in the soil quality concept and its application. *Journal of Soil and Water Conservation* 58, 180–187.
National Research Council (United States) Committee on Long-Range Soil and Water Conservation Board on Agriculture (1993) *Soil and Water Quality: an Agenda for Agriculture.* National Academy Press, Washington, DC, 516 pp.
Parr, J.F., Papendick, R.I., Hornick, S.B. and Meyer, R.E. (1992) Soil quality: attributes and relationship to alternative and sustainable agriculture. *American Journal of Alternative Agriculture* 7, 5–10.
Pathak, P., Wani, S.P., Singh, P., Sudi, R. and Srinivasa Rao, Ch. (2002) *Hydrological Characterization of Benchmark Agricultural Watersheds in India, Thailand, and Vietnam.* Global Theme 3. Report no. 2. International Crops Research Institute for the Semi-Arid Tropics, Patancheru, India, 52 pp.
Pierce, F.J., Larson, W.E., Dowdy, R.H. and Graham, A.P. (1983) Productivity of soils: Assessing long-term changes due to erosion. *Journal of Soil and Water Conservation* 38, 39–44.
Rego, T.J. and Rao, V.N. (2000) Long-term effects of grain legumes on rainy-season sorghum productivity in a semi-arid tropical Vertisol. *Experimental Agriculture* 36, 205–221.
Ritchie, J.T. (1981) Soil water availability. *Plant and Soil* 58, 327–338.
Sahrawat, K.L. (2004) Organic matter accumulation in submerged soils. *Advances in Agronomy* 81, 169–201.
Sanchez, P.A., Palm, C.A. and Buol, S.W. (2003) Fertility capability soil classification: a tool to help assess soil quality in the tropics. *Geoderma* 114, 157–185.
Shepherd, K.D. and Walsh, M.G. (2002) Development of reflectance spectral libraries for characterization of soil properties. *Soil Science Society of America Journal* 66, 988–998.
Singh, P. and Vittal, K.P.R. (1997) Soil moisture measurement. In: Laryea, K.B., Pathak, P. and Katyal, J.C. (eds) *Measuring Soil Processes in Agricultural Research.* Technical Manual no. 3. International Crops Research Institute for the Semi-Arid Tropics (ICRISAT), Patancheru, India, pp. 65–82.
Sojka, R.E., Upchurch, D.R. and Borlaug, N.E. (2003) Quality soil management or soil quality management: performance versus semantics. *Advances in Agronomy* 79, 1–68.
Steer, A. (1998) Making development sustainable. *Advances in Geoecology* 31, 857–865.

Stephen, N. (2002) Standardisation of soil quality attributes. *Agriculture, Ecosystems and Environment* 88, 161–168.

Stott, D.E., Kennedy, A.C. and Cambardella, C.A. (1999) Impact of soil organisms and organic matter on soil structure. In: Lal, R. (ed.) *Soil Quality and Soil Erosion*. Soil and Water Conservation Society, CRC Press, Boca Raton, Florida, pp. 57–74.

Taylor, H.M. and Terrell, E.E. (1982) Rooting pattern and plant productivity. In: Rechcigl, Jr., M. (ed.) *CRC Handbook of Agricultural Productivity*, Vol. 1. CRC Press, Boca Raton, Florida, pp. 185–200.

Tian, G., Olimah, O.A., Adeoye, G.O. and Kang, B.T. (2000) Regeneration of earthworm population in a degraded soil by natural and planted fallows under humid tropical conditions. *Soil Science Society of America Journal* 64, 222–228.

Turco, R.F., Kennedy, A.C. and Jawson, M.D. (1994) Microbial indictors of soil quality. In: Doran, J.W., Coleman, D.C., Bezdicek, D.F. and Stewart, B.A. (eds) *Defining Soil Quality for a Sustainable Environment*. Soil Science Society of America (SSSA) Special Publication no. 35. SSSA and American Society of Agronomy, Madison, Wisconsin, pp. 73–90.

Wani, S.P., McGill, W.B., Haugen-Kozyra, K. and Juma, N.G. (1994) Increased proportion of active soil N in Breton loam under cropping systems with forages and green manures. *Canadian Journal of Soil Science* 74, 67–74.

Wani, S.P., Pathak, P., Jangawad, L.S., Eswaran, H. and Singh, P. (2003) Improved management of Vertisols in the semi-arid tropics for increased productivity and soil carbon sequestration. *Soil Use and Management* 19, 217–222.

Young, C.E., Crowder, B.M., Shortle, J.S. and Alwang, J.R. (1985) Nutrient management on dairy farms in southeastern Pennsylvania. *Journal of Soil and Water Conservation* 40, 443–445.

4 Measurable Biophysical Indicators for Impact Assessment: Changes in Water Availability and Quality

K.L. Sahrawat, K.V. Padmaja, P. Pathak and S.P. Wani

International Crops Research Institute for the Semi-Arid Tropics (ICRISAT) Patancheru, India

Introduction

It is estimated that 94% of global water is in oceans and seas and that freshwater accounts for a mere 6% of the total volume. Freshwater is a scarce resource in many regions of the world, particularly in arid and semi-arid areas and during dry seasons in many regions that may otherwise have a surplus during wet seasons. Global freshwater availability is not a limiting factor but it is increasingly becoming a development constraint in regions with low rainfall, and in places where it is not easily accessible for human use. Thus, maintaining high quality freshwater resources is important to human, domestic livestock, and wildlife health (van der Leeden *et al.*, 1990).

Increased population and demand for food, floriculture, livestock, feed and fibre production is leading to over exploitation of freshwater in areas with limited renewable supplies. It is estimated that irrigation accounts for about 72% of global and 90% of developing-country water withdrawal (Cai and Rosegrant, 2003). In the dry areas (e.g. in West Asia and North Africa), agricultural use accounts for about 80% of the total consumption of water (Oweis and Hachum, 2003). Population growth is also leading to increased demand for freshwater for other competing uses such as domestic, agricultural, industrial and recreational activities. Agricultural activities could have adverse effects on both the quantity and quality of surface and groundwaters. Excessive and over-exploitation of groundwater is resulting in the depletion of water resources. Groundwater resources are heavily exploited for agriculture, particularly where they provide cheap water supplies that do not require large capital investments and/or do not incur high pumping costs.

The adverse effects of agricultural activities on surface and groundwater quality occur in both extensive and intensive agricultural production systems. In extensive agricultural systems, the quality of surface and groundwater is affected by the soil erosion associated with inappropriate management and over-exploitation of soil resources. Adverse effects on water quality can also occur when shifting cultivation or subsistence agriculture are practised on marginal or fragile lands, or on lands in ecologically sensitive regions. In the early phases of extensive agriculture, the use of chemical fertilisers was low and fallow periods were long, allowing soil fertility to recuperate. Such agricultural production systems also allowed soil to be conserved, and maintained its physical, chemical and biological integrity. Hence the effects on water quality were limited. Under intensive production systems, water resources become contaminated due to the increased intensity of fertiliser and pesticide use. The intensification of agricultural production systems based on high inputs of chemicals, especially in environmentally sensitive regions dominated by light-textured soils such as the porous soils of the Punjab in India, has led to nitrate contamination of surface and groundwater resources (Bajwa *et al.*, 1993).

Natural resource management (NRM) interventions can have substantial impacts on agricultural productivity and system sustainability. Similarly, agricultural and NRM practices can greatly impact water availability and quality. Assessing the impacts of agricultural and NRM interventions on water quantity and quality requires the development of appropriate indicators for measuring and monitoring such effects.

In this chapter the impact of agricultural and NRM practices on water quantity and quality are examined. The various biophysical indicators proposed to assess surface and groundwater quantity and quality impacts of agricultural and NRM interventions are discussed with examples drawn from recent literature and case studies from watersheds in the semi-arid tropics. Future research needs for developing more effective and measurable indicators of water quantity and quality for the purpose of monitoring the biophysical impacts of technological and resource management interventions are highlighted.

Agricultural Practices and Water Quantity

Water availability indicators

The water available for agricultural production includes soil moisture or water stored in the soil profile, surface water, and groundwater. Water stored in the soil profile is a function of rainfall quantity and intensity and its distribution, the storage capacity of the soil, bedrock contact, and water infiltration as influenced by ground slope and soil surface configuration and cover conditions. The available water in a watershed can be manipulated through harvesting excess rainwater and by directing the harvested water to storage in water tanks for future use.

NRM interventions can have impacts on water stored in the soil profile. For example, long-term experiments by the International Crops Research Institute for the Semi-Arid Tropics (ICRISAT) on Vertisols and Vertic Inceptisols on a watershed scale in India showed that a broadbed-and-furrow (BBF) land configuration compared to flat land treatment on average stored 40–50 mm more water in the soil profile and reduced runoff (from 45 to 25% of rainfall), soil loss (from 6.5 to 1.5 t/ha) and nitrate-N loss (from 15 to 10 kg/ha) (Singh *et al.*, 1999; Wani *et al.*, 2002, 2003). Similar results were also reported by Srivastava and Jangawad (1988) and Gupta and Sharma (1994) who showed that the BBF landform system compared to a flat land configuration reduced water runoff, soil loss and nitrate loss in runoff water during the rainy season on Vertisols and associated soils. Recent research on a watershed (500–1000 ha) scale in India has also shown that NRM interventions (the use of improved varieties along with soil fertility management and soil and conservation practices) reduced soil loss and increased groundwater recharge and storage in surface tanks (Wani *et al.*, 2002).

Various indicators can be used to monitor the changes in water availability that result from NRM interventions. The indicators commonly used to characterise surface and groundwater availability are summarised in Table 4.1. The indicators cover soil moisture, surface water flow, surface water availability and groundwater availability; each of them is discussed in the following sections.

Table 4.1. Selected indicators commonly used to characterise water availability.

Impact outcome	Indicator used	How measured
Soil moisture	Total water in soil profile Plant available water	Gravimetric method Moisture meters (neutron probes) Pressure membrane method
Surface water flow	Runoff volume	Stage level runoff recorder with hydraulic structure
Surface water	Number of water storage structures and their capacities	Through surveys and topographic maps
	Water levels in storage structures	Staff gauge readings Remote sensing
Groundwater	Water levels in open wells Water levels in tube wells and piezometers	Water level recorders' readings at regular intervals
	Water recovery rate after the pumping	Time in h or days to recover the water level
	Duration of water pumping	Pumping time in h or days

Indicators for available surface water

Available surface water constitutes water stored in water storage structures (introduced as part of an NRM intervention) such as tanks, check dams, ponds and streams. The indicators used to measure changes in surface water quantity on a watershed scale are based on the estimation of water available from tanks, check dams and streams together with their utilisation and seasonal and long-term trends (El-Ashry, 1991; Rao *et al.*, 1996). These indicators are, however, difficult to measure. To assess surface water quantity, it may therefore be useful to consider the use of such proxy indicators as:

- Total area irrigated from surface storage structures or reservoirs
- Number of reservoirs of different capacities
- Number of reservoirs that contain water at the middle and end of the cropping season
- Number and/or length of perennial rivers
- Duration of flows for ephemeral rivers.

The data required to measure the total available surface water in a watershed include the total water storage capacity of all water storage structures in the watershed, weekly or monthly observations on the quantity of available surface water, and its use. Long-term measurements are essential to develop trends of water availability that in turn are critical for the development of accurate surface water availability indicators (Hazell *et al.*, 2001).

Indicators for surface water outflow (runoff)

Surface water outflow (runoff) as an indicator is used to measure the extent of water outflow through runoff from a given hydrological unit (e.g. a watershed). The three runoff indicators commonly used are runoff depth, runoff volume, and peak runoff rate. They indicate runoff in terms of runoff water depth, runoff water volume, and the peak runoff water rate during a given rainfall event or averaged over the entire season. These indicators are useful in determining the effectiveness of various measures and/or watershed technologies in conserving water in a watershed (Farroukhi, 1995). The surface water outflow indicator provides a useful signal of the general quality of watershed management. Equally important, the three runoff indicators can also be used to assess the long-term effects of watershed management technologies on watershed hydrology (Pathak *et al.*, 2004). The loss of soil through soil erosion that has implications for short- and long-term agricultural productivity is also directly related to this measure of surface water loss.

Water runoff can be directly measured using a suitable runoff recorder (Pathak *et al.*, 2002), or by using runoff simulation models that incorporate data on soil, slope, vegetative cover, rainfall and other climatic parameters (Littleboy *et al.*, 1989; Pathak *et al.*, 1989; Rose, 2002). For example, in India in the Adarsha watershed, Kothapally, Andhra Pradesh, and Lalatora watershed, Madhya Pradesh, where ICRISAT is conducting on-farm trials for integrated community-based watershed management, runoff was used as an indicator to assess the impact of watershed management interventions in reducing water losses. The runoffs from treated and untreated sub-watersheds were measured and compared using digital runoff recorders. The results showed a significant reduction in runoff from the treated sub-watershed compared to that from the untreated sub-watershed. Results also showed that the peak runoff rates in treated and untreated watershed were similar, suggesting that the runoff volume is the main variable that changes between treated and untreated watersheds. During the 2000 rainy season, during which higher than the average rainfall was received, the runoff in the treated sub-watershed of Adarsha was 45% lower than that in the untreated sub-watershed. The same was true for Lalatora watershed in 1999. Even during years of low rainfall, the runoff in treated sub-watersheds was about 30% lower than that observed in the untreated counterpart. Results also showed that the peak runoff rates in treated and untreated watersheds were similar, suggesting that runoff volume is the main variable that changes with treatment (Table 4.2). These empirical results demonstrate how NRM interventions affect water availability and surface water flow. The difference in selected indicators between the two management regimes can be used to measure the impact of the new technologies on surface water flow.

Table 4.2. The impact of watershed management interventions on runoff and peak runoff rate at Kothapally and Lalatora watersheds (1999–2001) (ICRISAT, unpublished).

Location/ Year	Rainfall (mm)	Runoff[a] (mm)		Peak runoff rate (m^3/second per ha)	
		Untreated	Treated	Untreated	Treated
Kothapally					
1999	584	16	NR[b]	0.013	NR
2000	1161	118	65	0.235	0.230
2001	612	31	22	0.022	0.027
Lalatora					
1999	1203	296	224	0.218	0.065
2000	932	234	NR	0.019	NR
2001	1002	290	55	0.040	0.027

[a]Untreated = control, with no development work; treated = with improved soil, water, and crop management technologies.
[b]NR = not recorded.

Runoff depth, volume and peak runoff rate indictors are useful in measuring the effectiveness of improved soil and water conservation and other NRM technologies (Samra, 1998) and to determine whether or not additional interventions in the upstream parts of watersheds are needed. Such runoff indicators can be easily measured using recorders installed in a watershed. Pathak *et al.* (2002) used data on seasonal runoff and peak runoff rates to measure runoff from treated (with water harvesting structures) and untreated (without land treatment) sub-watersheds in Madhya Pradesh. The empirical results from runoff hydrograph measurements are shown in Fig. 4.1. For a period of 10 days (5–14 September 1999), the runoff from the

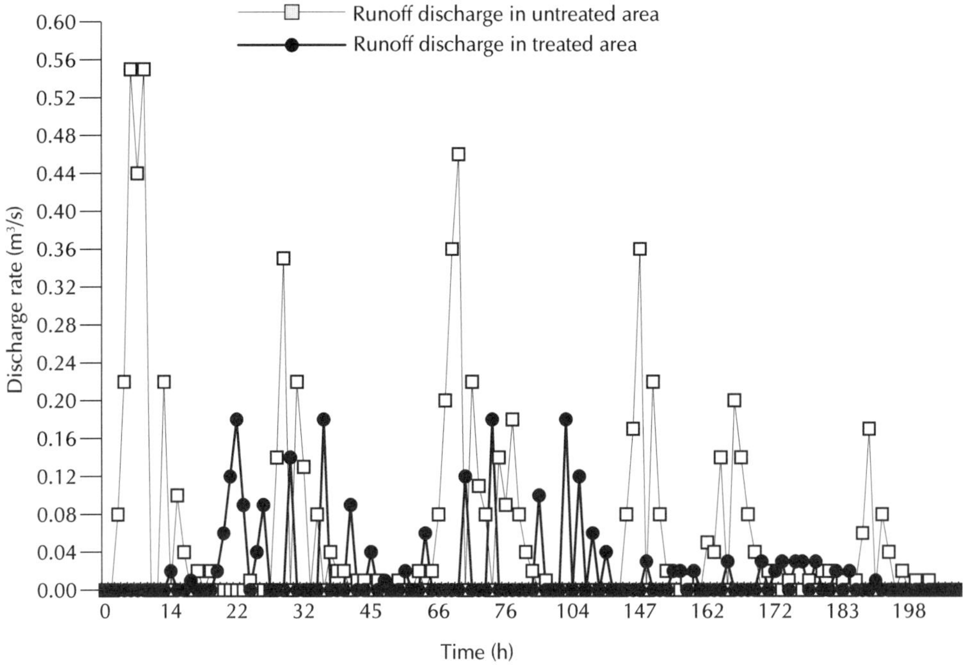

Fig. 4.1. The impact of integrated watershed management interventions on runoff as measured by a runoff hydrograph from untreated and treated sub-watersheds at Lalatora watershed, India, during 5–14 September 1999 (Pathak *et al.*, 2002).

treated sub-watershed was 130 mm compared to 150 mm in the untreated counterpart. Clearly, during the period under investigation the runoff discharge rate in the treated sub-watershed was lower than that in the untreated watershed. The majority of farmers from the treated sub-watershed reported that seasonal flooding (both frequency and the area affected by the floods) have significantly reduced. Their perception is that the construction of large check dams and other water-harvesting structures has helped to reduce flash floods. These results were influenced by the size of the sub-watersheds. This approach is designed for watersheds on a 500–1000 ha scale. However, results from this study show that treatment effects on water discharge rates are dynamic, even though they do not indicate whether the effects are sustainable.

Indicators for upstream and downstream temporary flood frequency and area affected

Flooding is caused by several factors. *In situ* flooding is caused by high rainfall on ground with low slope and soils with low infiltration (Vertisols) or with an impermeable layer (Planosols). Flooding in plains, known as induced waterlogging, is caused when a river bursts its banks or by flood irrigation. Main flooding indicators include the area affected, frequency, and duration of flooding; these indicators are important for decision-making and for assessing the impacts from NRM interventions.

Flooding indicators are used to characterise and measure the extent to which temporary or seasonal flooding upstream affects downstream parts (reaches) of streams and their tributaries. Apart from the human miseries and loss of property, seasonal flooding causes destruction of standing crops and loss of agricultural productivity, silting of lands in the course of rivers, and waste of rainwater (McCracken, 1990; Wasson, 2003). Temporary flooding or waterlogging is of major concern because it results in decreased crop productivity and/or complete destruction of crops and excess sedimentation (McCracken, 1990). For example, Vertisols in medium to high rainfall areas are very prone to severe damage as a result of temporary or seasonal flooding, particularly in downstream areas. This is mainly due to the low water infiltration rates associated with their high clay content and shrink–swell characteristics.

Data requirements for flood indicators include upstream, middle and downstream flood frequency records and estimates of damage, the extent to which land and water management practices are implemented, the number of water storage structures in a given area, and the implementation of other vegetative control measures (Sharma *et al.*, 1991). For large watersheds, aerial photographs taken during periods of temporary flooding and the use of other types of periodic remote-sensing tools are useful. These can be complemented by interviews with local farmers to assess short-term flood frequency and damage (Rao *et al.*, 1993). For small- and medium-sized watersheds (500–1000 ha), the peak runoff rate and total runoff volume can be used as indicators of temporary flooding and the area affected by such flooding (Pathak *et al.*, 2004).

Indicators for groundwater availability

The part of rainfall water that percolates deep into the ground strata, beyond shallow depth (due to a perched water-table), becomes part of groundwater. It is essential that rainfall recharges groundwater to a desirable level each season to ensure the sustained maintenance of available groundwater. Groundwater levels in many areas are declining despite the implementation of several measures to improve groundwater recharge because of excessive withdrawal of water (Moore, 1984; Khepar *et al.*, 2001). However, NRM interventions can be used to improve groundwater levels by changing the level of recharge. For example, this problem can be addressed by reducing runoff water through bunding and by increasing the percolation of rainwater to recharge the groundwater-table through check dams, percolation tanks, ponds and other water-harvesting and soil-conservation structures. However, in most locations off-take of water for irrigation and domestic use is increasing, resulting in a 'smaller than desired' effect of interventions on the groundwater-table. This trend has become more important over time despite the implementation of various practices to harvest, conserve and use rainwater.

Indicators of groundwater availability include depth of groundwater, safe yield (sustainable level of harvest), number of wells, spatial and temporal availability, and yield. To increase land productivity it is important that the use of available groundwater in a given hydrological unit is optimised. For the sustainable management of groundwater resources, it is necessary to have information on how much water can be stored, and how much can be taken off for irrigation and domestic use. The potential or permissible withdrawal of water is a function of groundwater recharge that in turn is a function of rainfall, runoff, evapotranspiration, percolation, and geological thresholds. The concept of safe yield needs to be evaluated on a watershed scale so that there is a balance between groundwater recharge and outflow (including pumping). To put the concept of safe yield into practice, the total numbers of open wells, tubewells and their depths and spacing need to be estimated and monitored for water status.

The depth of groundwater in wells is the most widely used parameter by researchers, development agencies and farmers for estimating the level and availability of groundwater (Moore, 1984; Khepar *et al.*, 2001; Wani *et al.*, 2003). But, several development agencies also use the number of operating or dry wells, and the area under irrigation as indicators of the water-table and quantity of available groundwater (Rao *et al.*, 1996).

Groundwater level measurements are often used as indicators to assess the impact of various soil and water conservation interventions on groundwater status. For example, in Adarsha watershed, Ranga Reddy district, Andhra Pradesh, ICRISAT monitored the water level in 62 open wells situated at different distances from water recharging facilities at fortnightly intervals. The results showed that after the construction of check dams and other soil and water conservation structures, the water level and yield in the open wells during the study period (1999–2002) improved significantly, particularly in open wells located near water-harvesting structures. The differences in groundwater levels in open wells near or away from check dams were relatively smaller during years of relatively low rainfall, but this difference grew during years of high rainfall, indicating the positive contribution of water-harvesting and recharging structures to increasing groundwater levels. This indicator showed a consistent pattern in groundwater levels during relatively low (1999, 2001 and 2002) and high rainfall (2000) years (Fig. 4.2). The effect of seasonal rainfall on groundwater levels in treated and untreated sub-watersheds is shown in Fig. 4.3. The groundwater level measured in the treated sub-watershed was higher than that in the untreated sub-watershed, where it fell steeply during low rainfall years. However, despite increased water withdrawal as farmers drilled more wells in the area, the treated sub-watershed maintained a higher groundwater level during the 2000–2002 seasons. This example shows how the selected indicator can be monitored at regular intervals to evaluate how improved catchment management contributes to increasing the availability of groundwater. The difference in groundwater levels between the two treatments can be used to estimate the impact of improved water management practices on groundwater availability.

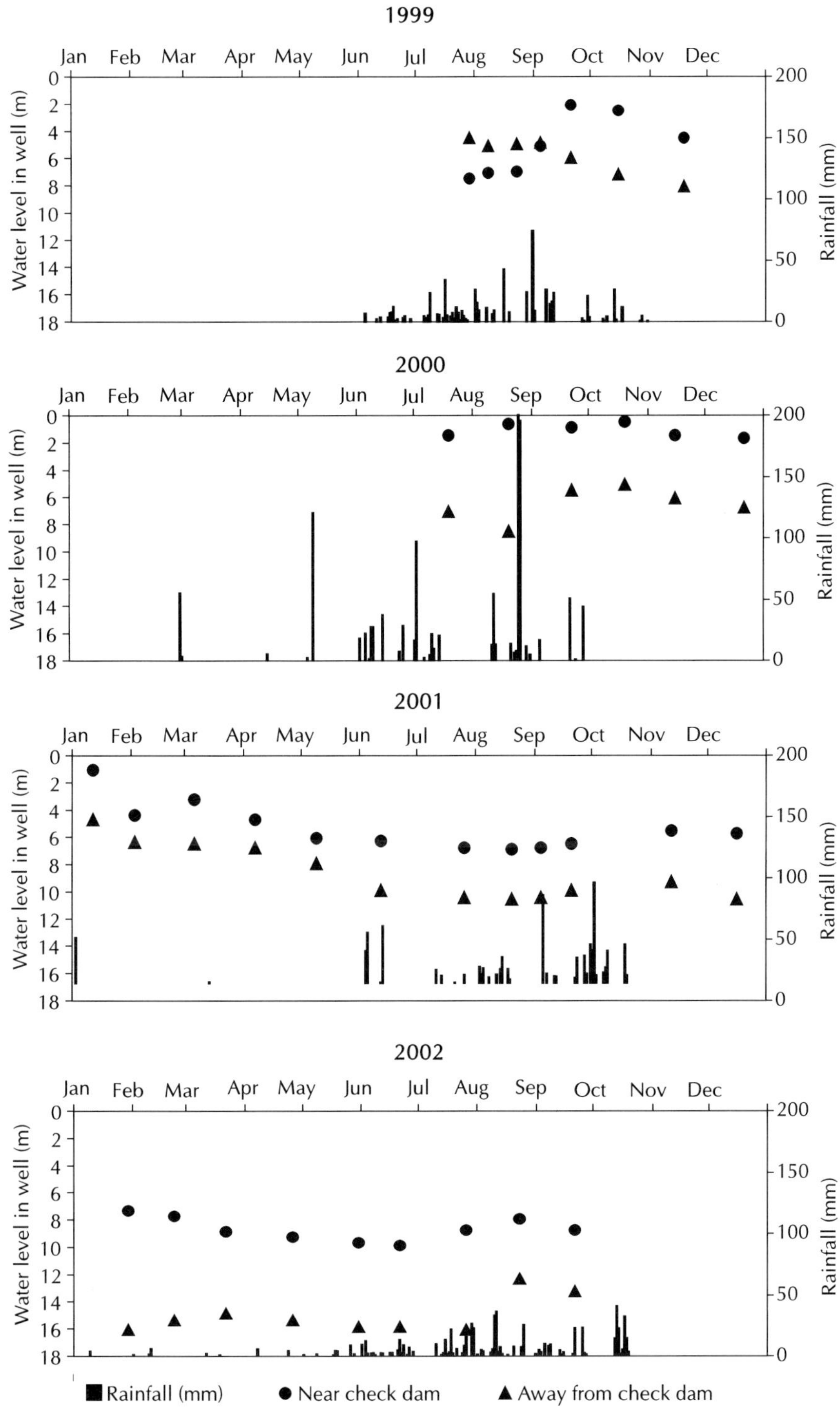

Fig. 4.2. The impact of check dam construction and soil and water conservation practices on groundwater levels at Adarsha watershed, Kothapally, India, 1999–2002 (ICRISAT, unpublished data).

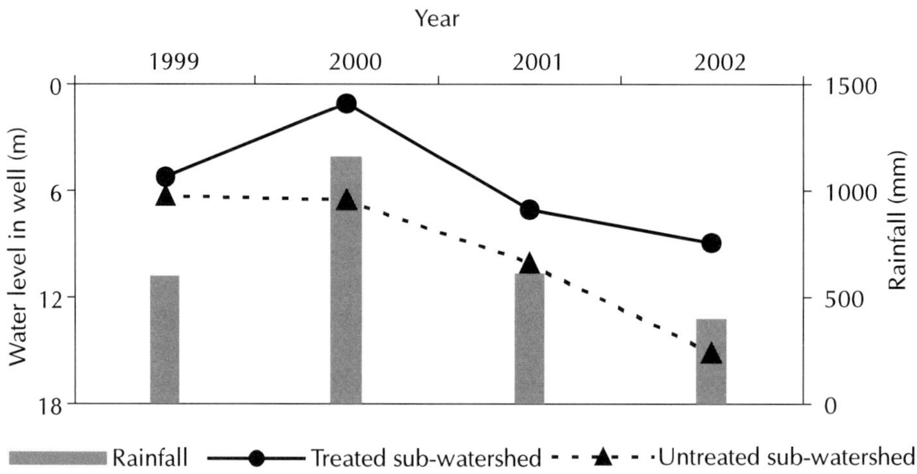

Fig. 4.3. The impact of integrated watershed management on groundwater levels at Adarsha watershed, India, 1999–2002 (ICRISAT, unpublished data).

Most of the existing groundwater indicators do not provide adequate information for planning and judicious management of groundwater resources. Moreover, simply monitoring changes in the water level in open wells or bore wells does not explain the extent to which changes in water levels are attributable to one or more of the following:

- Annual variations in rainfall and their effect on groundwater recharge and reduced runoff
- Increased off-take for irrigation resulting from increasing numbers of bore wells or deeper drilling of wells
- Increased off-take for domestic use.

The effect of variation in annual rainfall on groundwater recharge makes the relationship between annual or seasonal rainfall and groundwater levels quite complex. This requires a better understanding of the pattern of multi-annual fluctuations in the water-table and its relationship with variation in rainfall (Hazell *et al.*, 2001).

There is a clear need for more appropriate indicators of groundwater availability that can provide accurate information about its status. Such indicators need to provide enhanced information for management and planning, and adequate signals for tracking the long-term sustainability of groundwater resources (Farroukhi, 1995).

Recently watershed programmes have been adopting participatory methods to develop more effective indicators of groundwater availability. Farmers are being closely involved in monitoring groundwater levels and in deciding the equitable distribution of surface and groundwater. In some instances, participatory groundwater monitoring experiences in India have contributed towards the sustainable management of groundwater resources (APWELL, 2003). Preliminary survey results suggest that the participatory monitoring can be an effective way to equitably manage groundwater at the community level (Kerr, 2002). Most of the participatory groundwater monitoring research work is in the initial stages of testing. Its usefulness will depend on the outcome of such research.

Indicators for rainfall use efficiency

In this chapter, rainfall efficiency is defined as the economic yield or economic returns per millimeter of rainfall (for detailed reviews see Molden *et al.*, 2003). The underlying concept is to produce 'more crop per drop' of water or 'producing more with less water'. In addition to rainfall use efficiency as an indicator, other proposed sub-indicators include:

- The amount of water stored in the root zone divided by the total rainfall per growing season
- Crop transpiration divided by total rainfall
- Crop yield divided by total rainfall in a given growing season
- Gross margins divided by total rainfall (Barker *et al.*, 2003; Molden *et al.*, 2003).

Increasing rainfall use efficiency is crucial for rainfed farming and can be effected by the judicious use of external inputs such as fertilisers and by implementing soil and water conservation practices.

Rainfed production systems that do not use water efficiently result in irrecoverable loss of water resources, lost opportunities for higher crop yields, and the possible degradation of water quality (Samra, 1998). For example, in a water-deficit situation it is very important to use rainfall use efficiency as an indicator to assess the efficiencies of various NRM technologies. The data required to compute rainfall water use efficiency include: data on daily and annual rainfall; runoff; crop yields; evapotranspiration (measured or simulated value); outflow and inflow of surface and groundwater; and volume of water withdrawn for irrigation.

Water Quality Indicators

Water quality is generally defined by its physical, chemical, biological and aesthetic (smell or odour and appearance) characteristics. These quality parameters may differ with use (drinking, recreation, wildlife, industrial, agricultural or domestic). Like water availability, water quality is greatly influenced by NRM-based agricultural activities. Land and water management practices, tillage, and the use of fertilisers and plant protection chemicals all affect water quality. Several indicators have been proposed to characterise and monitor the physical, chemical, and biological characteristics that relate to water quality in its various uses (Table 4.3).

Water quality is high in undisturbed or natural ecosystems. Several soil processes are adversely affected by the conversion of lands under natural vegetation to agricultural production. Among these, the hydrologic cycle and cycles of carbon and plant nutrients are most relevant to the determination of water quality. The conversion of natural systems (under forest or grass) to agricultural land use reduces water quality due to the

Table 4.3. Selected water quality indicators for monitoring and impact assessment of natural resource management interventions.

Criteria	Water quality indicators
Physical/aesthetic quality	Odour Floating matter Colour Turbidity and clarity Dissolved solids Sediment load Suspended organic and inorganic materials
Chemical quality	pH (acidity/alkalinity) Salinity, electrical conductivity Dissolved oxygen Chemical oxygen demand Dissolved organic matter and organic nitrogen Dissolved load of chemical constituents (nitrate, phosphorus, fluoride, pesticides, toxic compounds, etc.) Heavy metals (copper, nickel, mercury, lead, chromium, cadmium, etc.)
Biological quality	Biomass Microorganisms Biological oxygen demand Pathogens (bacteria, algae, etc.) Phytoplankton and zooplankton Cyanobacteria

contamination of water with sediments, plant nutrients, and agricultural chemicals used in production systems. Studies in the humid tropical regions of Nigeria suggest that the quality of surface water is greatly influenced by agricultural operations (Lal, 1994). Water quality is significantly affected by land use and farming systems. The principal agricultural management practices that affect the quality of surface and groundwater include:

- Soil surface management including tillage methods and ground cover
- Crop residue management and the use of such crop residues as mulch, ploughing under, burning, or grazing
- Fertility management including type of fertiliser (inorganic or organic, soluble or slow-release), method of placement and time of application
- Crop rotations including cropping intensity, crop type, type of farming (commercial or subsistence) and use of chemicals to control insects and plant diseases
- Weed management including use of chemicals, cultivation and manual weeding (Angle *et al.*, 1984, 1993; Lal, 1994).

In general, farming practices that affect soil erosion also affect surface and groundwater quality (Lal, 1994; Evans, 1996).

The movement of sediment and associated agricultural pollutants (fertilisers, pesticides and amendments) into watercourses is the major offsite impact resulting from soil erosion. This not only results in the silting-up of dams, and disruption of wetland ecosystems, but also leads to the contamination of drinking water (Evans, 1996). It has been observed that pollution of surface and groundwater takes place even if the rate of soil erosion is not high, because significant amounts of agricultural chemicals can be transported off-site (Favis-Mortlock, 2002).

Water quality indicators associated with agricultural practices include: sediment load in runoff water, quality of runoff water, nitrogen (N) and phosphorus (P) concentrations and amounts in runoff water, and nitrate pollution of groundwater (Lal, 1994; Jones *et al.*, 1999; Thorburn *et al.*, 2003).

High levels of water pollution resulting from intensification of agriculture have negative effects on human and animal health that need to be accounted for in assessing the impact of agricultural practices and other NRM interventions. The World Health Organization guidelines for nitrate in drinking water recommended that the nitrate concentration be less than 50 mg nitrate/l or 11.3 mg nitrate-N/l (WHO, 1970). According to this recommendation, nitrate concentration in the range of 50–100 mg/l is acceptable, but a concentration of greater than 100 mg nitrate/l can be harmful. In 1980, the European Economic Community (EEC) recommended a maximum acceptable concentration of 50 mg nitrate or 11.3 mg nitrate-N/l unless waivers were granted by the member-state of the Union (EEC, 1980).

Among the plant nutrients, added N is of great concern because it is required in large amounts for crop production. Nitrogen is generally transported from soils into surface and groundwater by water runoff, erosion and leaching (mainly nitrate) (Foster *et al.*, 1982; Follett, 1989). In arable crop production systems, the nitrification of soil and fertiliser ammonium converts relatively immobile ammonium into highly mobile nitrate. That explains why the control or regulation of nitrification retards the contamination of surface and groundwater with nitrate by reducing the movement of nitrate in runoff water and through leaching (Sahrawat, 1989).

Singh and Sekhon (1976) studied the nitrate pollution of groundwater from N fertilisers and animal wastes on light-textured soils in Punjab where N fertilisers are intensively used to grow such cereal crops as maize and wheat. They found that in the Ludhiana district, 90% of the well water samples contained less than 10 mg/l nitrate-N. More importantly, the nitrate concentration of well water decreased significantly with depth, and correlated positively with the amount of fertiliser N added annually per unit area.

Monitoring the nitrate-N concentrations in shallow well water in Ludhiana in 1982 and 1988 revealed that the increase in fertiliser N consumption was associated with an increase of nitrate-N of almost 2 mg/l (Singh *et al.*, 1991). Bajwa *et al.* (1993) analysed 236 water samples from 21 to 38 m deep tube wells in different blocks of the Punjab where annual fertiliser-N consumption ranged from 151 to 249 kg N/ha. They found that 17% of the tube-wells in vegetable-growing areas contained more than 5 mg NO_3-N/l compared to 3% in the tube-wells located in rice–wheat and 6% in potato–wheat rotation areas. These results suggest that excess N not used by the crops moved to the groundwater with rainwater during the rainy season. These results drew attention to the need for rational use of fertiliser N to avoid nitrate pollution of surface and groundwater in porous soils.

Soil conservation practices such as landform configuration also help to conserve soil and reduce loss of N in runoff. For example, a study on Vertic Insceptisol at the ICRISAT farm in Patancheru, India (Table 4.4) showed that the BBF landform had less water runoff, soil loss and nitrate-N loss in water runoff than a flat landform during the 1998 rainy season (ICRISAT, unpublished).

Table 4.4. Impacts of improved land management (flat vs. broadbed-and-furrow (BBF)) on water runoff, soil and nitrate loss in Vertic Inceptisols, ICRISAT farm, Patancheru, India, 1998 (ICRISAT, unpublished data).

Parameter measured	Land management treatments	
	Flat	BBF
Water runoff (mm)	287	226
Soil loss (t/ha)	5.4	3.1
Nitrate-N loss (kg/ha)	13.3	9.3

Among the water quality indicators used to assess the impact of agricultural practices (Table 4.3), the most important and practical indicators of surface and groundwater quality include sediment load, odour or smell, dissolved load of chemical constituents (nitrate, P, pesticides, etc), turbidity and colour. These indicators are also simple and useful in decision-making. For example, waters with high proportions of suspended materials and foul smell are not considered suitable for domestic use, especially for drinking.

The contamination of groundwater with such chemicals as nitrate, phosphate, fluoride, basic cations (potassium, calcium, magnesium and sodium) and heavy metals (mercury, copper, nickel, lead, cadmium, chromium, etc.) is a problem. This contamination can be determined by chemical analysis of surface, shallow, or deep groundwater. Measurements of concentrations of the polluting chemical serve as quality indicators. The suitability of water for drinking, agricultural or other domestic use depends on several physical, chemical and biological properties and their acceptable concentrations or presence in the water (Lal, 1994). For example, long-term chemical analysis of rainwater samples from three locations on the ICRISAT farm showed that rainwater annually added significant amounts of N, sulphur, potassium, magnesium and calcium nutrients to the soil. This input of nutrients through rainfall offsets, at least partially, their removal by crops (Murthy *et al.*, 2000). The changes in water quality resulting from NRM interventions can also be compared to the threshold levels specified by the international water quality standards for chemical contaminants (Table 4.5).

The presence of such pathogens as bacteria, cyanobacteria and other algae or microorganisms has been found to be highly undesirable for the use of surface and groundwater for various domestic purposes. Little research has been reported on the contamination of both surface and groundwater with pesticides, but pesticide contamination of surface and groundwater is of great concern to human health.

Table 4.5. International water quality standards for some chemical constituents for human and livestock consumption (Lal, 1994).

	Concentration (mg/1000 ml)	
Chemical constituent	Human	Livestock
Nitrate	< 45	< 200
Ammonium	< 0.05	NA[a]
Chloride	< 400	< 1000
Calcium	< 200	< 1000
Barium	< 1.0	NA
Zinc	< 15	< 20
Molybdenum	NA	0.01
Lead	< 0.1	0.05
Arsenic	< 0.05	0.05
Selenium	< 0.01	0.01
Cadmium	< 0.01	0.01
Mercury	< 0.01	0.002

[a]NA = not available.

Application of Simulation Modelling

Hydrological models have been extensively used to assess surface and groundwater availability (Pathak and Laryea, 1992; Allerd and Haan, 1996; Sireesha, 2003). The models have been used to provide evidence of trends in the long-term availability of surface and groundwater. Pathak and Laryea (1992) used a water-harvesting model to estimate the probability of runoff and water availability in a tank. They also ran simulations using long-term data on rainfall, evaporation, soil characteristics and catchment area, to estimate the chances of adequate stored water being available for supplemental irrigation during drought stress periods in a growing season (Pathak and Laryea, 1992).

There is a direct link between soil conservation and the enhancement of surface and groundwater quality. This implies that without soil conservation practices water quality cannot be maintained. Research on water quality has focused on developing simulation models to evaluate suitable soil management practices that maintain surface and groundwater quality (McCool and Renard, 1990). Simulation modelling has an important role to play in the development of water quality indicators for monitoring and assessing water quality. Several water quality models (McCool and Renard, 1990; Williams *et al.*, 1994) have been used to generate information on how to solve a variety of complex water quality problems. It has been suggested that simple screening simulation models may be sufficient to identify pollution sources in surface and groundwater. On the other hand, rather comprehensive models may be required to compare the effects of various agricultural management practices on the transport of chemicals and pollutants by water runoff and sediment (Williams *et al.*, 1994).

For example, simulation models have been used to estimate the amount of nitrate-N in runoff water from the soil surface layer. The decrease in nitrate-N concentration by the volume of water flowing through a soil layer is simulated using an exponential function. In this way, an average daily concentration of nitrate-N can be obtained by integrating the exponential function to give nitrate-N yield, and dividing this value by the volume of water leaving the soil layer in runoff, lateral flow, and percolation. The amount of nitrate-N in surface runoff is estimated as the product of the volume of water and the average nitrate-N concentration. A provision is made in the model for estimating production of nitrate via nitrification and loss of ammonium via ammonia volatilisation. The loss of nitrate produced via denitrification is also taken into account under partial anaerobic or anaerobic conditions created by the water regime.

Simulation models have also been used to evaluate the impact of agricultural practices on environmental quality. For example, Kelly *et al.* (1996) simulated the long-term (30-year) impacts of different cropping systems and such NRM interventions as no-till, manure application, and cover crops on the tradeoffs between net returns and different aspects of environmental quality. Their study showed that no-till rotations provided the greatest returns, followed by conventional rotations. In terms of environmental impacts, no-till rotations dominated all other rotations with lowest N loss, and cover crop rotations had the best results in terms of soil erosion and P loss. However, since herbicides were used to control weeds in the no-till system, the pesticide index was very high, suggesting a trade-off between pesticide hazard and other environmental considerations. The authors also constructed an environmental hazard index to provide decision-makers with better information for analysing the trade-offs between potential chemical contamination of water bodies and net returns.

Recently, the combined use of geographic information systems (GIS) and mathematical modelling has been used to develop decision-support systems for quantifying:

- Runoff and movement of sediment, pesticides and nutrients
- Percolation and leaching of pesticides and nutrients to shallow ground-water
- The economic impact associated with crop management, land use, and other policy changes to improve water quality at the watershed and river basin levels (Lovejoy *et al.*, 1997).

Gardi (2001) evaluated the impact of a new agronomic framework protocol in a small watershed using combined applications of GIS and a crop-simulation model (CropSyst). It was found that the greatest leaching of nitrate occurred on coarser-textured soils. Erosion and herbicide effects on water quality were higher in sloping areas sown to spring–summer crops. It was concluded that the increase in row-crop cultivation, determined by European Union (EU) agricultural policy, represented the main adverse impact on water quality of the site studied.

Summary and Conclusions

With the impending freshwater scarcity in many regions of the world, water availability and issues relating to water quality are assuming increasing importance. Agricultural activities can affect the quantity and quality of surface and groundwater resources. Improved NRM practices are being developed and implemented to reduce the negative environmental outcomes of agricultural practices and to increase water availability and quality. Information reviewed in this chapter indicates that the use of fertilisers, especially fertiliser N in excess of that utilised by plants in intensive production systems on porous soils, has the potential to contaminate shallow and deep groundwater resources. Little information is, however, available on the contamination of surface and groundwater resources with pesticides and other agricultural chemicals. There is lack of sufficient data on biophysical indicators from tropical regions to fully assess the impact of agricultural practices and soil processes on water availability and quality.

Because of their simplicity, cost and effectiveness, commonly used water availability indicators include:

- Measurement of soil moisture using the gravimetric method
- The number of storage structures and their water levels to assess surface water availability
- Water levels in open wells, tube wells and piezometers, and duration of water pumping to determine groundwater availability.

Commonly used water quality indicators include:

- Aesthetic (smell, appearance, floating matter)
- Physical (sediment load, turbidity)
- Chemical (chemical constituents such as nitrate, fluoride, etc.) and
- Biological (presence of bacteria and pathogens, etc.) characteristics.

More importantly, unlike soil quality that takes a long time for observable changes to occur, water quality is extremely dynamic and needs regular monitoring.

Recent watershed research results reviewed in this chapter indicate that improved NRM interventions have the potential to decrease runoff and soil loss and increase surface and groundwater availability. However, there is a need to generate more empirical data on the impact of NRM technologies on water availability and the quality of surface and groundwater in different ecoregions, because these relationships are likely to be context- and location-specific.

Another important research area is understanding the relationships between soil management and water quality, especially in tropical regions where there is a shortage of such information (Karlen, 1999). When minimal empirical data is available, simulation models can be used to understand this relationship, and to provide information useful in developing indicators that consistently track impacts over time. More attention is needed to link technological options for water harvesting and use to regular monitoring of impacts on water budgets and quality of groundwater resources. In addition, threshold or tolerable limits in terms of the concentrations of major pollutants in natural waters need to be standardised.

Priority should be given to developing and applying simulation models that can effectively predict nitrate movements in surface water and its leaching into groundwater, and how this will be affected by agricultural and resource management practices. Such research can be helpful in developing ecofriendly and environmentally sound N management practices for intensive and high input-based agriculture (Moreels *et al.*, 2003).

Progress in generating information required to monitor the impacts of agricultural and management practices on water availability and quality in the developing regions has been slow and limited. The use of simulation modelling and remote sensing and GIS tools could help to bridge this gap and to develop useful decision-support systems. In addition to such biophysical factors as soils, climate, and land use, socio-economic and institutional factors and agricultural policies often play an important role in the management of water resources. Greater emphasis should therefore be given to integrated approaches that link socio-economic and biophysical information when assessing the impacts of NRM interventions on water quantity and quality (Faeth, 1993; Lal and Stewart, 1994; Shiferaw and Holden, Chapter 12, this volume).

Acknowledgements

We are grateful to the two anonymous reviewers and the editors for their critical comments and helpful suggestions on earlier versions of the manuscript for improving the presentation and focus of this chapter.

References

Allerd, B. and Haan, G.T. (1996) SWMHMS – Small watershed monthly hydrological modelling system. *Water Resources Bulletin* 32, 541–552.

Angle, J.S., McClung, G., McIntosh, M.S., Thomas, P.M. and Wolf, D.C. (1984) Nutrient losses in runoff from conventional and no-till corn watersheds. *Journal of Environmental Quality* 13, 431–435.

Angle, J.S., Gross, C.M., Hill, R.L. and McIntosh, M.C. (1993) Soil nitrate concentrations under corn as affected by tillage, manure and fertiliser applications. *Journal of Environmental Quality* 22, 141–147.

APWELL (Andhra Pradesh Groundwater Borewell Irrigation Schemes) (2003) *Judicious Management of Groundwater Through Participatory Hydrological Monitoring.* APWELL Project. PR (Panchayat Raj) and RD (Rural Development) Departments, Government of Andhra Pradesh, India, 60 pp.

Bajwa, M.S., Singh, B. and Singh, P. (1993) Nitrate pollution of groundwater under different systems of land management in the Punjab. In: *Proceedings of First Agricultural Congress.* National Academy of Agricultural Sciences, New Delhi, India, pp. 223–230.

Barker, R., Dawe, D. and Inocencio, A. (2003) Economics of water productivity in managing water for agriculture. In: Kijne, J.W., Barker, R. and Molden, D. (eds) *Water Productivity in Agriculture: Limits and Opportunities for Improvement.* CAB International, Wallingford, UK, pp. 19–35.

Cai, X. and Rosegrant, M.W. (2003) World water productivity: Current situation and future options. In: Kijne, J.W., Barker, R. and Molden, D. (eds) *Water Productivity in Agriculture: Limits and Opportunities for Improvement*. CAB International, Wallingford, UK, pp. 163–178.

El-Ashry, M.T. (1991) Policies for water resource management in semi-arid regions. *Water Resources Development* 7, 230–234.

EEC (European Economic Community) (1980) Council directive on the quality of water for human consumption. *Official Journal* 23 (80/778 EEC L 229), EEC, Brussels, Belgium, pp. 11–29.

Evans, R. (1996) *Soil Erosion and its Impact in England and Wales*. Friends of the Earth, London, UK, 126 pp.

Faeth, P. (ed.) (1993) *Agricultural Policy and Sustainability: Case Studies from India, Chile, the Philippines and the United States*. World Resources Institute, Washington, DC, 75 pp.

Farroukhi, L. (1995) Rainwater harvesting and groundwater recharge techniques: a case study of Moti Rayan Project in the salt affected coastal area of Southern Kachchh district, Gujarat State, India, *Working Paper* 271. International Rural Development Centre, Uppsala, Sweden, 71 pp.

Favis-Mortlock, D. (2002) Erosion by water. In: Lal, R. (ed.) *Encyclopedia of Soil Science*. Marcel Dekker, New York, pp. 452–456.

Follett, R.F. (ed.) (1989) *Nitrogen Management and Groundwater Protection*. Elsevier, New York, 395 pp.

Foster, S.S.D., Cripps, A.C. and Smith-Carington, A. (1982) Nitrate leaching to ground water. *Philosophical Transactions of the Royal Society of London Series B* 296, 477–489.

Gardi, C. (2001) Land use, agronomic management and water quality in a small northern Italian watershed. *Agriculture, Ecosystems and Environment* 87, 1–12.

Gupta, R.K. and Sharma, R.A. (1994) Influence of different land configurations on in-situ conservation of rainwater, soil and nutrients. *Crop Research* 8, 276–282.

Hazell, P., Chakravorty, U., Dixon, J. and Rafael, C. (2001) Monitoring system for managing natural resources: economics, indicators and environmental externalities in a Costa Rican watershed. *EPTD Discussion Paper No. 73*. Environment and Production Technology Division, International Food Policy Research Institute and Environment Department, World Bank, Washington, DC, 145 pp.

Jones, O.R., Southwick, L.M., Smith, S.J. and Hauser, V.L. (1999) Soil quality and environmental impacts of dryland residue management systems. In: Lal, R. (ed.) *Soil Quality and Soil Erosion*. CRC Press, Boca Raton, Florida, pp. 153–167.

Karlen, D. (1999) Opportunities and challenges associated with watershed-scale assessment of soil and water quality. *Journal of Soil and Water Conservation* 54, 626–627.

Kelly, T.C., Lu, Y.C. and Teasdale, J. (1996) Economic–environmental tradeoffs among alternative crop rotations. *Agriculture, Ecosystems and Environment* 60, 17–28.

Kerr, J. (2002) Watershed development, environmental services, and poverty alleviation in India. *World Development* 30, 1387–1400.

Khepar, S.D., Sondhi, S.K., Chawla, J.K. and Singh, M. (2001) Impact of soil and water conservation works on ground water regime in Kandi area of Punjab. *Journal of Soil and Water Conservation (India)* 45, 182–189.

Lal, R. (1994) Water quality effects of tropical deforestation and farming system on agricultural watersheds in Western Nigeria. In: Lal, R. and Stewart, B.A. (eds) *Soil Processes and Water Quality. Advances in Soil Science*. Lewis Publishers, Boca Raton, Florida, pp. 273–301.

Lal, R. and Stewart, B.A. (1994) Research priorities for soil processes and water quality in 21st century. In: Lal, R. and Stewart, B.A. (eds) *Soil Processes and Water Quality. Advances in Soil Science.* Lewis Publishers, Boca Raton, Florida, pp. 383–391.

van der Leeden, F., Troise, F.L. and Todd, D.K. (1990) *The Water Encyclopedia.* Lewis Publishers, Chelsea, Michigan, 808 pp.

Littleboy, M., Silburn, D.M., Freebairn, D.M., Woodruff, D.R. and Hammer, G.L. (1989) PERFECT: a computer simulation model of productivity, erosion and runoff functions to evaluate conservation techniques. *Bulletin, Queensland Department of Primary Industries,* Brisbane, Australia, 135 pp.

Lovejoy, S.B., Lee, J.G., Randhir, T.O. and Engel, B.A. (1997) Research needs for water quality management in the 21st century: A spatial decision support system. *Journal of Soil and Water Conservation* 52, 18–22.

McCool, D.K. and Renard, K.G. (1990) Water erosion and water quality. In: Singh, R.P., Parr, J.F. and Stewart, B.A. (eds) *Dryland Agriculture: Strategies for Sustainability. Advances in Soil Science,* Volume 13. Springer-Verlag, New York, pp. 175–185.

McCracken, R.J. (1990) *Indicators for assessing changes in natural resources in developing countries.* United States Agency for International Development (USAID), Washington, DC, 50 pp.

Molden, D., Murray-Rust, H., Sakthivadivel, R. and Makin, I. (2003) A water productivity framework for understanding and action. In: Kijne, J.W., Barker, R. and Molden, D. (eds) *Water Productivity in Agriculture: Limits and Opportunities for Improvement.* CAB International, Wallingford, UK, pp. 1–18.

Moore, C.V. (1984) Groundwater overdraft management: some suggested guidelines. *Gianninni Foundation Information Series* 84(1), Division of Agriculture and Natural Resources, University of California, Davis, California, 12 pp.

Moreels, E., De Neve, S., Hofman, G. and van Meirvenne, M. (2003) Simulating nitrate leaching in bare fallow soils: a model comparison. *Nutrient Cycling in Agroecosystems* 67, 137–144.

Murthy, K.V.S., Sahrawat, K.L. and Pardhasaradhi, G. (2000) Plant nutrient contribution by rainfall in the highly industrialised and polluted Patancheru area in Andhra Pradesh. *Journal of the Indian Society of Soil Science* 48, 803–808.

Oweis, T.Y. and Hachum, A.Y. (2003) Improving water productivity in the dry areas of West Asia and North Africa. In: Kijne, J.W., Barker, R. and Molden, D. (eds) *Water Productivity in Agriculture: Limits and Opportunities for Improvement.* CAB International, Wallingford, UK, pp. 179–198.

Pathak, P. and Laryea, K.B. (1992) Prospects of water harvesting and its utilization for agriculture in the semi-arid tropics. In: Gollifer, D.E. and Kronen, M. (eds) *Proceedings of the Symposium of the Southern African Development Coordination Conference* (SADCC) *Land and Water Management Research Program (L & WMRP) Scientific Conference, Gaborone, Botswana, 8–10 October 1990.* SADCC – L & WMRP, Gaborone, Botswana, pp. 266–278.

Pathak, P., Laryea, K.B. and Sudi, R. (1989) A runoff model for small watersheds in the semi-arid tropics. *Transactions of American Society of Agricultural Engineers* 32, 1619–1624.

Pathak, P., Wani, S.P., Singh, P., Sudi, R. and Srinivasa Rao, Ch. (2002) Hydrological characterization of benchmark agricultural watersheds in India, Thailand, and Vietnam. *Agroecosystem Global Theme, Report no. 2.* International Crops Research Institute for the Semi-Arid Tropics, Patancheru, India, 52 pp.

Pathak, P., Wani, S.P., Singh, P. and Sudi, R. (2004) Sediment flow behaviour from small agricultural watersheds. *Agricultural Water Management* 67, 105–117.

Rao, M.S.R.M., Adhikari, R.N., Chittaranjan, S. and Chandrappa, M. (1996) Influence of conservation measures on groundwater regime in a semi-arid tract of South India. *Agricultural Water Management* 30, 301–312.

Rao, R.S., Venkataswamy, M., Rao, C.M. and Krishna, G.V.A.R. (1993) Identification of overdeveloped zones of groundwater and the location of rainwater harvesting structures using an integrated remote sensing based approach – a case study in part of the Anantapur district, Andhra Pradesh, India. *International Journal of Remote Sensing* 14, 3231–3237.

Rose, C.W. (2002) Erosion by water, modeling. In: Lal, R. (ed.) *Encyclopedia of Soil Science*. Marcel Dekker, New York, pp. 468–472.

Sahrawat, K.L. (1989) Effects of nitrification inhibitors on nitrogen transformation, other than nitrification, in soils. *Advances in Agronomy* 42, 279–309.

Samra, J.S. (1998) Watershed management for sustainable agriculture. In: Dhaliwal, G.S., Arora, R., Randhawa, N.S. and Dhawan, A.K. (eds) *Ecological Agriculture and Sustainable Development*, Volume 1. Centre for Research in Rural and Industrial Development, Chandigarh, India, pp. 147–155.

Sharma, M.L., Barron, R.J.W. and Craig, A.B. (1991) Land use effects on groundwater recharge to an unconfined aquifer. *Divisional Report 91(1)*. Division of Water Resources Research, Commonwealth Scientific and Industrial Research Organisation, Canberra, Australia, 45 pp.

Singh, B. and Sekhon, G.S. (1976) Nitrate pollution of groundwater from fertilizers and animal wastes in the Punjab, India. *Agriculture and Environment* 3, 57–67.

Singh, B., Sadana, U.S. and Arora, B.R. (1991) Nitrate pollution of groundwater with increasing use of nitrogen fertilizers in Punjab, India. *Indian Journal of Environmental Health* 33, 516–518.

Singh, P., Alagarswamy, G., Pathak, P., Wani, S.P., Hoogenboom, G. and Virmani, S.M. (1999) Soybean–chickpea rotation on Vertic Incesptisols. I. Effect of soil depth and landform on light interception, water balance and crop yields. *Field Crops Research* 63, 211–224.

Sireesha, P. (2003) Prospects of water harvesting in three districts of Andhra Pradesh. M.Tech. Thesis, Centre for Water Resources, Jawaharlal Nehru Technological University, Hyderabad, India, 95 pp.

Srivastava, K.L. and Jangawad, L.S. (1988) Water balance and erosion rates of Vertisol watersheds under different management. *Indian Journal of Dryland Agricultural Research and Development* 3, 137–144.

Thorburn, P.J., Biggs, J.S., Weier, K.L. and Keating, B.A. (2003) Nitrate in groundwaters of intensive agricultural areas in coastal northeastern Australia. *Agriculture, Ecosystems and Environment* 94, 49–58.

Wani, S.P., Pathak, P., Tam, H.M., Ramakrishna, A., Singh, P. and Sreedevi, T.K. (2002) Integrated watershed management for minimizing land degradation and sustaining productivity in Asia. In: Adeel, Z. (ed.) *Integrated Land Management in Dry Areas: Proceedings of a Joint United Nations University–Chinese Academy of Sciences (UNU–CAS) International Workshop, Beijing, China, 8–13 September 2001*. United Nations University, Tokyo, Japan, pp. 207–230.

Wani, S.P., Pathak, P., Sreedevi, T.K., Singh, H.P. and Singh, P. (2003) Efficient management of rainwater for increased crop productivity and groundwater recharge in Asia. In: Kijne, J.W., Barker, R. and Molden, D. (eds) *Water Productivity in Agriculture: Limits and Opportunities for Improvement*. CAB International, Wallingford, UK, pp. 199–215.

Wasson, R.J. (2003) A sediment budget for the Ganga–Brahmaputra catchment. *Current Science* 84, 1041–1047.

Williams, J.R., Arnold, J.G., Jones, C.A., Benson, V.W. and Griggs, R.H. (1994) Water quality models for developing soil management practices. In: Lal, R. and Stewart, B.A. (eds) *Soil Processes and Water Quality. Advances in Soil Science.* Lewis Publishers, Boca Raton, Florida, pp. 349–382.

WHO (World Health Organization) (1970) *European Standards for Drinking Water*, Second Edition. World Health Organization (WHO), Geneva, Switzerland, 58 pp.

5 Biophysical Indicators of Agro-ecosystem Services and Methods for Monitoring the Impacts of NRM Technologies at Different Scales

S.P. Wani[1], Piara Singh[1], R.S. Dwivedi[2], R.R. Navalgund[2] and A. Ramakrishna[1]

[1] *International Crops Research Institute for the Semi-Arid Tropics (ICRISAT) Patancheru, India*
[2] *National Remote Sensing Agency (NRSA), Hyderabad, India*

Introduction

Natural resources such as soil, water, air, and vegetation that provide the vital needs of humans and animals are in a perpetually dynamic state. Agricultural interventions typically involve opening closed natural systems that may have attained a certain equilibrium. Such products as food, feed, fuel, etc. are exported from the system resulting in more outflows than inflows. When this happens, unless outflows are complemented by external inputs, resource productivity will gradually decline. Land degradation is a commonly used term to describe this situation and refers to the productivity loss and/or diminishing ability of land to provide such essential ecological services as groundwater recharging, carbon fixation and storage, detoxification of harmful compounds, and water purification.

In order to minimise the process of degradation and to maintain productive capacity and ability to provide ecosystem services for present and future generations, various natural resource management (NRM) options have been developed and implemented.

Socio-economists and natural resource experts have long struggled to assess the broader economic and environmental impacts of NRM technologies. This has been a difficult task because such technologies are not separately developed and marketed as divisible component inputs like seeds. Typically NRM practices are developed in an integrated approach to improve biophysical conditions and are used in conjunction with other yield-enhancing inputs. Hence, the direct economic benefits derived from such technologies are not always evident and are generally attributed to such other inputs as improved seeds. The new paradigm of integrated natural resource management (INRM) aims to provide multi-disciplinary solutions in a coordinated manner to achieve livelihood and sustainability objectives. However, the full social impact of INRM cannot be measured directly using conventional methods of economic evaluation (Shiferaw *et al.*, Chapter 2, this volume).

Therefore, appropriate qualitative and quantitative indicators of biophysical impact on varying spatial and time scales are needed. A good indicator must be sensitive enough to show temporal and spatial changes, predictable, measurable and interactive. Assessing NRM impacts will need new methods, tools and multidisciplinary teams of experts to understand and accurately quantify the benefits. Some non-marketed agro-ecosystem services are especially difficult to recognise and quantify. Such tools as simulation modelling, geographic information systems (GIS), and satellite imaging, used in conjunction with traditional productivity-based techniques, are vital in estimating some NRM impacts. Productivity-based indicators (e.g. biomass and crop yields) at micro levels need to be complemented by indicators like the vegetation index at ecoregional levels using satellite images and GIS tools. Simulation modelling is also useful for verifying and extrapolating results to larger scales and for studying long-term effects.

Chapters 3 and 4, this volume, dealt with biophysical indicators for assessing soil quality and water availability and quality. This chapter presents indicators used to monitor changes in the flow of such other ecosystem services as biodiversity conservation, carbon sequestration and ecosystem regulation and describes tools and methods available to monitor and estimate the impacts associated with adoption of NRM technologies on various scales. This chapter first presents the criteria and indicators for monitoring NRM impacts related to various ecological functions and ecosystem services. The use of simulation models to estimate biophysical changes is then discussed. Following is a discussion of how remote sensing and GIS tools can be used to monitor spatial and temporal changes. The key issues and areas for future research are highlighted.

Indicators of NRM Impact

An indicator is a sign or signal that relays a complex message, potentially from numerous sources, in a simplified and useful manner. It can reflect the biological, chemical or physical attributes of ecological conditions. The primary uses of an indicator are to characterise current status and to track or predict significant change. With a foundation of diagnostic research, an ecological indicator may also be used to identify major ecosystem stress. Glave and Escobal (1995) proposed a set of verifiable and replicable indicators to assess changes in natural resource conditions, the ecological and economic structure, and ecological, economic and social benefits in the Andes. Munasinghe and McNeely (1995) suggested the index of biophysical sustainability, soil and water conservation, efficiency of fertiliser use, efficiency of energy use, and the productive performance of forests as important NRM indicators. Ramakrishnan (1995) introduced such additional indicators as management practices, biodiversity and nutrient cycles. Smyth and Dumanski (1993) stated that good indicators are measurable and quantifiable, such as the environmental statistics that measure or reflect environmental status or changes in resource conditions. Agricultural systems can be analysed at various hierarchical levels. For land evaluation and farming systems analysis, FAO (1992) distinguishes between cropping, farming, sub-regional, regional, and national systems. The precision level and the purpose of a given indicator will change if it is extrapolated to a higher scale and time step.

Indicators for assessing NRM technology impacts are selected according to data availability, data sensitivity to temporal and spatial change, and the capacity of the data to quantify the behaviour of given agricultural systems. Table 5.1 presents commonly used and potential indicators for monitoring NRM impacts.

Table 5.1. Indicators for monitoring biophysical and sustainability impacts of NRM interventions.

Criteria	Indicators
1. Biodiversity	• Species richness • Species diversity • Species risk index
2. Agro-biodiversity	• Index of surface percentage of crops (ISPC) • Crop agro-biodiversity factor (CAF) • Genetic variability • Surface variability
3. Agro-ecosystem efficiency	• Productivity change • Cost–benefit ratio • Parity index
4. Environmental services	• Greenery cover/vegetation index • Carbon sequestered • Reduced emissions of greenhouse gases • Reduced land degradation/rehabilitation of degraded lands
5. Soil quality	• Soil physical indicators (e.g. bulk density, water infiltration rate, water holding capacity, water logging, soil loss, etc.) • Soil chemical indicators (e.g. soil pH, organic C, inorganic C, total and available N, P and other nutrients, CEC, salinity, etc.) • Soil biological indicators (e.g. soil microbial biomass, soil respiration, soil enzymes, biomass N, diversity of microbial species, etc.)
6. Water availability and quality	• Quantity of fresh surface water available • Fluctuations in groundwater level • Quality of surface water and groundwater

Biodiversity indicators

Natural resource management affects biodiversity on various scales. Indicators are required to assess the impacts of NRM interventions on natural and managed ecosystems. Biodiversity has been most generally defined as the 'full variety of life on Earth' (Takacs, 1996). It is the sum total of different kinds of diversities such as species diversity within communities, genetic diversity, i.e. the variety of individuals within populations, and life-form, floristic, and functional diversities. Some believe that it has simply replaced the terms 'nature' or 'wilderness' (Chadwick, 1993). In fact, 'biodiversity' is now sometimes used to mean 'life' or 'wilderness' and has served on occasion as a catch-all for 'conservation' itself. Biodiversity provides many benefits. Its reduction influences the structure, stability and function of ecosystems and diminishes the flow of valuable ecosystem goods and services to humans (Erlich and Erlich, 1992). Some of these benefits come in the form of goods that can be directly valued and costed while other critical indirect benefits to humans are difficult to value and quantify (Freeman *et al.*, Chapter 1, this volume; Shiferaw *et al.*, Chapter 2, this volume). These benefits include such ecosystem services as air and water purification, climate regulation, soil formation, and the generation of moisture and oxygen.

When exploring indicators that might shed light on the conservation of biodiversity, it is essential to identify the types of indicators needed on various scales to determine whether conservation objectives are being met. Reid *et al.* (1993) provide a summary of 22 biodiversity indicators defined on three levels: genetic, species, and community diversities.

Biodiversity on any scale can be measured with flora, fauna and species diversity of different types. The term species diversity or biodiversity at first instance means the number of different species found in a given area, but it must take into account the relative abundance of all the different species. Indicators are needed to measure the outcomes related to such effects. Changes in biodiversity can be measured in terms of indicators for species richness, diversity, and risk index. Species richness and species diversity are often confused and used interchangeably, but mean different things (Spellerberg and Fedor, 2003).

Species richness

This refers to the total number of species per site or habitat and can be estimated by counting all species within the target area (Simpson, 1949). Although species richness is a measure of the variety of species, it should be used to refer to the number of species in a given area of sample (Spellerberg and Fedor, 2003).

Species diversity

This measures the total number of species (abundance) and their relative distribution, i.e. as the index of some relationship between number of species and number of individuals. Diversity indices that take the relative abundances of different species into account, therefore provide more information about community composition than simply species richness. Species diversity can be measured in several ways; commonly used indices are the Shannon Index, the Simpson Index and the Species Risk Index.

The Shannon Index (H) is based on probabilities of occurrence. It measures the average degree of uncertainty in predicting the species of a given individual selected at random from a community (Shannon and Weaver, 1963):

$$H = -\sum_{i=1}^{K}[P_i \, ln(P_i)] \qquad (1)$$

where $P_i = n_i/N$ is the number of sample observations in category i, n_i is the number of individuals in category i, and N is the total number of individuals in the sample.

The index varies from a value of 0 (for communities composed of a single species) to high values (for communities with many species). The larger the index, the greater the diversity. This index, based on communication theory, is also referred to as the Shannon–Wiener Index (in recognition of the work of Norbert Wiener from which Shannon built the index) and the Shannon–Weaver Index (in recognition of the mathematician Warren Weaver with whom Shannon co-authored his original book in 1949). The index combines the number of species (species richness) with the distribution of individuals among species to provide a quantitative measure of diversity in any habitat.

The Simpson Index (SI) measures the probability that two individuals randomly selected from a sample will belong to the same species (or some category other than species) (Simpson, 1949). The index can be computed as:

$$SI = \sum_{i=1}^{K} n_i(n_i - 1)/N(N-1) \qquad (2)$$

where $0 \leq SI \leq 1$, n_i is individuals in species i and N is sample size (total number of individuals). With this index, 0 represents infinite diversity and 1, no diversity. In order to make the index more intuitive, it has been suggested to use *1–SI* or *1/SI* so that diversity increases with the index.

Moreover, when it is necessary to compare the degree of similarity in the abundance of different species in a given habitat, the evenness index (EI) can be calculated using *H* and *S* (Shannon and Weaver, 1963) as:

$$EI = H/ln(S) \qquad (3)$$

where *S* is an index of species richness (defined above).

When there are similar proportions of all species, *EI* will have a value of 1. When the abundances are very dissimilar, the value of *EI* increases to greater than 1.

The Species Risk Index combines information on endemic species within a community and on the status of that community in order to provide insights into the risk status of species, even in the absence of detailed threatened species lists. The index is calculated by multiplying the number of endemic species (per unit area) in a community by the percentage of the natural community that has been lost. Thus, an ecological community with many endemics that has lost a high proportion of its area would be ranked at high risk, while a community with few endemics or one that has experienced little conversion would be ranked at low risk (MacKinnon and MacKinnon, 1986; Reid *et al.*, 1993).

Agro-biodiversity indicators

Agricultural biodiversity or agro-biodiversity embodies cultural and spiritual dimensions of biodiversity together with the practical and economic values of gaining sustainable rural livelihoods for poor people (Altieri, 1999). Agro-biodiversity can be defined much more broadly as the many ways in which farmers use the natural diversity of the environment for production. It includes farmers' choice of crops, and management of land, water, and biota (Brookfield and Padoch, 1994). It goes beyond the concept of species and genetic diversity of plants and animals to incorporate other aspects of the farming system related to agriculture, namely: genetic resources, crops and non-cultivated edible and non-edible beneficial plants, livestock, freshwater fish, beneficial soil organisms; and naturally occurring biological pest and disease control agents (insects, bacteria, and fungi). The concept also includes habitats and species outside farming systems that benefit agriculture and enhance ecosystem functions.

Natural resource management interventions can engender significant changes in the state of agro-biodiversity (Thrupp, 1998). Agro-biodiversity has therefore been used as an important criterion for agricultural sustainability (Table 5.1). There are no universally accepted indicators of agro-biodiversity. Some practitioners suggest that the index of surface percentage of crops (ISPC), crop agro-biodiversity factor (CAF), genetic variability, and surface variability factors can all be used as useful indicators to monitor changes in agro-biodiversity (McLaughlin and Mineau, 1995). The ISPC expresses the ratio between the number of crops that represent 50% of the cultivated area and the number of crops commercially cultivated. The CAF indicates the relationship between the number of major crops in a given area and the crops that are agroecologically adapted to the prevailing management systems. Genetic variability or diversity refers to variation in the genetic composition within or among species. Traditional Mendelian methods are insufficient to provide a detailed estimation of genetic variability. The process is too time-consuming and is restricted to phenotypic characters. Today this can be overcome by using DNA-based molecular techniques that provide more precise information on genetic variability (Noss, 1990). To some extent, genetic variability in agro-ecosystems can also be inferred qualitatively from the proportional area of a given cultivar within the total cultivated area of that crop. For example, agro-ecosystems where single varieties or hybrids occupy a large share of the cultivated area indicate limited genetic variation. Surface variability refers to the area covered by agricultural crops in a given agro-ecosystem (Merrick, 1990). For example, regions with a large number of crops with similar areal coverage will have higher surface variability than those dominated by only a few crops. How changes in agro-biodiversity can be used to monitor the sustainability related impacts of NRM technologies is illustrated using information on crop diversity and surface percentage of crops that represent aspects of the stability and balance of agricultural systems at the watershed level (Box 5.1). The examples given for two watersheds, Thanh Ha (Vietnam) and Kothapally (India), show how such quantitative indicators as ISPC, CAF, and surface variability of main crops have changed as a result of integrated watershed management interventions (Wani *et al.*, 2003b).

Box 5.1. The impact of watershed management on agro-biodiversity.

In an operational scale watershed of the International Crops Research Institute for the Semi-Arid Tropics (ICRISAT) at Thanh Ha, Hoa Binh Province, northern Vietnam, a total of four different crops cover the agricultural surface, which represents a low diversity of commercially cultivated species grown. The CAF for the watershed is 0.25 indicating that only one-fourth of the potentially useful species is exploited. Cereals such as maize and rice together constitute 84% of the agricultural surface. These crops are largely cultivated as monocrops generating a very low ISPC.

Maize is the most extensive crop (83% of cropped area) and its production is based on hybrids bred from exotic or introduced genetic materials. In northern Vietnam fewer than five hybrids have produced more than 80% of maize in the last 15–20 years. Not only the number of prevailing hybrids in the ecosystem needs to be considered but also the parentage of such hybrids. In many cases few parental lines, particularly the male-sterile lines (female parent) are used in producing such hybrids, resulting in a narrow genetic diversity of cultivated hybrids, in contrast to the high genetic diversity found in traditional systems. Due to various NRM interventions in this watershed, the area under maize has declined from 380 ha to 148 ha while the area under groundnut, mungbean and soybean has increased from 18 ha to 250 ha changing the CAF from 0.25 in 1989 to 0.6 in 2002.

During 1998–2002, more pronounced impacts in terms of increasing agro-biodiversity were observed in a 500-ha micro-watershed at Kothapally, Ranga Reddy district, Andhra Pradesh, India. In this watershed the farmers grow a total of 22 crops, and a remarkable shift has occurred in the cropping patterns from cotton (200 ha in 1998 to 100 ha in 2002) to a maize/pigeonpea intercrop (40 ha in 1998 to 180 ha in 2002); thereby changing the CAF from 0.41 in 1998 to 0.73 in 2002.

Agro-ecosystem efficiency indicators

Agro-ecosystem efficiency can be approximated through various productivity and economic efficiency indicators. Crop yield is a land productivity indicator that reflects the efficiency of the system (soil, solar energy, water, etc.), with regard to genetic potential, ecological conditions, management, capital investment and labour use. It denotes the production of economic yield and total plant biomass from application of various inputs from a given parcel of land during a given period. It is used as a biological parameter for the evaluation of a system's behaviour and reflects its state at any given time. It is perhaps the best-known functional characteristic of agro-ecosystems and is widely used as a criterion for the assessment of both the biological and economic sustainability of agricultural systems. To assess the impact of NRM technologies, yield parameters sometimes converted in terms of economic returns serve as important indicators. Further, since yield is a final product that takes into account soil and other growing conditions, time-series yield data from a given system can directly indicate the dynamics and sustainability of the system.

At ICRISAT, Patancheru, operational watersheds have been maintained over the last 26 years and scientists have compared the productivity impacts of different NRM options on Vertisols (Wani *et al.*, 2003a). The best practice included: improved soil and water conservation options such as grassed waterways; land configuration (broadbed-and-furrow (BBF) on grade); integrated nutrient and pest management options; recommended varieties of maize intercropped with pigeonpea; plant population and crop husbandry. The farmers' traditional management practice included: rainy-season fallow; and flat-land cultivation with postrainy-season sorghum grown on stored soil moisture with application of 10 t/ha farmyard manure once in 2 years.

The productivity and sustainability impacts of NRM options were tested using time series yield data during 1977–2002 (Fig. 5.1) along with soil quality parameters. Crop yields increased under both management practices, but the annual productivity growth under improved management (77 kg/ha) is significantly higher than that under traditional management (26 kg/ha). The improved system with an average productivity of 4.7 t/ha has a higher carrying capacity (18 persons/ha) than the traditional system with 0.95 t/ha (4 persons/ha). Improved management is better able to respond to increasing population pressure while higher incomes enhance farmers' capacity to invest in more-sustainable practices.

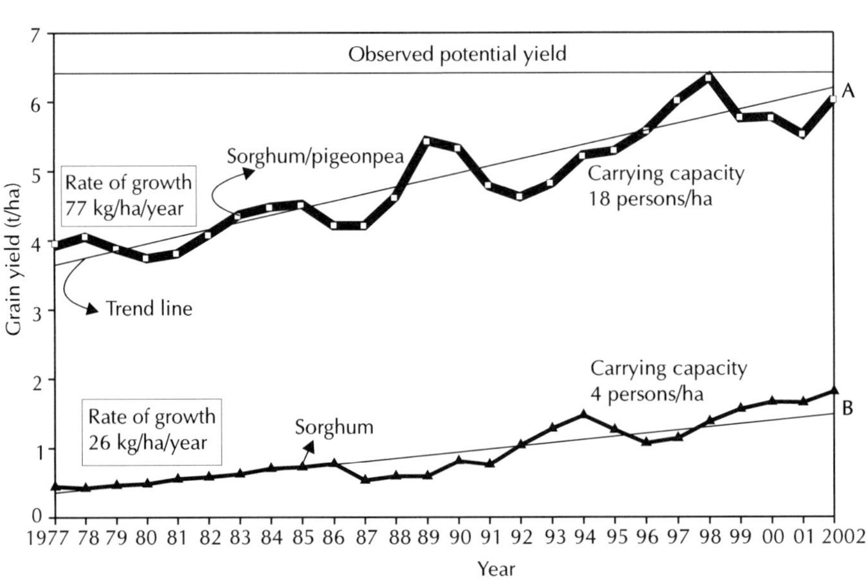

Fig. 5.1. Average grain yields under improved (A) and traditional (B) technologies on a Vertisol watershed at ICRISAT (1977–2002).

The potential yield can also be estimated for a fully optimised production situation using crop simulation models with a fixed limiting constraint such as soil-water availability. The gap between the potential yield that is often greater than that attainable under experimental conditions, and yields under farmers' growing conditions is often referred to as a 'yield gap'. In this sense, NRM impact can also be estimated in terms of the extent to which improved NRM succeeds in reducing the yield gap. The larger the reduction in the yield gap, the higher the success of the intervention in optimising production. Singh *et al.* (2002) used this approach to identify the soybean-growing districts where high yield gaps existed and to identify locations where the yield gaps could be bridged using improved NRM interventions to increase soybean productivity at the district level (Table 5.2). A similar approach was also applied in an operational-scale watershed to assess the potential of improved soil, water, nutrients and crop management options for soybean-based systems at ICRISAT (Singh *et al.*, 1999).

Table 5.2. Simulated soybean yields and yield gap for the selected locations in India.

Location	Mean sowing date	Harvest date	Simulated yields (kg/ha)		Mean observed yield[a] (kg/ha)	Yield gap (kg/ha)
			Mean	SD		
Raisen	22 Jun	11 Oct	2,882	1,269	–	–
Betul	19 Jun	08 Oct	2,141	603	858	1,283
Guna	30 Jun	14 Oct	1,633	907	840	793
Bhopal	16 Jun	08 Oct	2,310	615	1,000	1,310
Indore	22 Jun	10 Oct	2,273	939	1,122	1,151
Kota	03 Jul	16 Oct	1,165	936	1,014	151
Wardha	17 Jun	06 Oct	3,040	640	1,042	1,998
Jabalpur	23 Jun	11 Oct	2,079	382	896	1,183
Amaravathi	18 Jun	08 Oct	1,552	713	942	610
Belgaum	17 Jun	30 Sep	1,844	629	570	1,274

[a] Mean of reported yields during 1990–95.

Related to the productivity measure, various economic efficiency indicators like the benefit–cost ratio can also be computed to evaluate the efficiency of agroecosystems. Such indicators can be used to evaluate the economic feasibility of various cropping systems and sustainability enhancing NRM options (Lynam and Herdt, 1989; Tisdell, 1996). A simple economic productivity indicator like the benefit–cost ratio can be computed at the farm level to determine the economic benefits to farmers of adopting new management practices.

Another related economic indicator is the Parity Index that compares the relative efficiency of different crops or income-generating options in response to a given intervention. The relative index is computed as a percentage or ratio of the option that provides the highest net return. When data on benefits and costs are available, such simple agro-ecosystem efficiency indicators can be computed relatively easily. The challenge is in estimating the parity indices when some of the non-market benefits and costs are difficult to value. Application of environmental valuation methods can be useful approaches to estimate the efficiency of the system in such situations.

Environmental services indicators

Various environmental services such as groundwater recharging, reducing silt load and nitrate concentrations in the runoff water, carbon (C) sequestration in vegetation and in the soil, soil formation, reducing levels of greenhouse gases in the environment, etc. generated through NRM are very important but generally difficult to assess using conventional economic methods. Moreover, the benefits of the environmental services may occur off-site, i.e. far away from the point of NRM interventions.

Existing policies and legal frameworks in many developing countries are not able to properly value the environmental services provided by land-use systems and such ecosystem services as those generated by NRM investments. For example, the effects of deforestation, land degradation or environmental degradation on global warming and climate change are difficult to quantify and assess. Similarly, it is difficult to assess the effects of environmental improvements associated with NRM investment practices. Measurement problems and off-site effects complicate the process of monitoring such changes. However, with the advancement of science and technology, new methods and tools are evolving to quantify these environmental benefits. A good example is the measurement of C sequestration benefits from improved NRM, where some progress is being made at the global level. In 1997, the Kyoto Protocol to the United Nations Framework Convention on Climate Change established an international policy context for reduction of carbon emissions and increased carbon sinks in order to reduce global warming and effects on climate change. This has drawn attention to NRM practices that sequester more carbon from the atmosphere.

C sequestration in soils not only reduces atmospheric CO_2 concentrations but also improves the organic matter status and overall fertility of soils. There is great interest in C sequestration in soils and numerous strategies including technical and policy issues for increasing C in cultivated land have been identified (Bruce *et al.*, 1999; Izaurralde *et al.*, 2001; Pretty and Ball, 2001; Wani *et al.*, 2003a; Smith, 2004). The application of nutritive amendments required for biomass production, including the chemical fertilisers that provide N, P, S, etc. (Vlek, 1990; Wani *et al.*, 2003a) and organic amendments, and diversification of monocropped cereal systems through inclusion of legumes, all favour build-up of soil C and the improvement of soil quality (Wani *et al.*, 1994, 2003a; Paustian *et al.*, 1997). It is clear that soils can sequester C and reduce the atmospheric concentration of CO_2.

Several soil and crop management practices affect C sequestration in soil. Lal (1999) reviewed the role of various practices on C sequestration potential in soil (Table 5.3). According to him conservation tillage, regular application of compost at high rates, integrated nutrient management, restoration of eroded soils, and water conservation management all have a relatively high potential for sequestering C and enhancing and restoring soil fertility.

The level of C sequestered by agricultural, agroforestry, and agrihorticultural systems can be quantified using suitable biochemical methods based on data collected from long-term experiments. The amount of C sequestered

Table 5.3. Carbon sequestration potential of various land management practices under dryland conditions.

Management practice	C sequestration potential (t C/ha/year)
Conservation tillage	0.10–0.20
Mulch farming (4–6 t/ha/year)	0.05–0.10
Compost application (20 t/ha/year)	0.10–0.20
Integrated nutrient management	0.10–0.20
Restoration of eroded soils	0.10–0.20
Restoration of salt-affected soils	0.05–0.10
Water conservation management	0.10–0.30
Afforestation	0.05–0.10

Source: Lal (1999)

by vegetation is quantified by assessing biomass accumulation and the C content of the biomass using standard methods of C estimation. Carbon sequestered in soils is estimated by analysing samples from different soil profiles and calculating the stocks in the profile using the bulk density for a given depth and the area covered by a particular system under study. Following the Kyoto Protocol, C sequestered by agricultural and NRM systems, once quantified in C units, can now be valued in economic terms.

Using this approach, Bruce *et al.* (1999) recorded an annual soil C gain of 0.2 t/ha on pasture and rangelands in the USA following adoption of best management practices. In the SAT of India, Wani *et al.* (2003a) evaluated the effect of long-term (24 years) improved management of Vertisols on C sequestration and reported a difference of 0.3 t C/ha/year attributable to NRM. Under improved soil fertility (60 kg N and 20 kg P/ha/year) and land management (BBF to drain excess water) and cropping systems (maize/pigeonpea intercrop), the soils contained 46.8 t C/ha in 120 cm soil profile as compared to farmers' traditional management practices that contained 39.5 t C/ha. This amounts to a gain of about 7.3 t C/ha over the 24-year period.

Growing knowledge on the C-sequestration benefits of NRM options and the possibilities for C trading have opened new opportunities for C-based rural development in many poor regions where the relative returns to agricultural land use are low. However, several hurdles remain in harnessing such initiatives for community development. For other environmental services, more work is needed in the area of quantification and policy development.

Simulation modelling for the estimation of biophysical changes

Simulation models are mathematical representations of various processes of soil, plant and climate systems in the form of computer programs that describe the dynamics of crop growth in relation to the biophysical environment. These models usually operate in daily time steps. They require soil, climate, crop, and management data as inputs and produce output variables describing the state of the crop and the soil at different points in time. The models are used to evaluate soil and crop management options for a given environment, to extrapolate the results of management strategies over time and space, and to study the long-term effects of NRM on productivity, soil quality, and the environment. Before the models are used to do this, they must be validated with observed field data for the specific soil–plant processes to be evaluated. There are several kinds of simulation models available in the literature, each with its own strengths and weaknesses. Selection of a model depends on its strengths, the purpose for which it is used, and the availability of input data in a given environment for model operation. Table 5.4 provides a summary of different types of simulation models.

Table 5.4. Simulation models and their potential application.

Acronym	Extended name	Purpose/simulation
APSIM	Agricultural production systems simulator	Effect of agronomic management practices on crop productivity and changes in soil properties
APSIM–SWIM	Agricultural production systems simulator – soil water infiltration and movement	Effect of agronomic practices on crop productivity and soil processes using SWIM module
CENTURY	-	Change in nitrogen (N), organic carbon (C), phosphorus (P), and sulphur (S) in the soil due to changes in agronomic management of various land-use systems
CERES–RICE	Crop estimation through resource environment synthesis for rice	A component model of DSSAT v3.5
DSSAT v3.5	Decision support systems for agrotechnology transfer, version 3.5	Effect of agronomic management practices on crop productivity and changes in soil properties
PERFECT	Productivity, erosion, runoff functions to evaluate conservation techniques	Effect of various conservation techniques on runoff, soil erosion and crop productivity
RothC–26.3	Rothamsted Carbon model, version 26.3	Carbon changes in the soil in response to various land and crop-residue management practices
SCUAF	Soil changes under agroforestory	C and N changes in soils in response to land clearing and agronomic management of agroforestory systems
SIMOPT2–MAIZE	A simulation-multi-criteria optimisation software for maize	Optimise productivity and N losses using CERES–MAIZE model
WATBAL	A simple water balance model	Estimate the soil moisture regimes of a site from readily available climatic data

The sustainability of production, soil quality and other environment resources are the major impact factors of NRM. Detailed empirical research over a period of time and space is required to quantify the impacts of improved management on these desirable outcomes. However, such long-term studies are costly and time-consuming; simulation models provide a cost-effective and efficient complementary approach to long-term field experimentation for *ex ante* analysis of the long-term impacts of NRM options. These models have often been validated on a plot or field scale. On a watershed scale, the models can be integrated with GIS to study spatial variability effects on crop production and the state of natural resources, enhancing their capability for up-scaling and user-friendly mapping. Thus, the models are useful when undertaking temporal trend analyses, and when incorporating a spatial component to assess the NRM impact on various processes governing sustainability. For example, considering past trends and current management practices using simulation models, Fisher *et al.* (2002) assessed the long-term (25–50 year) impact on crop yields of climatic change including the occurrence of droughts. In the following section, examples and approaches for assessing the impact of NRM using simulation models and GIS are discussed.

Impacts of land surface management on runoff, soil erosion and productivity

Runoff, soil loss and nutrient depletion are the major agents of human-induced land degradation (Pathak *et al.*, Chapter 3, this volume; Sahrawat *et al.* Chapter 4, this volume). Freebarin *et al.* (1991) used the results of two long-term field experiments to develop coefficients for soil processes and to validate the PERFECT model for two sites in Australia. Then they used the model to assess the impact of various management practices such as crop/fallow sequences, tillage, and effects of various amendments that modify soil physical processes. Long-term (100^{+} years) simulated results showed the decline in yields associated with soil erosion and removal of the previous season's crop stubble from the field. Singh *et al.* (1999) used DSSAT v3.5 to assesses the impact of two land surface configurations on surface runoff and yields of soybean and chickpea using experimental data (2 years) and historical weather data (22 years). It was found that in most years BBF decreased runoff from the soil, but had a marginal effect on yields of soybean and chickpea. The decreased runoff was associated with an increase in deep drainage and reduced soil loss. Wani *et al.* (2002) used a simple WATBAL model (Keig and McAlpine, 1974) along with GIS to assess the available soil moisture and excess runoff water available for harvest at the district level.

Nelson *et al.* (1998) used the APSIM model to evaluate the sustainability of maize crop management practices in the Philippines using hedgerows to minimise land degradation. Intercropping maize with hedgerows was used to assess the long-term sustainability of maize production due to reduced soil erosion. In the absence of hedgerows, continuous maize cultivation turned out to be unsustainable in the long term, although the inclusion of a fallow period slowed the productivity decline by spreading the effect of erosion over a larger cropping area.

Impact of nitrogen management on leaching

Field experiments conducted in environments with highly variable climates may give misleading results, as the years in which they are conducted might not represent the long-term average. In such cases, simulation models provide a rigorous mechanism to assess the long-term risks of specific management options. Verburg *et al.* (1996) using the APSIM–SWIM model assessed the long-term (33 years) impact of different irrigation management strategies and N application on sugarcane yield and nitrate leaching. Alocilja and Ritchie (1993) used the SIMOPT2–MAIZE model to investigate the trade-offs between maximised profits and minimised nitrate leaching. Thornton *et al.* (1995) took the analysis a step further by linking it to GIS with spatial databases of soils and weather to analyse the influence of N management on crop yield and leaching at the regional level. Such a linkage not only allowed an analysis of the spatial variability due to different soil types and weather across the region, but also the temporal variation associated with changes in weather.

Singh and Thornton (1992) simulated the effects of various nutrient management strategies on N leaching from rice fields in Thailand using the CERES–RICE model. The results obtained from a 25-year simulation suggested that on well-managed clayey soils, medium- to high-input agriculture can be highly productive and environmentally sustainable. Leaching losses were considerably higher on sandy soils than on clay soils. The N loss was inversely related to the depth of urea incorporation and could be minimised by deep placement.

Production systems and soil quality

A number of cropping systems simulation models incorporate the simulation of soil processes such as soil water dynamics, decomposition and mineralisation of added crop residues and organics, with simulation of N fixation by legumes, thus providing the opportunity to evaluate yield responses to application of organic matter and the integration of legumes. Probert *et al.* (1998) used the APSIM for simulating the performance of hypothetical chickpea–wheat rotations on clay soils in Queensland, Australia. The simulation results indicated that soil organic matter (SOM) and N steadily declined over 25 years under continuous wheat cropping without N fertiliser application, whereas the integration of chickpea into the rotation considerably reduced the soil fertility decline. Similar results were obtained by Bowen and Baethgen (1998) using the DSSAT models to assess the long-term sustainability impacts of various cropping systems in Brazil. A continuous maize–fallow system without fertiliser application caused maize yields to decline gradually over 50 years, whereas a green-manure–maize–fallow system was able to sustain yields over the same period.

Menz and Grist (1998) applied the SCUAF model to evaluate the impact of vegetation burning and changing the length of the fallow period in shifting cultivation systems in Indonesia. The results were used to assess the economic viability of different management options in terms of returns from rice cultivation. It was concluded that although more-intensive cultivation carried a future yield penalty, systems with extended fallow periods were unable to overcome the more immediate economic gains to be made from intensive cropping.

Shepherd and Soule (1998) developed a farm simulation model to assess the long-term impact of existing soil management strategies on productivity, profitability, and sustainability of farms in western Kenya. The model linked soil management practices with nutrient availability, crop and livestock productivity, and farm economics. A wide range of soil management options was simulated, including crop residue and manure management, soil erosion control measures, green manuring, crop rotations, and N and P fertiliser application. The dynamic model was applied for Vihiga district in western Kenya, and was used to assess the sustainability of the existing systems using three household groups (farms) in the area. It was shown that the low and medium resource endowment farms had declining SOM, negative C, N and P budgets, and low productivity and profitability. The high resource endowment farms, on the other hand, had increasing SOM, low soil nutrient losses and were productive and profitable. This approach showed the dangers of relying on nutrient balances of an 'average' farm-type. The authors concluded that when the required capital is available, farmers can invest in NRM options that improve profitability without sacrificing long-term sustainability.

Carbon sequestration

Conducting long-term experiments could also be used to monitor the changes in soil C contents associated with NRM investments. Alternatively, soil C simulation models can also be used to simulate the impact of NRM interventions on C sequestration in soils on farm and catchment scales. The most commonly used models are RothC–26.3 (Coleman and Jenkinson, 1996) and CENTURY (Parton *et al.*, 1987). More recently DSSAT v3.5 (Gijsman *et al.*, 2002) and APSIM softwares have also incorporated soil C balance subroutines to simulate soil C change along with analysis of crop productivity. The simulation approach avoids long-term experimentation and the models can be validated using empirical data along with known biochemical relationships in the soils. Probert *et al.* (1998) used the CENTURY and APSIM models to examine the effects of tillage, stubble management and N fertiliser on the productivity of a winter-cereal–summer-fallow cropping system in Australia. Both models predicted that for this continuous cereal cropping system there would be a decline in SOM (organic C = SOM/1.72).

Furthermore, the C stocks at regional or ecoregional levels can be calculated using GIS and measurements of C at benchmark sites for a given soil series and management system. Velayutham *et al.* (2000) calculated C stocks in India using information on soil series and measurements at benchmark locations that were extrapolated using GIS techniques.

Monitoring Spatial and Temporal Dynamics of Agro-ecosystems

Natural resource management interventions result in multi-faceted biophysical impacts including the establishment of vegetation cover, reduction in soil loss, increase in the number and spatial coverage of water bodies, changes in water quality, and groundwater recharge. These changes can be monitored over space and time. Remote sensing and GIS are the most suitable tools for monitoring such spatial and temporal dynamics. By providing synoptic and repetitive coverage at regular intervals, remote sensing offers high potential for monitoring observable changes. Remote sensing refers to making an observation on a feature or phenomenon without being in physical contact with it. In nature, every object reflects and/or emits a fraction of incident radiant energy that makes it possible to derive coded information that will help to remotely sense the condition of the resource under study. *In situ* air and/or spaceborne spectral measurements are made to detect various natural and/or cultural features. GIS is a tool used to store, retrieve, analyse and integrate spatial and attribute data. The system helps to generate development plans by integrating information on natural resources with the ancillary information, and to develop a decision-support system.

Impact assessment of NRM technologies/interventions often involves the evaluation and monitoring of changes in selected indicators at a reference site. For this purpose, the reference site needs to be characterised in terms of its natural resources and environmental conditions. Remote sensing holds very good promise for providing information on changes in land use/land cover, quality of surface water, vegetation cover and dynamics of degraded land, which can in turn be used as indicators of agricultural sustainability. Since NRM is implemented on various scales ranging from plot/farm to watershed and river basin, impact assessments also need to be made using a database with a matching spatial scale. In this context, spaceborne/airborne spectral measurements with varying spatial resolution, ranging from about 1 km (geo-stationary satellites) to the sub-metre level (Quickbird–II mission), provide the desired details of terrain features that enable assessment of the impact of diverse biophysical NRM impacts. How spaceborne multispectral data could be used to monitor the spatial and temporal dynamics of agro-ecosystems is discussed below. A synthesis of different satellite systems used in monitoring biophysical dynamics of agro-ecosystems is given in Appendix 5.1.

Land-use change and intensification

Gemini and Apollo space photographs were used to map land use/land cover in the late 1960s (Aldrich, 1971), but operational use of spaceborne multispectral measurements for land use/land cover mapping only began with the launching of the Earth Resources Technology Satellite (ERTS–1), later named Landsat–1, in July 1972 (Anderson *et al.*, 1976). Subsequently, data from other satellites in the Landsat series, along with the Satellite pour observation de la terre (SPOT) and the Indian Remote Sensing Satellite (IRS–1A/–1B/–1C/–1D) have been operationally used to collect information on land use/land cover on various scales ranging from regional to micro-watershed level (Landgrebe, 1979). The utility of spaceborne multispectral data in the detection of changes in land-use patterns is illustrated by an example from a micro-watershed of Ghod catchment in Maharashtra, India. The Linear Imaging Scanning Sensor (LISS–III) aboard IRS–1C/–1D, and Landsat–5 Thematic Mapper (TM) data for the period 1985/86 and 1999/2000 were used to generate agricultural land use maps (not shown) and data (Table 5.5). The area estimated from analysis of satellite data revealed that compared with 166 ha during the period 1985/86, the area under postrainy-season cropping had increased to 251 ha during 1999/2000. A similar trend was observed in the spatial extent of other land uses.

Table 5.5. Impact of NRM on land use in gd24 micro-watershed, Ghod catchment, Maharashtra, India, during 1985/86 to 1999/2000.

	Area (ha)	
Land use	1985/86	1999/2000
Rainy season *(Kharif)*	192	193
Postrainy season *(Rabi)*	166	251
Double crop	144	243
Fallow	158	99
Forest	6	6
Scrubland	256	177
Barren/rocky	411	360
Water bodies	0	4
Built-up	0	0
Total	1,333	1,333

Vegetation cover

Amongst various biophysical parameters relevant to NRM impact assessment, vegetation density and vigour, and above ground biomass can be detected from spaceborne spectral measurements. Higher reflection in the near-infrared region (NIR) and considerable absorption in the red region (*R*) of the spectrum of green plants enables their detection using remote-sensing techniques. Absorption in the red region is due to the presence of chlorophyll in plant leaves, while reflection in the NIR region results from the inter-cellular space of plant leaves. Various vegetation indices – normalised difference vegetation index (NDVI), transformed vegetation index (TVI), and soil-adjusted vegetation index (SAVI) – can be derived from spectral measurements that are related to biomass, vegetation density and vigour, and crop yield. The NDVI is most commonly used as a surrogate measure of the vigour and density of vegetation, and is computed from spectral measurements in the red (0.63–0.69 µm) and near-infrared (0.76–0.90 µm) region as follows:

$$\text{NDVI} = \frac{\text{NIR} - \text{R}}{\text{NIR} + \text{R}} \tag{4}$$

where NIR is spectral responses of vegetation in the near infrared and R for red regions of the spectrum. Index values can range from – 1.0 to 1.0, but vegetation values typically range between 0.1 and 0.7. Higher index values are associated with higher levels of healthy vegetation cover. NDVI can be used as an indicator of change in relative biomass and greenness.

The utility of NDVI for assessment of vegetation development is illustrated in Table 5.6 for a micro-watershed in the Ghod catchment, Maharashtra, India. Soil and water conservation interventions resulted in the establishment of vegetation cover during the period 1985/86 to 1999/2000, that could be monitored through temporal NDVI images. As is evident from Table 5.6, the area under the three NDVI ranges (0.20–0.39, 0.40–0.59 and 0.6) has increased substantially (National Remote Sensing Agency, 2001a). This shows that the area under various levels of vegetation cover has increased from 1985 to 1999.

Table 5.6. Vegetation dynamics in gc3b micro watershed in Ghod catchment, Maharashtra.

	Area (ha)	
NDVI range	1985/86	1999/2000
<0.0	1,519	1,312
0.00 – 0.19	936	859
0.20 – 0.39	329	469
0.40 – 0.59	117	227
>0.60	96	130
Total	2,997	2,997

Monitoring changes in surface water resources

Because of its characteristic absorption feature in the near-infrared region of the electromagnetic radiation, surface water is easily detected in remotely sensed images. The high transmittance of incident radiation in the blue region (0.45–0.52 µm) enables the discrimination of clear water from turbid water. The turbidity causes most of the incident radiation in the blue region to reflect, resulting in a higher spectral response. Moore and North (1974) and Adam *et al.* (1998) used optical and microwave sensor data to delineate floodwater boundaries. Lathrop and Lillesand (1986) used Landsat–TM data to assess water quality in Southern Great Bay and West Central Lake, Michigan,

USA. The temporal change in the spatial coverage of reservoirs after NRM interventions has been studied in the Ghod catchment (Fig. 5.2). While the water spread in the reservoir was about 3 ha in 1985, it increased to 16 ha by 1999 following the implementation of soil and water conservation measures (National Remote Sensing Agency, 2001a).

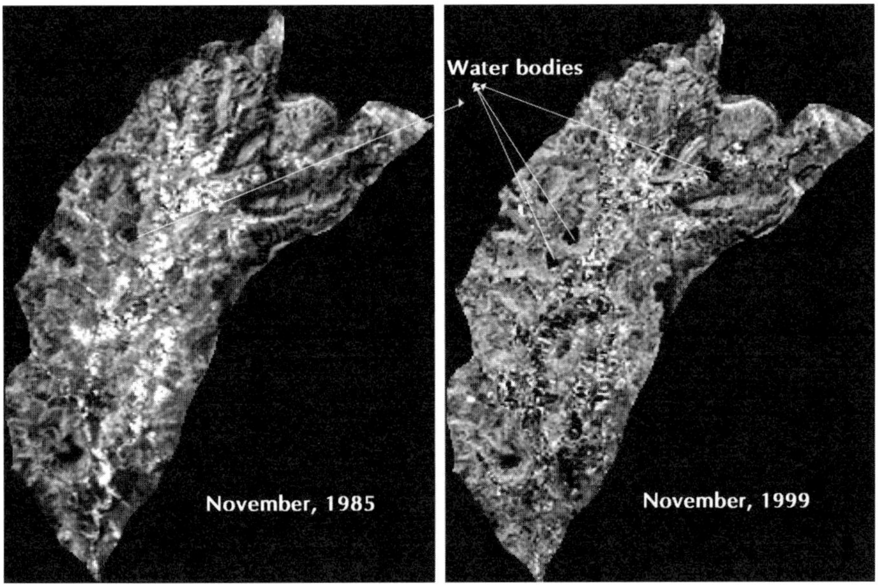

Fig. 5.2. Changes in surface water resources in gc1f micro-watershed, Ghod catchment, Maharashtra, India.

Monitoring the dynamics of degraded lands

Natural resource management interventions in degraded land areas often result in improvements in soil quality and gradual improvement in vegetation cover. Spaceborne multispectral images have been extensively used to inventory and study the dynamics of eroded lands (Wu *et al.*, 1997), salt-affected soils (Dwivedi *et al.*, 2001), waterlogged areas (Wallace *et al.*, 1993), areas of shifting cultivation (Dwivedi and Ravi Sankar, 1991) and the land affected by tanneries' effluents (National Remote Sensing Agency, 1999). The following examples illustrate the use of Earth Observation Satellite data in this endeavour.

Eroded lands

Investment in soil conservation measures in a given area, generally, results in reduced soil loss, reduced soil erosion, and improved soil moisture status, and vegetation cover/biomass. The extent of land degradation is directly related to ground cover that can be quantified using remote sensing data. An illustrative example of eroded lands in the 'rg2h' mini-watershed of the Ramganga catchment, Uttaranchal Pradesh, northern India, during the periods 1985/86 and 1999/2000 is shown in Fig. 5.3. The figure shows that

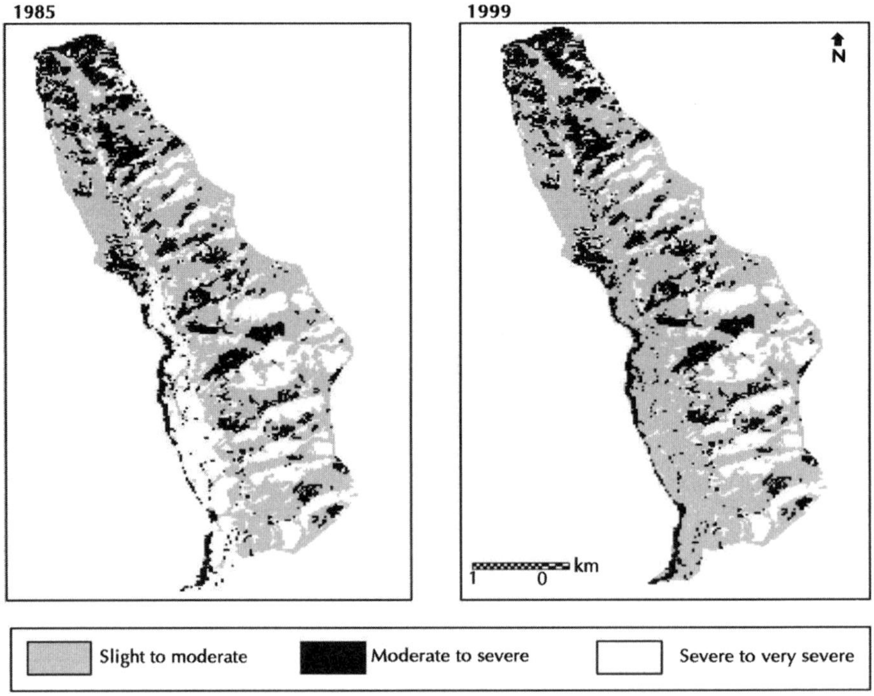

Fig. 5.3. Change in spatial extent and distribution of eroded lands in rg2h micro-watershed, Ramganga catchment, Uttaranchal, northern India.

there has been substantial shrinkage in the spatial extent of moderately eroded lands with concomitant increase in the slightly eroded category (National Remote Sensing Agency, 2001b). In 1985 an estimated 691 ha of land suffered due to moderate soil erosion. By 1999, this had been reduced to 457 ha while the slightly eroded category expanded to 1128 ha from 901 ha in 1985.

Waterlogged areas

Waterlogging in arid and semi-arid regions with alternate wet and dry periods leads to the development of soil salinity. By virtue of the very low response of water in the near-infrared region of the spectrum, the detection of waterlogged areas, especially those with surface ponding or a thin film of water at the surface from remote sensing images is easy. Figure 5.4 shows an example from Mahanadi Stage-1 command area in Kendrapara district, Orissa, eastern India. Gentle slopes and the presence of lenses of clay that act as a hydrological barrier, and irrigation by flooding have contributed to the development of waterlogging. There has been an appreciable increase in the spatial extent of both seasonally and perennially waterlogged areas. Whereas an estimated 389 ha of land were found to be subject to seasonal waterlogging in 1985, by 1999 this had risen to 442 ha.

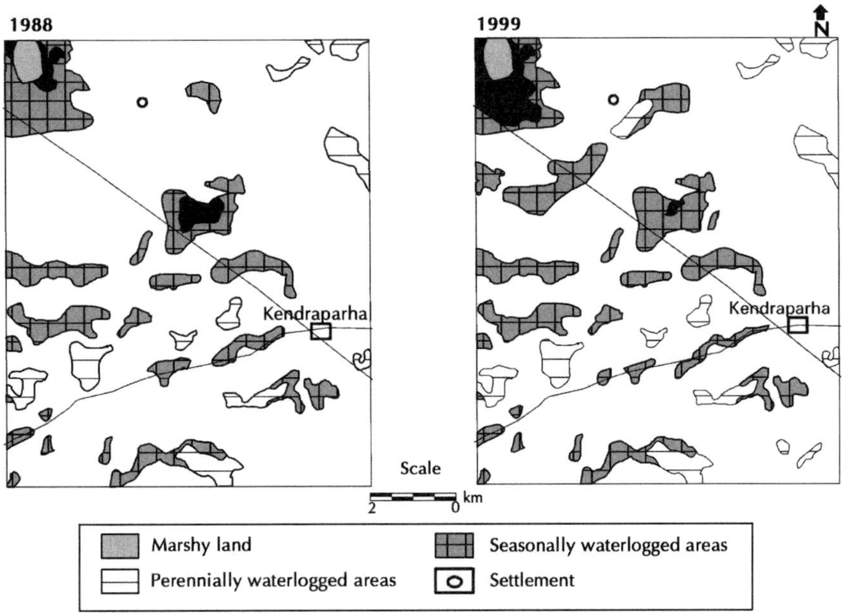

Fig. 5.4. Dynamics of waterlogging in part of Kendrapartha district, Mahanadi Stage-I command, Orissa, India.

Summary and Conclusions

Assessing the multi-dimensional impacts of NRM interventions – especially in non-tangible environmental services – is not an easy task. Monitoring selected indicators through direct observation during and after project implementation or through simulation modelling is a useful approach that will enhance options for evaluating the impacts of NRM interventions. Difficulties on various scales could be overcome through the application of such available tools as GIS and remote sensing. Off-site impacts on ecological functions and ecosystem services such as the effects on water quality, land quality, siltation, groundwater recharge, and C sequestration can also be assessed by systematic monitoring using remote sensing and ground-truthing measurements.

In this chapter various indicators and tools that can be used to monitor the impacts of NRM interventions were presented. They focused on biophysical indicators for ecosystem services and discussed various tools used to generate data on such indicators. Agro-biodiversity and agro-ecosystem efficiency indicators can be applied on different spatial scales. The impacts of NRM technologies on C sequestration and other ecosystem services can be either measured directly through long-term studies or simulated using agro-biological simulation models. The latter approach is becoming increasingly popular as long-term experimentation and monitoring become either impossible or highly costly. However, the approach requires climatic and agronomic data to estimate potential impacts by calibrating the models to specific local conditions.

Remote sensing in conjunction with *in situ* observations/measurements (ground-truthing) offers tremendous potential in providing timely information on the spatial extent and temporal behaviour of various indicators on scales ranging from micro-watersheds to regional/ecoregional levels. Remote sensing methods are being used to monitor changes in land resource conditions, vegetation dynamics, surface water resources, and to assess changes in levels of land degradation. In the future, the impact of NRM on such environmental services as C sequestration and groundwater recharging could also be monitored or derived from satellite images as new satellites equipped with an array of sensors are launched. On a watershed scale, crop simulation models and water balance models can be important tools for evaluating the biophysical impacts of proposed interventions. Several indicators including those for agro-biodiversity and agro-ecosystem efficiency could also be useful at the micro-watershed level.

Such recently launched satellites as Resourcesat–1(IRS–P6) with varying spatial resolution ranging from 56 m from Advanced Wide Field Sensor (AWiFS) to 23 m from LISS–III to 5.8 m from LISS–IV offer unique opportunities to monitor biophysical impact indicators on different spatial scales. Integrating panchromatic data with 2.5-m and 1-m spatial resolution from such future Earth observation missions as Cartosat–1 and Cartosat–2, will further enhance the value of data from the Resourcesat–1 satellite.

Despite the technological advances and the impressive progress made in the last few years, there will be a need for future research to enhance and develop methods and indicators to assess NRM impacts on ecoregional scales. Such indicators will complement and enhance economic approaches for evaluating the impacts of NRM interventions, especially on larger spatial scales. Methods and indicators for the quantification of various difficult-to-quantify environmental services and for monitoring such non-quantitative impacts as effects on implementation processes, policies and institutional arrangements, changes in social capital, and capacity building and empowerment of local communities will also need attention in future research.

References

Adam, S., Wiebe, J., Collins, M. and Pietroniro, A. (1998) RADARSAT flood mapping in the Peace-Athabasca Delta, Canada. *Canadian Journal of Remote Sensing* 24, 69–79.

Aldrich, R. (1971) Space photos for land use and forestry. *Photogrammetric Engineering and Remote Sensing* 37, 389–401.

Alocilja, E.C. and Ritchie, J.T. (1993) Multicriteria optimization for a sustainable agriculture. In: Penning de Vries, F.W.T., Teng, P. and Metselaar, K. (eds) *Systems Approaches for Agricultural Development*. Kluwer Academic Publishers, Dordrecht, The Netherlands, pp. 381–396.

Altieri, M.A. (1999) The ecological role of biodiversity in agroecosystems. *Agriculture, Ecosystems and Environment* 74, 19–31.

Anderson, J.R., Hardy, E.E., Roach, J.T. and Witmer, R.E. (1976) A land use and land cover classification for use with remote sensor data. *U.S. Geological Survey Professional Paper no. 964*. U.S. Government Printing Office, Washington, DC, 28 pp.
Bowen, W.T. and Baethgen, W.E. (1998) Simulation as a tool for improving nitrogen management. In: Tsuji, G.Y., Hoogenboom, G. and Thornton, P.K. (eds) *Understanding Options for Agricultural Production. Systems Approaches for Sustainable Agricultural Development*. Kluwer Academic Publishers, Dordrecht, The Netherlands, pp. 189–204.
Brookfield, H. and Padoch, C. (1994) Appreciating agro-biodiversity: a look at the dynamism and diversity of indigenous farming practices. *Environment* 36(5), 6–11, 37–43.
Bruce, J.P., Frome, M., Haites, E., Janzen, H., Lal, R. and Paustian K. (1999) Carbon sequestration in soils. *Journal of Soil and Water Conservation* 54, 382–389.
Chadwick, D.H. (1993) Seeking meanings. *DEFENDERS Magazine Winter* 1992/1993, 3, 166–168.
Coleman, K. and Jenkinson, D.S. (1996) RothC-26.3 – A model for the turnover of carbon in soil. In: Powlson, D.S., Smith P. and Smith, J.U. (eds) *Evaluation of Soil Organic Matter Models Using Existing Long-term Datasets*. NATO ASI series I, Vol. 38. Springer-Verlag, Heidelberg, Germany, pp. 237–246.
Dwivedi, R.S. and Ravi Sankar, T. (1991) Monitoring shifting cultivation areas using spaceborne multispectral and multitemporal data. *International Journal of Remote Sensing* 12(3), 427–433.
Dwivedi, R.S., Ramana, K.V., Thammappa, S.S. and Singh, A.N. (2001) The utility of IRS-1C LISS-III and PAN-merged data for mapping salt-affected soils, 2001. *Photgrammetric Engineering and Remote Sensing* 67(10), 1167–1175.
Erlich, P.R. and Erlich, A.H. (1992) The value of biodiversity. *Ambio* 21(3), 219–226.
FAO (Food and Agriculture Organization of the United Nations) (1992) *Sustainable Development and the Environment. FAO Policies and Actions, Stockholm 1972–Rio 1992*. FAO, Rome, Italy, 88 pp.
Fisher, G., Shah, M. and van Velthuizen, H. (2002) *Climate Change and Agricultural Vulnerability. A Special Report*. IIASA (International Institute for Applied Systems Analysis), Laxenburg, Austria, 152 pp.
Freebarin, D.M., Littleboy, M., Smith, G.D. and Coughlan, K.J. (1991) Optimizing soil surface management in response to climate risk. In: Muchow, R.C. and Bellamy, J.A. (eds) *Climatic Risk in Crop Production – Models and Management for the Semi-Arid Tropics and Sub-Tropics*. CAB International, Wallingford, UK, pp. 283–305.
Gijsman, A.J., Hoogenboom, G., Parton, W.J. and Kerridge, P.C. (2002) Modifying DSSAT crop models for low-input agricultural systems using a soil organic matter–residue module from century. *Agronomy Journal* 94, 462–474.
Glave, M. and Escobal, J. (1995) Indicatores de sostenibilidad para la agricultura Andina. *Debate Agrario* 23, 69–112.
Izaurralde, R.C., Rosenberg, N.J. and Lal, R. (2001) Mitigation of climate change by soil carbon sequestration: Issues of science, monitoring and degraded lands. *Advances in Agronomy* 70, 1–75.
Keig, B. and McAlpine, J.R. (1974) WATBAL: A computer system for the estimation and analysis of soil moisture regimes from simple climatic data (2nd edn). *Technical Memorandum no. 74/4*. Division of Land Use Research, Commonwealth Scientific and Industrial Research Organisation (CSIRO), Canberra, Australia, 45 pp.

Lal, R. (1999) Soil management and restoration for C sequestration to mitigate the greenhouse effect. *Progress in Environmental Science* 1, 307–326.

Landgrebe, D.A. (1979) Monitoring the earth's resources from space – Can you really identify crops by satellites. In: *Proceedings of the National Computer Conference, New York, NY, 7 June 1979*. Laboratory for Applications of Remote Sensing (LARS) Technical Note 060779, pp. 233–241.

Lathrop, R.G. and Lillesand, T.M. (1986) Utility of thematic mapper data to assess water quality in Southern Green Bay and West-Central Lake, Michigan. *Photogrammetric Engineering and Remote Sensing* 52, 671–680.

Lynam, J. and Herdt, R. (1989) Sense and sustainability as an objective in international agricultural research. *Agricultural Economics* 3, 381–398.

MacKinnon, J. and MacKinnon, K. (1986) *Review of the Protected Areas System in the Indo-Malayan Realm*. IUCN (The World Conservation Union) and the United Nations Environment Programme, Gland, Switzerland.

McLaughlin, A. and Mineau, P. (1995) The impact of agricultural practices on biodiversity. *Agriculture, Ecosystems and Environment* 55, 201–212.

Menz, K. and Grist, P. (1998) Bio-economic modelling for analyzing soil conservation policy issues. In: Penning de Vries, F.W.T., Agus, F. and Kerr, J. (eds) *Soil Erosion at Multiple Scales*. CAB International, Wallingford, UK, pp. 39–49.

Merrick, L. (1990) Crop genetic diversity and its conservation in traditional agroecosystems. In: Altieri, M. and Hecht, S. (eds) *Agroecology and Small Farm Development*. CRC Press, Boca Raton, Florida, pp. 3–11.

Moore, G.K. and North, G.W. (1974) Flood inundation in the southeastern United States from aircraft and satellite imaging. *Water Resources Bulletin* 10(5), 1082–1097.

Munasinghe, M. and McNeely, J. (1995) Key concepts terminology of sustainable development. In: Munasinghe, M. and Shearer, W. (eds) *Defining and Measuring Sustainability. The Biogeophysical Foundations*. The United Nations University (UNU), Tokyo, Japan, and The World Bank, Washington, DC, pp. 20–56.

Nelson, R.A., Dimes, J.P., Silburn, D.M., Paningbatan, E.P. and Cramb, R.A. (1998) Erosion/productivity modelling of maize farming in the Philippine uplands. II. Simulation of alternative farming methods. *Agricultural Systems* 58, 147–163.

Noss, R.F. (1990) Indicators for monitoring biodiversity: a hierarchical approach. *Conservation Biology* 4, 355–364.

NRSA (National Remote Sensing Agency) (1999) Assessment of land degradation due to tanneries: A remote sensing approach. *Technical Report*. NRSA, Hyderabad, India, 43 pp.

NRSA (National Remote Sensing Agency) (2001a) Evaluation of the impact of implementation of soil conservation measures in Ghod catchment, Maharashtra. *Technical Report*. NRSA, Hyderabad, India, 29 pp.

NRSA (National Remote Sensing Agency) (2001b) Evaluation of the impact of implementation of soil conservation measures in Ramganga catchment, Uttar Pradesh. *Technical Report*. NRSA, Hyderabad, India, 20 pp.

Parton, W.J., Stewart, J.W.B. and Cole, C.V. (1987) Dynamics of C, N, S and P in grassland soils: a model. *Biogeochemistry* 5, 109–131.

Paustian, K., Agren, G. and Bosatta, E. (1997) Modelling litter quality effects on decomposition and soil organic matter dynamics. In: Cadische, G. and Giller, K.E. (eds) *Driven by Nature: Plant Litter Quality and Decomposition*. CAB International, Wallingford, UK, pp. 313–336.

Pretty, J. and Ball, A. (2001) Agricultural influences on carbon emissions and sequestration. A review of evidence and the emerging trading options. *Occasional Paper no. 2001-03*. Centre for Environment and Society, University of Essex, Colchester, UK, 31 pp.

Probert, M.E., Carberry, P.S., McCown, R.L. and Turpin, J.E. (1998) Simulation of legume-cereal system using APSIM. *Australian Journal of Agricultural Research* 49, 317–327.

Ramakrishnan, P. (1995) Currencies for measuring sustainability: Case studies from Asian highlands. In: Munasinghe, M. and Shearer, W. (eds) *Defining and Measuring Sustainability. The Biogeophysical Foundations*. The United Nations University (UNU), Tokyo, Japan and The World Bank, Washington, DC, pp. 193–206.

Reid, W.V., McNeely, J.A., Tunstall, D.B., Bryant, D.A. and Winograd, M. (1993) *Biodiversity Indicators for Policy-Makers*. World Resources Institute, Washington, DC, 42 pp.

Shannon, C.E. and Weaver, W. (1963) *The Mathematical Theory of Communication*, 2nd edn. University of Illinois Press, Urbana, Illinois, 117 pp.

Shepherd, K.D. and Soule, M.J. (1998) Soil fertility management in West Kenya: a dynamic simulation of productivity, profitability and sustainability at different resource endowment levels. *Agricultural, Ecosystems and Environment* 71, 131–145.

Simpson, H.E. (1949) Measurement of diversity. *Nature* 163, 668.

Singh, P., Alagarswamy, G., Pathak, P., Wani, S.P., Hoogenboom, G. and Virmani, S.M. (1999) Soybean–chickpea rotation on Vertic Inceptisols I. Effect of soil depth and landform on light interception, water balance and crop yields. *Field Crops Research* 63, 211–224.

Singh, P., Vijaya, D., Srinivas, K. and Wani, S.P. (2002) Potential productivity, yield gap, and water balance of soybean–chickpea sequential system at selected benchmark sites in India. *Research Report No. 1. Global Theme 3: Water, Soil and Agrobiodiversity Management for Ecosystem Health*. International Crops Research Institute for the Semi-Arid Tropics (ICRISAT), Patancheru, India, 52 pp.

Singh, U. and Thornton, P.K. (1992) Using crop models for sustainability and environmental quality assessment. *Outlook on Agriculture* 21, 209–218.

Smith, P. (2004) Carbon sequestration in croplands: the potential in Europe and the global context. *European Journal of Agronomy* 20, 229–236.

Smyth, A. and Dumanski, J. (1993) FESLM: An international framework for evaluating sustainable land management. *FAO World Soil Resources Reports No. 73*. Food and Agriculture Organization of the United Nations (FAO), Rome, Italy, 76 pp.

Spellerberg, I.F. and Fedor, P.J. (2003) A tribute to Claude Shannon (1916–2001) and a plea for more rigorous use of species richness, species diversity and the 'Shannon-Wiener' Index. *Global Ecology & Biogeography* 12, 177–179.

Takacs, D. (1996) *The Idea of Biodiversity: Philosophies of Paradise*. The Johns Hopkins University Press, Baltimore, Maryland, 393 pp.

Thornton, P.K., Saka, A.R., Singh, U., Kumwenda, J.D.T., Brink, J.E. and Dent, J.B. (1995) Application of a maize crop simulation model in the central region of Malawi. *Experimental Agriculture* 31, 213–226.

Thrupp, L.A. (1998) *Cultivating Diversity: Agrobiodiversity and Food Security*. World Research Institute, Washington, DC, 70 pp.

Tisdell, C. (1996) Economic indicators to assess the sustainability of conservation farming projects: An evaluation. *Agriculture, Ecosystems and Environment* 57, 117–131.

Velayutham, M., Pal, D.K. and Bhattacharyya, T. (2000) Organic carbon stock in soils of India. In: Lal, R., Kible, J.M. and Stewart, B.A. (eds) *Advances in Soil Science. Global Climate Change and Tropical Ecosystems*. CRC Press, Boca Raton, Florida, pp. 71–95.

Verburg, K., Keating, B.A., Bristow, K.K., Huth, N.I., Ross, P.J. and Catchpoole, V.R. (1996) Evaluation of nitrogen fertilizer management strategies in sugarcane using APSIM-SWIM. In: Wilson, J.R., Hogarth, D.M., Campbell, J.A. and Garside, A.K. (eds) Sugarcane: Research Towards Efficient and Sustainable Production. Division of Tropical Crops and Pastures, Commonwealth Scientific and Industrial Research Organisation (CSIRO), Brisbane, Queensland, Australia, pp. 200–202.

Vlek, P.L.G. (1990) The role of fertilizers in sustaining agriculture in sub-Saharan Africa. *Fertilizer Research* 26, 327–339.

Wallace, J., Campbell, N.A., Wheaton, G.A. and McFarlane, D.J. (1993) Spectral discrimination and mapping of waterlogged cereal crops in Western Australia. *International Journal of Remote Sensing* 14(14), 2731–2743.

Wani, S.P., McGill, W.B., Robertson, J.A., Haugen Kozyra, K.L. and Thurston, J.J. (1994) Improved soil quality, increased barley yields with faba beans and returned manure in a crop rotation on a gray Luvisol. *Canadian Journal of Soil Science* 74, 75–84.

Wani, S.P., Pathak, P., Tam, H.M., Ramakrishna, A., Singh, P. and Sreedevi, T.K. (2002) Integrated watershed management for minimizing land degradation and sustaining productivity in Asia. In: Adeel, Z. (ed.) *Integrated Land Management in Dry Areas: Proceedings of a Joint UNU–CAS International Workshop, 8–13 September 2001, Beijing, China.* United Nations University, Tokyo, Japan, pp. 207–230.

Wani, S.P., Pathak, P., Jangawad, L.S., Eswaran, H. and Singh, P. (2003a) Improved management of Vertisols in the semi-arid tropics for increased productivity and soil carbon sequestration. *Soil Use and Management* 19, 217–222.

Wani, S.P., Maglinao, A.R., Ramakrishna, A. and Rego, T.J. (eds) (2003b) *Integrated Watershed Management for Land and Water Conservation and Sustainable Agricultural Production in Asia: Proceedings of the ADB–ICRISAT–IWMI Project Review and Planning Meeting, 10–14 December 2001, Hanoi, Vietnam.* International Crops Research Institute for the Semi-Arid Tropics, Patancheru, India, 268 pp.

Wu, J., Nellis, M.D., Ransom, M.D., Price, K.P. and Egbert, S.L. (1997) Evaluating soil properties of CRP land using remote sensing and GIS in Finny county, Kansas. *Journal of Soil and Water Conservation* 52(5), 352–358.

Appendix 5.1. Brief summary of remote sensing satellites and their characteristics.

Satellite	Owner	Launch	Sensors[a]	Spectral range (mm)	Resolution (m)	Swath (km)	Revisit (days)
Cartosat–1	India	Expected launch 2004	PAN	0.55–0.75	2.5	30	6–7
Cartosat–2	India	Expected launch 2006	PAN	0.55–0.75	<1	10	
IKONOS–II	USA	1999	PAN	0.45–0.9	1	11	1–4
			MSS	0.45–0.52, 0.52–0.6 0.63–0.69, 0.76–0.9	4		
IRS–1A and 1B	India	1988 and 1991	LISS–I	0.45–0.52, 0.52–0.59 0.62–0.68, 0.77–0.86	72.5	148	22
			LISS–II	0.45–0.52, 0.52–0.59, 0.62–0.68, 0.77–0.86	36.25	74	
IRS–1C and 1D	India	1995 and 1998	WiFS	0.62–0.68, 0.77–0.86	189	810	5
			LISS–III	0.52–0.59, 0.62–0.68 0.77–0.86	23.6	142	24–25
				1.55–1.70	70.8	148	
			PAN	0.50–0.75	5.8	70	
Landsat–1	USA	1972	MSS	0.5–0.6, 0.6–0.7, 0.7–0.8, 0.8–1.1	82	185	18
Landsat–5	USA	1984	MSS	Same as Land-sat–1	Same as Land-sat–1		
			TM	0.45–0.52, 0.52–0.6 0.63–0.69, 0.76–0.9 1.55–1.75, 2.08–2.35	30	185	16
				10.4–12.5	120		
Quickbird–II	USA	2001	MSS	0.45–0.52, 0.52–0.6 0.63–0.69, 0.76–0.89	2.5	17	1–3.5
			PAN	0.45–0.9	0.61		

Appendix 1. Continued.

Satellite	Owner	Launch	Sensors	Spectral range (mm)	Resolution (m)	Swath (km)	Revisit (days)
Resourcesat–1	India	2003	LISS–IV	0.52–0.59,			
				0.62–0.68			
				0.77–0.86	5.8	23.9 (MX)	5
						70 (mono)	
			LISS–III	0.52–0.59,			
				0.62–0.68			
				0.77–0.86,			
				1.55–1.70	23.5	141 24	
			AWiFS	0.52–0.59,	56	740 5	
				0.62–0.68	(nadir)	(combined)	
				0.77–0.86,			
				1.55–1.70	70	370	
					(pixel	(each	
					end)	head)	
SPOT–4	France	1998	MLA	0.5–0.59,			
				0.61–0.68			
				0.79–0.89,			
				1.58–1.75	20	60 26	
			PLA	0.61–0.68	10		

[a] Sensors: AWiFS = Advanced Wide Field Sensor, LISS = Linear Self-Scanning Sensor, MSS = Multi Spectral Scanner, MLA= MSS Linear Array, PAN = Panchromatic, PLA = Panchromatic Linear Array, TM = Thematic Mapper, WiFS = Wide Field Sensor.

Part III.

Methodological Advances for a Comprehensive Impact Assessment

6 Econometric Methods for Measuring Natural Resource Management Impacts: Theoretical Issues and Illustrations from Uganda

J. Pender

International Food Policy Research Institute (IFPRI), Washington, DC, USA

Introduction

Research and development efforts related to natural resource management (NRM) in developing countries have increased substantially in the decade since the Earth Summit in Rio de Janeiro in 1992. Within the Consultative Group on International Agricultural Research (CGIAR), research on NRM is viewed as one of the two major pillars of the CGIAR's research agenda (the other being the development of improved germplasm), and is using an increasing share of the Group's resources. A large body of literature has focused on measuring the impacts of agricultural research in general and crop variety improvement in particular on agricultural productivity (Alston *et al.*, 2000); but very little is focused on the impacts of NRM research on productivity or on other such outcomes of interest as poverty, food security, ecosystem resilience, or other aspects of the sustainable use of natural resources (NR) (Barrett, 2003).

Evaluating the impacts of NRM research requires two kinds of assessments: 1. assessing the impacts of NRM research on farmers' NRM practices and other aspects of their behaviour (since changes in NRM practices may cause farmers to alter other decisions); and 2. assessing the impacts of farmers' NRM practices (and other decisions affected by NRM research) on the outcomes of interest. This chapter focuses on the second type of assessment, illustrating it with an example from recent research in Uganda.

Evaluating the impacts of NRM practices on such outcomes as productivity, poverty and NR degradation (or improvement) is challenging, for several reasons. First, both NRM and outcomes like poverty and resource degradation are complex, multidimensional concepts that are not easy to conceptualise or measure in terms of simple indicators measured on a cardinal scale. Natural resource management includes many interrelated decisions about the use and management of land (such land uses as cropping vs. pasture vs. fallowing; land investments such as terraces or planting trees; choice of crops to plant; soil fertility management; tillage and crop rotation practices, etc.); water (investments in irrigation, water harvesting, drainage); natural vegetation (weed control methods, deforestation) and other aspects of biodiversity (control of vermin, hunting). Poverty may mean much more than simply low income or consumption: it may be related to limited assets, limited education, poor health and other human capital constraints; limited access to infrastructure and services; or other limitations in the portfolio of physical, human, natural, financial or social capital that households are able to draw upon to ensure their livelihoods (Carney, 1998). Natural resource degradation can include physical or chemical degradation of the soil, depletion or pollution of water sources, depletion or degradation of natural vegetation and wildlife, or more generally, reduction in the services provided by natural ecosystems.

The nature and extent of NRM impacts on these outcomes may be very context-dependent, largely influenced by biophysical and socio-economic factors that vary widely from place to place, from one household to the next, from micro- to macro-scales, and over time. For example, the effects of investments in soil and water conservation structures such as terraces may be very different in humid environments compared to semi-arid environments, or on shallow vs. steep slopes, since the benefits of conserving soil moisture may vary widely across such environments, as may the opportunity costs of such investments (Herweg, 1993; Shaxson, 1999). The benefits and opportunity costs of NRM investments may also vary as a result of differences in the socio-economic environment; for example, differences in population density (and hence resource scarcity) (Boserup, 1965; Tiffen *et al.*, 1994; Pender, 2001), access to markets and infrastructure (affecting local prices of outputs and factors of production) (Binswanger and McIntire, 1987; Pender *et al.*, 2001c), social and information networks (affecting access to information and ability to cope with risks, finance investments or achieve collective action) (Baland and Platteau, 1996; Barrett *et al.*, 2002), and land tenure (affecting incentives to invest and ability to finance investments) (Feder *et al.*, 1988; Place and Hazell, 1993; Otsuka and Place, 2001).

The impacts of NRM practices often do not occur immediately, and they vary over time and space. For example, investments in soil conservation measures may take several years to show their full impact on productivity (Shiferaw and Holden, 1998a). Furthermore, farmers may adapt, or disadopt as well as adopting NRM practices and investments, and understanding the factors leading to adaptation and disadoption and their impacts can be quite important (Shiferaw and Holden, 1998b; Adesina and Chianu, 2002). NRM practices often involve spatial externalities and collective action issues, further complicating the assessment of factors affecting adoption and the impacts of such practices (Knox *et al.*, 2002). Thus, investigation of the pattern of adoption, adaptation and disadoption and the flow of benefits and costs of NRM investments and practices over time and space are needed to fully assess their impacts.

Finally, the relationships between NRM and outcomes are complex, and may occur through many different mechanisms. For example, investment in irrigation may increase the value of agricultural production and income by increasing the productivity of existing crops, by promoting more intensive use of inputs (if irrigation enables multiple crops per year to be grown), by promoting complementary investments or land management practices (e.g. use of fertiliser or manure), or by promoting the adoption of higher-value (e.g. horticultural) crops. Such multiple mechanisms must be considered if the full impacts of NRM practices (or other factors) are to be accounted for.

The next section discusses econometric approaches to NRM impact assessment in general, their merits and drawbacks, and the types of problems that must be addressed. This is followed by an exploration of some of the key econometric issues in more detail using simple mathematical notation, illustrations of the econometric approach, using data collected from Uganda, and a final concluding section.

Econometric Approaches to NRM Impact Assessment

This chapter reviews econometric approaches to NRM impact assessment using household survey data. Econometric analysis is certainly not the only tool that can, or should, be used in evaluation of NRM impacts. For example, experimental approaches that apply NRM practices under carefully controlled treatments subject to random assignment have many advantages over econometric approaches based on analysis of survey data, since confounding factors can be more reliably controlled for and outcomes carefully measured. However, this strength of experiments can also be a weakness if the control conditions used in experiments do not closely approximate those occurring in farmers' fields. It is difficult and costly to set up experiments that represent the wide range of biophysical and socio-economic contexts in which farmers adopt NRM practices, and this is one reason that technologies and practices found to be promising in experimental trials are often not widely adopted.

One response to this problem has been to increase the emphasis on participatory research with farmers, including farmer-managed (and sometimes also farmer-designed) experiments. This is a promising approach that is helping to improve the impact of agricultural research and technology development (see Pound *et al.*, 2003). With the right programme design and monitoring system, the impacts of such programmes can be rigorously investigated. Using a quasi-experimental programme design and analysis of survey data on intervention and impact indicators measured for both programme households and non-programme households before, during, and after the intervention, researchers can distinguish the impacts of the programme interventions and NRM practices from outcomes resulting from differences between participating and non-participating households, changes in the biophysical or socio-economic environment, or other factors that may confound programme impacts. Careful quasi-experimental designs can outperform cross-sectional 'with–without programme' or 'before–after programme' type comparisons (Ravallion, 2001). Unfortunately, such quasi-experimental implementation and evaluation designs are rare among programmes promoting improved NRM practices, but are becoming more common for other types of programmes.[1]

Although formal experiments and quasi-experiments promoting NRM practices are rare, many farmers are conducting their own 'experiments' with NRM practices every day in a wide variety of contexts without the involvement of research or development projects, and much can be learned from also assessing the impacts of such 'experiments'. Econometric analysis of survey data is one useful way to assess the impacts of such farmer-initiated experiments. Even for this purpose, econometric analysis of survey data is not the only useful approach, and faces numerous problems that will be discussed further. Such analysis can be even more valuable if complemented by more qualitative research approaches used to investigate farmers' assessments of their reasons for adoption or non-adoption of specific practices, perceptions about the impacts of different practices, or the dynamics of impacts.

Econometric methods can be used to address many of the challenges encountered when assessing NRM impacts. The multidimensional nature of NRM can be addressed by incorporating measures of multiple types of NRM practices as explanatory variables in the analysis. The multidimensional nature of poverty and resource degradation indicators can also be handled by using separate indicators of these different dimensions, and by conducting separate analyses with different indicators as dependent variables, or analysing such models as a system of dependent variables when possible.[2] Such data-reduction methods as principal components analysis and factor analysis can also be used to construct indexes of such multidimensional factors (Stevens, 2002).

Adoption or intensity of use of NRM practices does not have to be measured on a cardinal or even interval scale; quite often NRM adoption is measured as a binary categorical (yes or no) variable and sometimes as an ordinal variable (indicating more or less of something, though not exactly how much), and there are econometric methods to deal with these types of indicators (see the studies of NRM adoption cited in Barrett *et al.*, 2002). Measurement problems with outcome variables can also be addressed by using categorical, ordinal or interval measures rather than cardinal measures. For example, perceived changes in particular welfare and NR conditions can be measured on an ordinal scale (indicating whether conditions have improved, not changed significantly, or worsened) (Pender *et al.*, 2001a, 2001b), and the share of households or land having a particular welfare or resource degradation indicator can be reasonably well estimated within interval ranges (<25%, 25–50%, etc.) even if it is unreasonable to expect a point estimate (Pender *et al.*, 2001a, 2001b).

Using categorical or ordinal variables results in loss of information if more precise and accurate measures are possible. For example, more information about the impacts of manure use on crop production can be obtained if the quantity and timing of manure use are measured and incorporated into the analysis, than if a simple binary measure is used measuring only whether or not manure was applied. However, there is always a trade-off between the precision and cost of measurement, and it may not always be feasible to use such cardinal measures, especially in a large survey. Furthermore, precise measures may be inaccurate if poorly measured, potentially contributing to spurious conclusions and illusory benefits of precision.

Using such types of non-continuous measures of dependent variables has implications for the type of econometric analysis that is appropriate. Ordinary least squares (OLS) estimation is generally not appropriate for such limited dependent variable models, but appropriate alternatives exist for most cases. For example, ordered probit or ordered logit can be used to estimate models with an ordinal dependent variable (Maddala, 1983; Amemiya, 1985), multinomial logit or nested logit can be used with polychotomous categorical dependent variables (Maddala, 1983; Amemiya, 1985), tobit (Maddala, 1983; Amemiya, 1985) or censored least absolute deviations models (Powell, 1984) can be used for censored dependent variables, and interval regression (a generalisation of tobit) can be used for dependent variables measured within interval ranges (StataCorp, 2003a).

The context-dependence of impacts can be addressed in econometric analysis by conducting separate analyses for different contexts (in highlands vs. lowlands), or by using interaction terms in the regressions to account for expected interaction relationships. For example, the impacts of fertiliser may depend upon the level of rainfall or the presence of soil and water conservation structures. Other factors may not interact with NRM practices, but should still be controlled for in an econometric analysis to avoid problems of omitted variable bias leading to incorrect conclusions. For example, land quality may be correlated with adoption of particular NRM practices (returns to fertiliser use or land investment may be greater on higher-quality land) and also contribute to agricultural production; thus, failing to adequately account for land quality in the regressions may cause improper attribution of productivity effects to NRM practices when they are actually due to land quality.

Closely related to the problem of omitted variables is the problem of endogenous explanatory variables. Variables that are endogenous choices of the household in the current year (such as many NRM practices) may be influenced by production conditions observed by the household, but unobserved by the econometrician, and thus be correlated with the error term in the regression and cause a bias. For example, use of fertiliser in the current year may be influenced by weather conditions early in the cropping season, causing a correlation between observed fertiliser use and unobserved (by the econometrician) weather conditions. One approach could be to collect detailed data on weather conditions affecting all households, but this is often not feasible. The more common approach is to use instrumental variables (IV) or two-stage least squares (2SLS) estimation. These issues are discussed further in the next section.

Addressing the time-varying nature of impacts requires time-series data, ideally panel data with repeated observations from the same households and plots over a period of many years so that the dynamics of these impacts and their feedback effects on household endowments and subsequent NRM decisions can be adequately assessed. Unfortunately, household and plot-level panel data sets with information on both NRM practices and causal factors and outcomes are quite rare. In the absence of such data, inferences about NRM impacts will remain limited to those possible based on available short-term experimental data and cross-sectional econometric studies. These can provide information on near-term impacts, for example, on current production, income and current rates of resource degradation or improvement, but do not reveal feedback effects such as how changes in income or resource conditions may lead to changes in future adoption, adaptation or disadoption of NRM practices. Neither can short-term, cross-sectional studies answer such important longer-term questions as whether promotion of new NRM practices helps to stimulate a pathway out of the poverty and resource degradation trap in which many farmers in developing countries appear to be caught (Barrett *et al.*, 2002; Place *et al.*, 2002).

Assessing the multiple and complex mechanisms by which NRM (and other factors) may affect outcomes is also an important issue, and one that is more difficult to address when limited dependent variable models (such as the probit, ordered probit, and tobit models mentioned above) or other non-linear models are estimated. In linear systems of structural equations, the total impacts of any variable on the outcomes can be determined by total differentiation of the system and by adding up the partial effects, as in Fan *et al.* (1999). But, with systems of limited dependent variable models or other non-linear models, this approach does not work. There will be no simple general relationship between the estimated coefficients of the structural model and the total impacts; these relationships all depend on the level of each variable in non-linear models.

An alternative approach to estimating total effects in non-linear models is to use predictions from the estimated model to simulate both indirect and direct impacts of changes in the explanatory variables. For example, using a set of regression results that predict the impact of irrigation on various NRM practices and on outcomes such as agricultural production, one can predict how a change in irrigation use would influence agricultural production, both directly (via the coefficient of irrigation in the production regression), and indirectly (by tracing the changes in NRM practices predicted by the change in irrigation, and then predicting the impact of those changes in NRM practices on production). This simulation approach is applied in the analysis of Uganda data discussed later in this chapter.

A Simple Exposition of Some of the Econometric Issues

In this section some of the issues involved in estimating the impacts of NRM practices are explained using mathematical notation. Here the focus is only on econometric issues, abstracting from the issues of measurement discussed in the previous sections.

Suppose that there is interest in the impacts of an NRM practice (P) on an outcome of interest (y), such as crop production.

$$y_i = \alpha + \beta P_i + u_i \qquad (1)$$

where:
α refers to the expected outcome if no practice is used,
β represents the impact of the practice on the expected outcome,
u_i refers to the effects of unobserved random factors (weather or soil quality) on the outcome,
the subscript i indexes separate replications of y resulting from different values of P_i and different draws of the random variable u_i; for example, results of experimental treatments on different plots.

There is interest in estimating the value of β based on observed values of y_i and P_i.

If P_i is a discrete variable, such as whether a specific practice is used or not (whether the plot was fallowed in the prior year), β represents the difference in the expected value of y_i between plots using the practice and those that do not use the practice. In this case, β can be consistently estimated by the difference in the sample means between plots with and without the practice, provided that the treatments are randomly assigned. [The consistency of an estimator means that the estimate tends toward the true parameter value as the sample size increases (Davidson and MacKinnon, 2004)]. If P_i is measured as a continuous variable, then β represents the marginal impact of a unit increase in P_i on y_i, assuming that the true relationship between P_i and y_i is linear, and can be estimated by ordinary linear least squares regression (OLS). Non-linear relationships can be detected by graphing the relationship between y_i (or the regression residuals) and P_i and/or statistically testing for non-linear relationships between the residuals and P_i (Mukherjee *et al.*, 1998). If non-linearity exists, it can be taken into account by specifying a suitable non-linear relationship between P_i and y_i; for example, by using higher order polynomials of P_i in the specified relationship such as $y_i = \alpha + \beta_1 P_i + \beta_2 P_i^2 + u_i$. Such a polynomial model is still linear in the parameters, the dependent variable and the error term, and is linear in the explanatory variables P_i and x_i if we define $x_i = P_i^2$. Thus this model can still be estimated by OLS. If the model is not linear in the parameters but linear in the dependent variable and error term (e.g. $y_i = \alpha + P_i^{\ \beta} + u_i$), non-linear least squares can be used (Davidson and MacKinnon, 2004). Sometimes an apparently non-linear model can be transformed to a linear model; e.g. $y_i = \alpha\, P_i^\beta\, exp(u_i)$ can be transformed by taking logarithms to obtain $ln(y_i) = ln(\alpha) + \beta ln(P_i) + u_i$, which is linear in the parameters, dependent variable, explanatory variable and error term, if the dependent variable is defined as $Y_i = ln(y_i)$, the explanatory variable defined as $p_i = ln(P_i)$, and the intercept is redefined as $A = ln(\alpha)$.

Assuming that the specification of the functional relationship in equation (1) is correct, correlation of P_i with u_i is the major concern in estimating β consistently. If P_i and u_i are correlated, then the estimated coefficient for β will 'pick up' some of the effect of u_i on the outcome. For impact analysis purposes, such a correlation could lead the analyst to err in estimating the true impact of an NRM practice. For example, if the plots where the practice is applied tend to be of higher soil quality than plots where the practice is not applied (in this case soil quality is part of u_i), the mean value of crop production is likely to be higher where the practice is applied, even if the practice itself has no impact. In an experimental study, such confounding influences are controlled for by randomly assigning plots with similar observable characteristics to different treatment groups.

Unfortunately, as discussed in the previous section, it is not always possible to determine impacts of NRM practices under carefully controlled experimental conditions using random assignment of treatments. If the intention is to determine the impacts of NRM practices under conditions that are actually faced by farmers, the problem that such practices are not randomly assigned to different plots is encountered. This implies that P_i may well be correlated with unobserved factors. It is important to note that this problem is not unique to econometric evaluation of NRM impacts; it is a potential problem for any method of evaluating the impacts of a programme or technology in which there is non-random assignment of the programme or practice to be evaluated. Thus use of simple descriptive statistics, qualitative comparative case studies, participatory evaluation, or other methods are also subject to this problem. Econometric approaches have an advantage over many other evaluation methods in that rigorous methods have been developed to address this problem.

One approach is to measure and control for other factors likely to determine y_i that may be correlated with P_i. For example, the different indicators of land quality, climate (if this varies across the sample), and other relevant factors can be measured and included in the regression model:

$$y_i = \alpha + \beta P_i + \gamma x_i + u_i \tag{2}$$

where:
x_i is a vector of factors expected to influence y_i,
γ is a vector representing the marginal impacts of each component of x_i on y_i.

Using a multivariate regression model, as in Equation 2, helps to 'control for' observable factors that could confound inferences about the relationship between P_i and y_i. It also allows investigation of the impacts of multiple types of NRM practices simultaneously, if P_i is taken as a vector of practices rather than as a single practice, and β as the vector of marginal impacts of those practices. Interactions between different NRM practices or between certain practices and components of x_i can also be specified, if the level of use of one practice or x_i is expected to affect the marginal impact of another practice (e.g. $y_i = \alpha + \beta P_i + \gamma x_i + \delta P_i x_i + u_i$).

One potential problem in estimating Equation 2 is that β may be difficult to estimate with reasonable precision if the correlations between P_i and components of x_i are very high. Intuitively, if there is high correlation among these variables, it is difficult to identify the independent influence of each on y_i. In the limiting case where there is perfect correlation, identification is impossible. Econometricians refer to this as the problem of multicollinearity. A useful measure of multicollinearity is the variance inflation factor (VIF), which measures the extent to which the variance of the estimate of a coefficient is increased by multicollinearity. The VIF for variable j is equal to $1/(1-R_j^2)$, where R_j^2 is the coefficient of determination in an auxiliary regression of variable j on the other explanatory variables in the regression (Mukherjee *et al.*, 1998). If there is perfect multicollinearity for a variable, $R_j^2 = 1$, the variance is infinite and the model is not estimable.

Of course, it is still possible that P_i is correlated with u_i, even after controlling for observable factors believed to be correlated with P_i and y_i. There is always the possibility that unobservable differences in land quality, rainfall, farmer ability, or other factors are responsible for differences in outcomes that are being attributed to NRM practices, if those unobserved factors are correlated with adoption of NRM practices. This is particularly likely to be a concern because farmers choose to adopt NRM practices based on information that is usually not available to the econometrician (and such information is likely correlated with u_i). This is referred to as the problem of endogeneity bias in the econometrics literature, since the choice of NRM practices is an endogenous decision of the farmer. How can this problem be addressed?

If panel data are available that include multiple observations per entity, one approach is to use fixed effects estimation. The basic idea behind fixed effects estimation is that, with panel data, it is possible to control the impacts of any fixed factors that are unique to each entity by estimating a separate intercept for each entity in the regression. In symbols, the fixed-effects model is a slight variation of Equation 2:

$$y_{ij} = \alpha_i + \beta P_{ij} + \gamma x_{ij} + u_{ij} \quad (3)$$

In Equation 3, the subscript i indexes the entities having multiple observations and j indexes the observations within each entity (for example, observations from different years for the same plot, from different plots for the same household, from different households for the same village, etc.). The intercept α_i is now subscripted by i, indicating that a separate intercept is estimated for each i, but this intercept is constant across j within each i. This intercept will pick up the effect of any 'fixed' (not varying across j) factors that are associated with the entity i, thus purging the error term of any such effects. For example, if i indexes plots and j indexes different years, then α_i will pick up the effects on y_i of any unobservable plot quality characteristics that do not vary across the years in the sample. It will also pick up the effects of more aggregated fixed factors, such as household or village characteristics that do not vary across the sample years for a given plot.

The ability to control for all fixed factors is both a strength and a weakness of fixed effects estimation. It is a strength because it controls for unobserved fixed factors that could confound the estimation of β, but is a weaknesses because α_i may pick up the effects of variables of interest, eliminating or weakening the ability to identify those effects. For example, if the panel consists of multiple observations from the same plots during different years, including plot-level fixed effects will wash out the effects of any plot characteristics that did not change during those years, which may include the presence of land investments or use of certain NRM practices. Even if there is variation in these practices across years, the estimation will not have much statistical power to identify β if there is not much variation within plots across years. Fixed effects also reduce statistical power by reducing the degrees of freedom (number of observations minus the number of explanatory variables) of the estimator. Such fixed effects also eliminate the ability to estimate the impacts of observed fixed household, village, or higher-level factors on y_i, which may not be desirable.

The power of fixed effects estimation can be optimised if a quasi-experimental design is used, in which data are collected for households and plots that are applying the NRM practices of interest and those that are not, both before and after adoption by the households that are adopting. In this case, all fixed differences in the nature of households who have adopted and those who have not can be controlled for (unlike cross-sectional studies only of 'with–without' households), as can changes in the biophysical or socio-economic environment that are affecting all households over time in the same way (unlike 'before–after' studies of impact). An extension of Equation 3 can illustrate the point:

$$y_{ijt} = \alpha_{ij} + \alpha_t + \beta P_{ijt} + \gamma x_{ijt} + u_{ijt} \quad (4)$$

In Equation 4, the subscript i indexes different groups of households depending on whether they have adopted some set of NRM practices, j subscripts households, and t subscripts the time period studied. α_{ij} is a vector of constants representing fixed productivity differences across households (unchanging over time) and α_t is a vector of constants reflecting average productivity of all households at different points in time, relative to the first year. These two sets of constants pick up the effects of all fixed differences across households and of general changes over time affecting all households equally (note, separate α_t's could be estimated for different sub-samples, such as households from different regions, if different patterns of exogenous change are occurring across regions). x_{ijt} then controls for observable household specific factors that affect y_{ijt} and that change over time (e.g. education and wealth), and β measures the marginal effect of P_{ijt}, controlling for all of these other effects. A suitable quasi-experimental design is needed to ensure that there is sufficient variation in P_{ijt} to enable identification of β; i.e. there needs to be sufficient variation in adoption of the practices within households over time and not just fixed differences in levels of adoption between adopters and non-adopters or common changes over time in adoption by all sample households (since those effects are picked up by α_{ij} and α_t).

Another approach to the problem of correlation between P_i and u_i is to use instrumental variables (IV) estimation to remove the correlation effect. IV estimation involves three major steps:

1. Finding variables that are good predictors of P_i, but are not correlated with u_i ('instrumental variables' in econometrics jargon),

2. Using those variables to obtain predicted P_i values that are uncorrelated with u_i,
3. Regressing predicted P_i on y_i.

Two-stage least squares (2SLS) regression is a special case of IV estimation, where P_i is a continuous variable predicted by a first stage least squares regression and the instruments are chosen 'optimally', to minimise the asymptotic variance of the estimator (Amemiya, 1985).[3] However, if P_i is not a continuous variable (e.g. if it is a binary response variable), then P_i may be predicted using a limited dependent variable model, such as probit or logit. Nevertheless, so long as y_i is a continuous variable, IV estimation can still be employed by using the predicted value of P_i as an instrumental variable (Dubin and McFadden, 1984).

Although conceptually simple, IV estimation often poses practical problems. Finding instrumental variables that are good predictors of P_i but uncorrelated with u_i is not easy. Farmers are likely to choose NRM practices because they are expected to have an impact on production. Thus, factors that affect the marginal impact of an NRM practice on production that are observed by the farmers, such as soil quality, topography, etc. (part of u_i in Equation 1) are likely to be determinants of P_i, but will not be valid instrumental variables if Equation 1 is used as the specification, since they will be correlated with u_i. If such factors are observed by the econometrician, they should be included in the regression equation, as in Equation 2, but this causes another problem. Using the same variables to predict P_i as are included as explanatory variables in the regression (x_i) implies that there will be multicollinearity between the value of predicted P_i and x_i in an IV regression. If a linear regression model is used to predict P_i using only x_i (or some subset of x_i), the model will not even be estimable because the impacts of predicted P_i cannot be distinguished from x_i. Econometricians call this the problem of identification in 2SLS models, but it is basically a problem of multicollinearity. If P_i is a binary choice variable that is predicted using a non-linear model, such as probit or logit, the IV model may be estimable even if x_i is used to predict P_i, so long as there is not perfect *linear* correlation between predicted P_i and x_i. Nevertheless, the degree of correlation is still often quite high, so that identification of β with reasonable precision still may be difficult.

Identification of β in a linear IV model requires that at least one of the instrumental variables used to predict P_i is not part of x_i and can validly be excluded from the regression model in Equation 2. The more such excluded instrumental variables that can be used to predict P_i and the better they are at predicting P_i, the more reliable the estimate of β will be. Where markets do not function well, community- and household-level socio-economic characteristics may influence households' decisions about NRM and use of agricultural inputs, and such factors often can be used as instrumental variables, since these may not influence productivity directly, controlling for such decisions. Variables such as access to markets, population density, household composition, non-agricultural assets, access to credit and land tenure are therefore good candidates to consider as instrumental variables in regressions predicting crop production. The validity of excluding such variables from the regression can be tested statistically in a variety of ways (see Davidson and MacKinnon, 2004, on tests of over-identifying restrictions). The analysis discussed below uses Wald tests of the joint hypothesis that the coefficients of the excluded variables are equal to zero, by first estimating an unrestricted model, and then only excluding such variables if this hypothesis cannot be rejected.[4]

It has been shown that in finite samples, 2SLS estimation using 'weak instrumental variables' (i.e. those which are poor predictors of P_i) can cause more biased results than using OLS (Bound *et al.*, 1995). Thus it is important to establish that the instrumental variables that are excluded from the regression are good predictors of the endogenous explanatory variables in the regression. This is addressed by testing the joint significance of the excluded instrumental variables in the regressions predicting P_i (referred to as a relevance test in the literature).

Even if the instrumental variables are relevant and the exclusion restrictions of the model are valid, IV estimation may be inferior to OLS if the endogenous explanatory variables of concern are not actually correlated with the error term in the regression. In this case, OLS would be preferred because it is consistent and more efficient (i.e. it has a smaller covariance matrix of the estimated parameters) than IV estimation. A test that can be used to compare OLS vs. IV models is the Hausman (1978) specification test, which compares two estimators, one of which is consistent and efficient under some null hypothesis (e.g. OLS under the hypothesis that all explanatory variables are uncorrelated with the error term), and one of which is consistent under both the null hypothesis and the alternative (e.g. IV if some of the explanatory variables are correlated with the error term). If the null hypothesis cannot be rejected, OLS is the preferred specification.

The following example presents the use of econometric methods to assess the impacts of NRM and other factors on crop production in Uganda. Many of the approaches discussed above were incorporated into the study, including tests of multicollinearity, exclusion restrictions, relevance, and the Hausman test of consistency of OLS vs. IV estimation.

An Example from Uganda

Empirical model

The following example applies econometric methods to explore the impact of land management and investments on the value of crop production. The empirical model is derived from a theoretical dynamic household model, presented in Nkonya *et al.* (2004). The analysis examines the proximate determinants of production, including household choices about crop choice, labour use, land management, land investment and other decisions, together with the underlying determinants of these choices. The exposition of the analysis begins with the structural model of crop output supply and input demand equations, next it explains the specific empirical variables used and characteristics of the associated data set, and finally, it explores results and interpretations.

Value of crop production

Crop production is by far the most important source of agricultural income in Uganda. The value of crop production by household h on plot p (y_{hp}) is assumed to be determined by:

- the shares of area planted to different types of crops (C_{hp})
- the amount of labour used (L_{hp})
- the annual land management practices used (LM_{hp}) (use of manure, fertiliser, compost, etc.)
- the stock of prior land investments on the plot (LI_{hp}) (presence of irrigation, trees, live barriers, etc.)
- the 'natural capital' of the plot (NC_{hp}) (slope, position on slope, soil type, other biophysical characteristics)
- the tenure characteristics of the plot (T_{hp}) (land rights category, how plot acquired, tenure security)
- the household's endowments of physical capital (PC_h) (land, livestock, equipment), human capital (HC_h) (education, age, and gender), and 'social capital' (SC_h) (indicated by participation in programmes and organisations)
- the household's income strategy (IS_h) (measured by primary income source)
- village-level factors that determine local comparative advantages (X_v) (agroecological conditions, access to markets and infrastructure, and population density)
- random factors (u_{yhp}).

$$y_{hp} = y(C_{hp}, L_{hp}, LM_{hp}, LI_{hp}, NC_{hp}, T_{hp}, PC_h, HC_h, SC_h, IS_h, X_v, u_{yhp}) \quad (5)$$

The total value of crop production depends on the choice of crops and farm-level prices of these crops, the inputs and land management practices used in producing them, prior investments on the plot, and the natural conditions of the plot. Many different crops are produced at different locations in Uganda, so crop prices are omitted as determinants of the value of crop production, because including them would result in many missing observations for prices. Instead, farm-level prices are captured by the village-level factors determining local supply, demand and transportation costs of commodities (X_v) and household-level factors affecting households' transactions costs and marketing abilities (HC_h, SC_h, IS_h). Household endowments of physical capital (PC_h) can also affect crop production if there are imperfect factor markets. In addition, agroecological conditions (part of X_v), and households' human and social capital may also influence agricultural productivity by affecting their knowledge about farming practices, even if these factors have no impact on local prices.

Crop choice, labour use and land management

In Equation 5, annual crop choice (C_{hp}), labour use (L_{hp}), and land management (LM_{hp}) are all choices in the current year,[5] determined: by the natural capital and tenure of the plot; by the household's endowments of physical, human, social, and financial capital and of family labour (L_{fh}) at the beginning of the year; by the household's income strategy; and by agroecological conditions, access to markets and infrastructure, and population density (X_v):

$$C_{hp} = C(LI_{hp}, NC_{hp}, T_{hp}, PC_h, HC_h, SC_h, FC_h, IS_h, L_{fh}, X_v, u_{chp}) \quad (6)$$

$$L_{hp} = L(LI_{hp}, NC_{hp}, T_{hp}, PC_h, HC_h, SC_h, FC_h, IS_h, L_{fh}, X_v, u_{lhp}) \quad (7)$$

$$LM_{hp} = LM(LI_{hp}, NC_{hp}, T_{hp}, PC_h, HC_h, SC_h, FC_h, IS_h, L_{fh}, X_v, u_{mhp}) \quad (8)$$

In order to meet the assumptions of OLS regression, the explanatory variables must be determined independently of the outcome variable to be explained. Independence may result either from complete exogeneity or else from temporal predetermination. An example of a predetermined explanatory variable would be land planted to a perennial crop like banana. Most of the determinant factors in Equations 6, 7 and 8 are either exogenous to the household (e.g. X_v) or else state variables that are predetermined at the beginning of the current year (e.g. LI_{hp}, NC_{hp}, T_{hp}, PC_h, HC_h, and FC_h). In theory, households' income sources (IS_h) may be partly determined in the current year. However, as no change was found in the reported primary income source of most households between 1990 and 2000 (Nkonya *et al.*, 2004), IS_h was treated as a predetermined variable relative to crop management in the current year.

Participation in programmes and organisations (SC_h) can also be partly determined in the current year, and hence partly endogenous to current decisions about crop choice, labour use and land management. It is assumed that participation in programmes and organisations is affected by fixed cultural factors, reflected by the ethnicity of the household (Eth_h), households' endowments of labour, human and natural capital, and village factors determining local comparative advantages:

$$SC_h = SC(Eth_h, L_{fh}, HC_h, NC_h, X_v, u_{sh}) \quad (9)$$

Dependent variables

The dependent variables in the econometric models include the logarithm of the value of crop production at the plot level (y_{hp}), the shares of plot area planted to different types of annual crops (cereals, legumes, root crops, and vegetables) (C_{hp}), the logarithm of pre-harvest labour used on the plot, in person-hours equivalent (L_{hp}), a vector of dummy variables for whether particular agricultural or land management practices (use of slash and burn, inorganic fertiliser, manure or compost, incorporation of crop residues, crop rotation, mulching, household residues, pesticides or integrated pest management) were used on the plot (LM_{hp}),[6] and a vector of dummy variables for whether the household participated in various programmes and organisations (agricultural extension, agricultural training, NGO programmes oriented towards agricultural and environment, various other types of organisations) (SC_h). Since many of these are limited dependent variables (censored or discrete), the system cannot be estimated by a system of linear equations, such as three-stage least squares. The estimation approach is discussed below.

Explanatory variables

The village-level explanatory variables (X_v) include the agroecological and market access zones, and the population density of the parish (the second lowest administrative unit, consisting of several villages). Ruecker *et al.* (2003) classified the agroclimatic potential for perennial crop (banana and coffee) production in Uganda, based upon the average length of growing period, rainfall pattern (bimodal vs. unimodal), maximum annual temperature, and altitude. Potential for maize production (the most important annual crop) was also mapped and the map was found to be very similar. Wood *et al.* (1999) classified Uganda into areas of low and high market access, using an index of 'potential market integration' based upon estimated travel time to the five nearest markets, weighted by their population.

Household-level factors include: income strategy (primary income source of the household); ownership of natural and physical capital (area of land, value of livestock and farm equipment); human capital (education, age, and gender of household head); the family labour endowment (size of household and proportion of dependents); social capital (participation in longer-term training and shorter-term extension programmes and in various types of organisations); and the ethnicity of the household. Plot-level factors include:

- Size, tenure and land rights status of the plot
- Whether the plot has a formal title
- Whether the household expects to have access to the plot in 10 years
- Altitude of the plot
- Distance of the plot from the farmer's residence, nearest road and nearest market
- Investments that have been made on the plot (presence of irrigation, trenches, grass strips, live barriers and planted trees; share of area planted to perennial crops)
- Various plot quality characteristics (slope, position on slope, soil depth, texture, colour and perceived fertility).

Data

The above model was estimated using econometric analysis of survey data collected from 451 households in 107 communities during 2001. The study region included most of Uganda, including more densely populated and more secure areas in the southwest, central, eastern and parts of northern Uganda, representing seven of the nine major farming systems of the country (Nkonya *et al.*, 2004). Within the study region, communities were selected using a stratified random sample, with the stratification based on development domains defined by the agroecological and market access zones and by population density (Pender *et al.*, 2001b). Sample weights were constant within each stratum but varied across strata based on the number of communities selected, with a minimum number of four villages selected from small strata and a maximum of 15 selected from the largest stratum, to ensure a minimum representation of each domain. One hundred villages were selected in this way. Additional communities were purposefully selected in areas of southwestern and central Uganda, where the African Highlands Initiative (AHI) and the Centro Internacional de Agricultura Tropical (CIAT) are conducting research. Within each community, a random sample of four households was selected in most cases (more in some cases). This extensive sampling approach was chosen to represent as adequately as possible the different development domains in the study region within the budgetary limits of the project, but is not an ideal approach to assessing impacts of particular NRM practices, since this was not the primary purpose of the study. If the purpose had been to assess NRM impacts, a more-focused sampling approach ensuring adequate representation of households using and not using the practices to be studied would have been preferable.

A community-level survey was conducted with representatives of each selected community to collect information on access to infrastructure and services, local markets and prices, and other community-level factors. For each household selected, a household-level questionnaire collected information about household endowments of assets, household demographic composition, income and expenditures, and adoption of agricultural and land management technologies. A plot-level survey was also conducted to collect information on all of the plots owned or operated by the household, including information about land tenure, plot quality characteristics, land management practices, use of inputs and outputs from the plot in the year 2000. The survey information was supplemented by secondary information collected from the 1991 population census and available geographic information to estimate population density.

Analysis

Equations 5–9 were estimated econometrically to analyse the determinants of crop choice, labour use, land management practices, and participation in programmes and organisations, and their impacts on the value of crop production. The analysis used a log–log specification (logarithm of the dependent variable and of all continuous explanatory variables) of Equation 5 because this reduces the non-normality of these variables and problems with outliers, improving the robustness of the regression results (Mukherjee *et al.*, 1998). Because there are zero values for some household assets (land, livestock, and equipment) for some households, it was not possible to use a simple logarithmic transformation for these variables. Instead, the analysis used the logarithm of assets for households with positive asset levels and set this variable equal to zero if the asset level was zero. To account for the impact of having zero level of an asset, a dummy variable for positive asset ownership was included, to allow for an intercept shift for households with zero assets.

A systems estimation approach, such as three-stage least squares or full information maximum likelihood would be ideal to deal with endogenous explanatory variables and account for correlation of error terms across the different equations (Davidson and MacKinnon, 2004). This is not feasible, however, due to the nature of many of the dependent variables. Several of the endogenous variables in this system are limited dependent variables (categorical or censored), for which a linear estimator (or even non-linear least squares) is not appropriate. Full information maximum likelihood estimation of this system is also infeasible, given the complex multidimensional joint probability distribution that would have to be integrated. Thus equations in the system were estimated separately.

C_{hp} are area shares under different crops and thus censored continuous variables (censored below at 0 and above at 1); a maximum likelihood tobit estimator (with left and right censoring) was used for Equation 6. LM_{hp} and SC_h are dichotomous choice variables (whether certain land management practices are used, whether the household participates in different types of programmes and organisations); probit models were used to estimate Equations 8 and 9. y_{hp} and L_{hp} are continuous uncensored variables; thus least squares regression can be used for Equations 5 and 7. In this chapter, the focus is on the estimation of Equation 5, since the interest is in estimating impacts of NRM practices on crop production.

Inclusion of endogenous explanatory variables in Equation 5 could result in biased estimates, as discussed in previous sections. Instrumental variables (IV) estimation was used to develop a model known to be consistent for testing against the OLS model. The vector of household ethnicity dummy variables in Equation 9 is assumed to determine participation in programmes and organisations, but does not enter in the other equations. Thus, this vector is assumed to provide instrumental variables useful in estimating the other equations. The predicted participation variables from Equation 9 are used as instrumental variables in the IV version of Equation 5. In addition, predicted crop choice, labour use and land management practices from Equations 6, 7 and 8 are also used as instruments in estimating Equation 5. Other candidates for instrumental variables in Equation 5 include market access, population density, household assets, household demographic composition, access to credit, and land tenure, since these may affect crop production only via their impacts on crop choice, labour use and land management decisions, and not directly. Hypothesis testing was used to select instrumental variables from this set of candidates; i.e. variables that were jointly statistically insignificant in the full version of the model (at the 0.50 probability level in both OLS and

IV models) were dropped from the IV regression and used as instrumental variables.

In addition to testing whether instrumental variable candidates could be excluded from the model, the relevance of the instrumental variables used in the IV estimation was tested and a Hausman test conducted to compare the IV and OLS models. Multicollinearity was tested but found not to be a serious problem (variance inflation factors <5) for almost all explanatory variables in the OLS regressions (except for some assets when the logarithmic specification with the intercept shift dummy variables were used). Since stratified random sampling was used, all parameters were corrected for sample stratification and sample weights (StataCorp, 2003b). Estimated standard errors are robust to heteroskedasticity and clustering (possible non-independence) of observations from different plots for the same household (StataCorp, 2003b). Outliers were detected using graphical methods (Mukherjee *et al.*, 1998) and any errors found were corrected.

Predicted impacts of selected variables

In a complex structural model, like the one estimated in this study, a change in a particular causal factor may have impacts on outcomes of interest through many different channels, given the many intervening response variables that may be affected. For example, improvements in education may affect the value of crop production directly by affecting farmers' production or marketing abilities. But they may also influence production indirectly by affecting households' choice of crops, labour use, land management practices, or participation in programmes and organisations. Such indirect effects must be accounted for if the full effect of causal factors on agricultural production and other outcomes is to be understood. To address this issue, the predicted responses implied by the estimated econometric relationships, under alternative assumptions about the values of the explanatory variables for the entire sample, were simulated and these predicted responses carried forward to determine their impact on subsequent relationships in the system (see details in Nkonya *et al.*, 2004).

Results

The results of estimating Equation 5 by OLS and IV estimation are presented in Table 6.1. The variables for market access, population density, ownership of equipment, gender of household head, size of household, proportion of dependents, access to credit in the village, land tenure and title of plot were jointly statistically insignificant in the unrestricted version of both the OLS and IV models (P=0.57 in the OLS model and P=0.99 in the IV model), and thus were dropped and used as instruments in the restricted version of the IV model. Relevance tests of the instruments found that the excluded instruments were significant at the P=0.01 level in regressions explaining

Table 6.1. Determinants of Output Value.

	Least squares regressions	
Variable[a]	Ordinary least squares (OLS)	Instrumental variables[b]
Crop choice (compared to cereals)		
Legumes	–0.068	0.752
Root crops	–0.468*[c]	1.553
Vegetables	0.525	2.523
Coffee	0.098	1.097
Bananas	0.988***	2.090***
Land management practices		
Slash and burn	–0.048	–0.140
Inorganic fertiliser	0.276	0.028
Manure and compost	0.103	–1.384*
Crop residues	0.043	0.483
Crop rotation	–0.201*	–0.892**
Mulch	–0.171	–0.152
Household residues	–0.093	0.103
Pesticides	0.059	0.620
Integrated pest management	0.158	–1.369
ln (pre-harvest labour use)	0.385***	0.563**
Primary income source (general agricultural production)		
Gifts/donations	0.230	–1.026
Wages/salary	0.169	0.348
Livestock	0.626**	0.457
Non-farm	0.549***	0.775***
Forestry/fishing	–0.732***	–0.720**
Brewing beer	0.279	0.244
Legumes	0.490**	0.600*
Horticultural crops	1.676***	1.159***
Bananas	0.164	0.105
Cereals	0.484***	0.575**
Root crops	0.117	–0.047
Export crops	0.483***	0.197
High market access	0.013	
Distance to (km)		
Residence	–0.093*	0.002
All-weather road	0.007	0.018*
Nearest market	–0.012	–0.015
ln (population density)	0.014	
Assets		
Own land (yes = 1, no = 0)	0.305	0.365
ln (area owned)	–0.097*	–0.260**
Own livestock (yes = 1, no = 0)	–0.828*	–0.437
ln (value of livestock)	0.068*	0.062
Own equipment (yes =1, no = 0)	0.010	
ln (value of equipment)	0.001	
Education of household head (none)		
Primary	–0.155	–0.276*
Secondary	0.129	0.071
Higher education	0.117	0.040

Continued

Table 6.1. Continued.

Variable[a]	Least squares regressions: Ordinary least squares (OLS)	Instrumental variables[b]
ln (age of head)	–0.359**	–0.044
Female household head	–0.152	
ln (size of household)	0.011	
Proportion of dependents	–0.266	
Participation in organisations		
Agriculture/environment	–0.168	
Credit	0.129	
Poverty reduction	0.229	
Community services	–0.038	
Participation in technical assistance programmes		
Training	0.271***	0.331
Extension	0.287***	0.629
Access to credit in village		
Formal credit	0.001	
Informal credit	0.055	
Land tenure (freehold)		
Leasehold	–0.436	
Mailo	0.217	
Customary	0.133	
Formal title to plot	–0.306	
How plot acquired (purchased)		
Leased in	–0.138	–0.403
Borrowed	–0.414	–0.663*
Inherited	–0.288***	–0.253*
Encroached	–0.331	–1.108**
Expect to operate plot in ten years? (no)		
- Yes	–0.008	
- Uncertain	0.213	
ln (area of plot)	0.580***	0.648***
Land investments on plot		
Irrigation	0.790	2.426**
Trenches	–0.009	0.115
Grass strips	0.046	0.499
Live barriers	–0.330	–0.376
Trees	0.030	0.096
Intercept	11.461***	6.986***
Number of observations	930	920
R^2	0.565	0.308

[a]Coefficients of agroclimatic zones and plot quality variables (slope, position on slope, soil depth, texture, colour and perceived fertility) not reported due to space limitations. Full regression results available upon request.

[b]Variables that were jointly statistically insignificant in the unrestricted OLS regression and IV regressions (P = 0.57 in OLS, P = 0.99 in IV) were excluded from the reported restricted IV regression. The excluded instrumental variables were significant in predicting all of the endogenous explanatory variables at the P = 0.01 level. A Hausman test failed to reject the OLS model as consistent (P = 1.000).

[c]Reported coefficient is statistically significant at * 10%, ** 5%, or *** 1% probability level. Coefficients and standard errors adjusted for stratification and probability weights, and are robust to heteroskedasticity and non-independence of errors across plots from the same household.

each of the endogenous explanatory variables. These tests establish that the instrumental variables used are suitable. A Hausman test of the restricted versions of the OLS vs. IV models failed to reject the OLS model (P=1.000). Hence the OLS model is preferable, since it is more efficient.

The results (both OLS and IV) indicate that the value of crop production is substantially higher on plots where bananas are grown than where cereals and many other types of crops are grown, controlling for labour use, land management, agroecological potential and other factors. Statistically significant differences in the value of production among other types of crops were not found.

Among land management practices, crop rotation reduces value of production significantly, at least in the short run. In the longer term, however, crop rotation may contribute to productivity by helping to restore soil fertility and by breaking cycles of pests, disease or weeds that may result from monocropping. No statistically significant and robust impacts of other land management practices on value of production, controlling for labour use and other factors were found.

Statistically significant impacts of land investments in the OLS (preferred) model were not found, although irrigation has a significant positive impact in the IV model. Since there were very few plots with irrigation in the sample (17 plots), the IV results for irrigation may be spurious. Thus, there is little evidence that the land investments found on the sample plots are having direct impacts on crop productivity (though they may have indirect impacts by affecting use of inputs, adoption of land management practices, and other factors).

Not surprisingly, the value of crop production on a plot increases significantly with both plot size and labour use. Other factors that significantly affect the value of crop production include: agroecological zone; primary income source of the household; age of the household head (negative effect); amount of land owned (negative effect); value of livestock owned (positive effect); participation in agricultural extension and training programmes (positive effect); and how the plot was acquired.

These results imply that other factors are having more impact on the value of crop production in Uganda than land management practices and most land investments, except investment in banana production. Other particularly important factors include the amount of labour used, access to agricultural extension and training, and the income strategy of the household.

Potential Impacts of Selected Interventions

The focus of this section is on the potential impacts of several policy-relevant changes on the value of crop production, including improved access to all-weather roads, improved access to education, participation in agricultural technical assistance programmes, participation in non-governmental organisations and access to irrigation. The potential impacts of such interventions on crop production are explored using the predicted relationships from the econometric model. Both the direct effects of such interventions based on the results reported in Table 6.1, and the indirect effects of such interventions, via their impacts on households' choice of participation in programmes and organisations, crops planted, land management practices and labour use are considered. Impacts for the full sample, and for highland and lowland zones are considered separately, in case there are differential impacts.

Population growth of 10% is predicted to have a small and statistically insignificant impact on the mean value of crop production in the full sample and the lowlands, but a significant negative impact in the densely populated highlands (Table 6.2). This suggests that priority should be given to reducing population pressure in the highlands.

Improved access to all-weather roads is predicted to have a small and statistically insignificant impact on the value of crop production in both the lowlands and the highlands.

Universal primary education is predicted to result in reduced crop production in the lowlands, but increased production in the highlands. In the lowlands, education can help households take advantage of better access to non-farm employment opportunities. In the highlands, better education may increase access to information and finance useful in agricultural production.

Government agricultural technical assistance programmes, whether through longer-term training programmes or short-term extension visits have positive and significant impacts on production in the lowlands. By contrast, NGOs focusing on agriculture and environment have more positive impacts in the highlands. These differences may be due to variation in the types of technologies promoted by these different types of programmes and organisations. For example, conservation technologies promoted by NGOs may have more beneficial immediate impacts on production in the highlands by helping to conserve both soil moisture and soil. In steeply sloping highland areas, soil moisture is usually a more important constraint on production than in lowland areas, so measures to conserve soil moisture may have more immediate impact (Shaxson, 1999).

Irrigation also has more positive predicted impacts on the value of crop production in the lowlands than in the highlands. This may be related to greater market access in the lowlands.

In general, these results demonstrate that the impacts of particular interventions are quite context-specific in Uganda. Interventions therefore need to be carefully targeted if they are to be most effective in increasing agricultural production.

Conclusions

Econometric approaches are well suited to address many of the challenges faced in assessing the impacts of NRM practices. The multidimensional nature of NRM and its impacts can be assessed using econometrics, and confounding factors influencing such impacts can be taken into account. Complex impacts arising through different mechanisms can be accounted for using simulations

Table 6.2. Simulated impacts of changes in selected variables on value of crop production.[a]

		Change in mean predicted values (%)					
		Mean of selected variable		Full sample value of crop production		Lowlands (< 1500 m.a.s.l)	Highlands (> 1500 m.a.s.l.)
Variable	Scenario	Before change	After change	Direct effects	Total effects	Total effects	Total effects
Population density (persons/km^2)	10% increase	220	242	+0.1%	+0.4%	+1.1%	−5.0%**
Distance to all-weather road (km)	All households next to an all-weather road	2.250	0.000	−2.2%$^{-}$	−0.9%	−0.9%	−2.9%
Primary education (proportion of households)	Universal primary education	0.480	1.000	−8.2%$^{-}$	−7.7%	−11.1%$^{**—}$	+42.1%$^{*++}$
Post-secondary education (proportion of households)	Higher education for all heads with secondary education	0.078	0.149	−0.1%	−0.7%	−0.7%	+0.3%
Agricultural training (proportion of households)	All households receive training	0.502	1.000	+13.1%***	+12.2%	+12.5%$^{***+++}$	−16.9%
Extension (proportion of households)	All households receive extension	0.311	1.000	+18.5%***	+13.7%	+10.8%***	+12.0%
Agricultural/environment NGOs (proportion of households)	All households participate	0.241	1.000	−11.8%	−8.7%	−10.7%**	+115.9%**
Irrigation (percentage of plots)	10% of plots irrigated	0.4%	10.0%	+6.0%$^{++}$	+0.6%	+3.7%***	−2.1%

[a]Simulation results for direct effects based upon predictions from OLS model regressions reported in Table 6.1. Results of regressions predicting choices of income sources, crops, land management practices and labour use were used to predict indirect impacts.

Direct effect is based on a coefficient that is statistically significant in the OLS regression at * 10%, ** 5%, or *** 1% probability level. Statistical significance of indirect effects not computed.

Direct effect is of the sign shown and statistically significant in the IV regression at $^{+,-}$ 10%, $^{++,—}$ 5% or $^{+++,—}$ 1% probability level.

based on a structural model, as demonstrated in the example from Uganda. If panel data on NRM and impact indicators are available, the dynamics of NRM impacts and responses can also be investigated using econometric methods.

Although econometric methods are useful in assessing NRM impacts, they are not without problems and limitations. Among the most important problems are the problems of endogeneity of NRM practices and omitted variable bias. These problems can be addressed through careful data collection, by analysis that accounts for as many important causal factors as possible, and by the use of instrumental variables estimators. It is always a challenge to identify instrumental variables that are strong predictors of the endogenous explanatory variables but which do not directly affect the outcomes of interest. Where markets do not function well, community and household-level socio-economic characteristics often can be used as instrumental variables for choices about NRM practices. For example, several of the human capital, social capital and land tenure variables did not significantly affect the value of crop production directly when controlling for NRM practices and input use, yet these same variables did affect adoption of NRM practices and inputs. Thus it was possible to use them as instrumental variables for NRM and technology adoption.

The Uganda study has several important limitations with regard to estimating impacts of NRM. The sample was not designed with impact assessment as the main objective. A better sample design for this purpose would have been more focused on otherwise similar households that were adopting and not adopting selected practices. Even better would have been a quasi-experimental panel design, in which adopters and non-adopters are followed over time.

The study did not address intertemporal or spatial impacts of NRM practices such as effects on future production or water flows downstream, which could be quite important in assessing their overall benefits and costs. A study design to address these impacts would have focused on sampling households likely to be affected and attempted to measure the relevant temporal and spatial externalities.

A final limitation of econometric analysis is that it doesn't provide explanations for the results obtained, some of which are usually puzzling. Sometimes puzzling results are simply statistical artefacts. Even a statistical significance level of 5% implies that the true value of 5% of the statistically significant coefficients may actually be zero. Confidence in econometric results can be increased by relating them to economic theory and to results of similar studies elsewhere. However, predictions based on theory are often ambiguous (especially when markets are imperfect). Furthermore, the impacts of selected variables may differ greatly in different contexts (as the Ugandan example demonstrates), so reference to findings of other studies may not help to increase confidence. The most effective way to clarify interpretations of econometric results is to vet them with farmers and other experts. Experiments designed to deepen understanding of empirical relationships identified through econometric analysis of survey data can also be very valuable. Complementing econometric analysis with other analytical methods can lead to more robust conclusions about the impacts of NRM, and how to improve these impacts in the future.

Endnotes

[1] An example of a large education and nutrition programme that incorporated a quasi-experimental design is the PROGRESA programme in Mexico, whose impacts on rural household welfare have been evaluated using the best available econometric approaches (Skoufias, 2003).

[2] With a linear system of equations, it is possible to estimate the model using system approaches such as seemingly unrelated regression or three stage least squares, in order to increase the efficiency of estimation (Davidson and MacKinnon, 2004). With limited dependent variable models (models with categorical, ordinal, or censored dependent variables) or other maximum likelihood models, however, it may not be feasible to estimate a system of equations, because this would require integrating over multivariate probability density functions, which is generally quite difficult with more than two dependent variables.

[3] Such a two-stage approach to estimate 2SLS is not much used any more, since this produces incorrect estimates of the covariance matrix of the coefficients (Davidson and MacKinnon 2004). Modern econometrics software packages that include 2SLS and IV estimators use the appropriate estimator of the covariance matrix.

[4] If the unrestricted IV model is poorly identified, such Wald tests may have low statistical power. As a check on the robustness of the approach in the analysis discussed here, similar Wald tests were also conducted using the unrestricted OLS model. The unrestricted OLS model did not have a serious problem of multicollinearity, and we only excluded variables that were highly statistically insignificant in both unrestricted regressions.

[5] Crop choice refers to choice of areas of annual crops to plant. Planting of perennial crops is treated as an investment, and the share of area already planted to perennial crops at the beginning of the current year is treated as part of the stock of land investments on the plot. The value of crop production in Equation 5 includes the value of both annual and perennial crops.

[6] These land management practices were measured using simple binary indicators of whether or not the practice was used, rather than continuous measures such as quantity or value of fertiliser or manure used, because of the difficulty of obtaining reliable quantity or value estimates. (An attempt was made to measure these but examination of the data led to serious questions about their reliability.) As mentioned previously, use of such categorical indicators involves a loss of information, but it was judged preferable to use less-precisely measured but more accurate indicators than more precisely measured but inaccurate indicators. Seed use was not included as part of LM_{hp} in Equations 5 and 8 for a similar reason; i.e. there was no reliable measure of the value of seed use, and a binary indicator of seed use is pointless since seeds are used for all annual crops.

References

Adesina, A.A. and Chianu, J. (2002) Farmers' use and adaptation of alley farming in Nigeria. In: Barrett, C.B., Place, F. and Aboud, A.A. (eds) *Natural Resources Management in African Agriculture.* CAB International, Wallingford, UK, pp. 51–64.

Alston, J.M., Chan-Kang, C., Marra, M.C., Pardey, P. and Wyatt, T.J. (2000) A meta-analysis of rates of return to agricultural R&D: ex pede Herculem? *IFPRI Research Report no. 113.* International Food Policy Research Institute (IFPRI), Washington, DC, 148 pp.

Amemiya, T. (1985) *Advanced Econometrics.* Harvard University Press, Cambridge, Massachussetts, 521 pp.

Baland, J.M. and Platteau, J.P. (1996) *Halting Degradation of Natural Resources: Is There a Role for Rural Community?* Clarendon Press, Oxford, UK, 423 pp.

Barrett, C.B. (2003) *Thematic Working Paper for the Operations Evaluation Department (OED) Report – The CGIAR at 31: An Independent Meta-Evaluation of the CGIAR.* World Bank, Washington, DC, 141 pp.

Barrett, C.B., Lynam, J., Place, F., Reardon, T. and Aboud A.A. (2002) Towards improved natural resource management in African agriculture. In: Barrett, C.B., Place F. and Aboud, A.A. (eds) *Natural Resources Management in African Agriculture.* CAB International, Wallingford, UK, pp. 287–296.

Binswanger, H.P. and McIntire, J. (1987) Behavioural and material determinants of production relations in land-abundant tropical agriculture. *Economic Development and Cultural Change* 36(1), 73–99.

Boserup, E. (1965) *The Conditions of Agricultural Growth.* Aldine Publishing Company, New York, 124 pp.

Bound, J., Jaeger, D.A. and Baker, R.M. (1995) Problems with instrumental variables estimation when the correlation between the instruments and the exogenous explanatory variable is weak. *Journal of the American Statistical Association* 90, 443–450.

Carney, D. (ed.) (1998) *Sustainable Rural Livelihoods: What Contribution Can We Make?* Department for International Development (DFID), London, UK, 213 pp.

Davidson, R. and MacKinnon, J.G. (2004) *Econometric Theory and Methods.* Oxford University Press, Oxford, UK, 750 pp.

Dubin, J.A. and McFadden, D. (1984) An econometric analysis of residential electric appliance holdings and consumption. *Econometrica* 52, 345–362.

Fan, S., Hazell, P. and Thorat, S. (1999) Linkages between government spending, growth, and poverty in rural India. *IFPRI Research Report no. 110.* International Food Policy Research Institute (IFPRI), Washington, DC, 81 pp.

Feder, G., Onchan, T., Chalamwong, Y. and Hongladaron, C. (1988) *Land Policies and Farm Productivity in Thailand.* The Johns Hopkins University Press, Baltimore, Maryland, 165 pp.

Hausman, J. (1978) Specification tests in econometrics. *Econometrica 46*, 1251–1272.

Herweg, K. (1993) Problems of acceptance and adaption of soil conservation in Ethiopia. Topics in Applied Resource Management 3: 391–411.

Knox, A., Meinzen-Dick, R. and Hazell, P. (2002) Property rights, collective action, and technologies for natural resource management: a conceptual framework. In: Meinzen-Dick, R., Knox, A., Place, F. and Swallow, B. (eds) *Innovation in Natural Resource Management: The Role of Property Rights and Collective Action in Developing Countries.* The Johns Hopkins University Press, Baltimore, Maryland, pp. 12–44.

Maddala, G.S. (1983) Limited-dependent and qualitative variables in econometrics. *Econometric Society Monographs no. 3.* Cambridge University Press, Cambridge, UK, 401 pp.

Mukherjee, C., White, H. and Wuyts, M. (1998) *Econometrics and Data Analysis for Developing Countries.* Routledge, London, UK, 496 pp.

Nkonya, E., Pender, J., Jagger, P., Sserunkuuma, D., Kaizzi, C.K. and Ssali, H. (2004) Strategies for sustainable land management and poverty reduction in Uganda. *IFPRI Research Report no. 133.* International Food Policy Research Institute (IFPRI), Washington, DC.

Otsuka, K. and Place, F. (2001) Issues and theoretical framework. In: Otsuka, K. and Place, F. (eds) *Land Tenure and Natural Resource Management: A Comparative Study of Agrarian Communities in Asia and Africa.* The Johns Hopkins University Press, Baltimore, Maryland, USA, pp. 3–33.

Pender, J. (2001) Rural population growth, agricultural change and natural resource management in developing countries: a review of hypotheses and some evidence from Honduras. In: Birdsall, N., Sinding, S. and Kelley, A. (eds) *Population Matters: Demographic Change, Poverty and Economic Growth in Developing Countries.* Oxford University Press, Oxford, UK, pp. 325–368.

Pender, J., Gebremedhin, B., Benin, S. and Ehui, S. (2001a) Strategies for sustainable development in the Ethiopian highlands. *American Journal of Agricultural Economics* 83(5), 1231–1240.

Pender, J., Jagger, P., Nkonya, E. and Sserunkuuma, D. (2001b) Development pathways and land management in Uganda: causes and implications. *Discussion Paper no. 85.* Environment and Production Technology Division, International Food Policy Research Institute (IFPRI), Washington, DC, 118 pp.

Pender, J., Scherr, S.J. and Durón, G. (2001c). Pathways of development in the hillsides of Honduras: causes and implications for agricultural production, poverty, and sustainable resource use. In: Lee, D.R. and Barrett, C.B. (eds) *Tradeoffs or Synergies? Agricultural Intensification, Economic Development and the Environment.* CAB International, Wallingford, UK, pp. 171–196.

Place, F. and Hazell, P. (1993) Productivity effects of indigenous land tenure systems in sub-Saharan Africa. *American Journal of Agricultural Economics* 75(1), 10–19.

Place, F., Swallow, B.M., Wangila, J. and Barrett, C.B. (2002) Lessons for natural resource management technology adoption and research. In: Barrett, C.B., Place, F. and Aboud, A.A. (eds) *Natural Resources Management in African Agriculture.* CAB International, Wallingford, UK, pp. 275–286.

Pound, B., Snapp, S., McDougall, C. and Braun, A. (eds) (2003) *Managing Natural Resources for Sustainable Livelihoods: Uniting Science and Participation.* Earthscan and International Development Research Centre (IDRC), London, UK, 252 pp.

Powell, J.L. (1984) Least absolute deviations estimation for the censored regression model. *Journal of Econometrics* 25, 303–325.

Ravallion, M. (2001) The mystery of the vanishing benefits: an introduction to impact evaluation. *World Bank Economic Review* 15(1), 115–140.

Ruecker, G., Park, S.J., Ssali, H. and Pender, J. (2003) Strategic targeting of development policies to a complex region: a GIS-based stratification applied to Uganda. *Discussion Papers on Development Policy no.* 69. Centre for Development Research (ZEF), University of Bonn, Germany, 41 pp.

Shaxson, T.F. (1999) *New Concepts and Approaches to Land Management in the Tropics with Emphasis on Steeplands.* Food and Agricultural Organization of the United Nations (FAO), Rome, Italy, 125 pp.

Shiferaw, B. and Holden, S. (1998a) Investment in soil conservation in the Ethiopian Highlands: Does it pay small farmers? *Discussion Paper no. D-32/1998.* Agricultural University of Norway, Ås, Norway, 37 pp.

Shiferaw, B. and Holden, S. (1998b) Resource degradation and adoption of land conservation technologies in the Ethiopian highlands: a case study in Andit Tid North Shewa. *Agricultural Economics* 18, 233–247.

Skoufias, E. (2003) *PROGRESA and its Impacts on the Welfare of Rural Households in Mexico.* International Food Policy Research Institute (IFPRI), Washington, DC, 118 pp. (mimeo)

StataCorp (2003a) *Stata Statistical Software: Release 8.0. Reference Manual. Volume 4.* Stata Corporation, College Station, Texas, 391 pp.

StataCorp (2003b) *Stata Statistical Software: Release 8.0. Users Guide.* Stata Corporation, College Station, Texas, 423 pp.

Stevens, J.P. (2002) *Applied Multivariate Statistics for the Social Sciences*, 4th edn. Lawrence Erlbaum Associates, London, UK, 699 pp.

Tiffen, M., Mortimore, M. and Gichuki, F. (1994) *More People – Less Erosion: Environmental Recovery in Kenya.* John Wiley & Sons, Chichester, UK, 311 pp.

Wood, S., Sebastian, K., Nachtergaele, F., Nielsen, D. and Dai, A. (1999). Spatial aspects of the design and targeting of agricultural development strategies. *Discussion Paper no. 44.* Environment and Production Technology Division, International Food Policy Research Institute (IFPRI), Washington, DC, 81 pp.

7 Assessing Economic Impacts of Natural Resource Management Using Economic Surplus

S.M. Swinton

Michigan State University, East Lansing, Michigan, USA

Introduction

Over the past three decades, major advances have been made in the economic assessment of agricultural research impacts. Yet in a recent appraisal of accomplishments in the Consultative Group on International Agricultural Research (CGIAR) international agricultural research centres (IARCs), Pingali notes that little progress has been made in measuring the impacts of research on natural resource management (NRM) (Pingali, 2001). In particular, although attempts have been made to conduct benefit–cost analysis of NRM projects, there have been scarcely any attempts to assess the economic impacts of new NRM practices using the economic surplus approach (Alston *et al.*, 1998).

In the aftermath of the Green Revolution, NRM research has seen a resurgence that is pushing beyond its historic focus on soil fertility and conservation. But assessing the economic impacts of technologies that are not embedded in an improved seed can be difficult. Particular challenges for NRM impact evaluation are attribution, measurement and valuation. First, like other types of cross-commodity research, NRM often defies easy attribution of its impacts. Clearly, yield gains from genetic research are attributable to the research investment. But many NRM research projects modify existing technologies or document benefits of established conservation practices (some of them very old). If soil is conserved because farmers built earthen terraces, is it attributable to public research that refined terracing techniques? Second, measurement of any research impact hinges on documenting the difference in output with and without the new technology. For NRM practices, establishing the counterfactual case (what would have happened without the NRM technology) is tricky because measuring biophysical impacts on natural resources can be costly, imprecise, and slow. How much soil, that would otherwise have eroded, was conserved by the earthen terraces? Would some other field receiving the eroded soil have realised offsetting

 Natural Resource Management in Agriculture: Methods for Assessing Economic and Environmental Impacts (eds B. Shiferaw, H.A. Freeman and S.M. Swinton)

gains? Third, assigning economic values to these outcomes is fraught with difficulty. Is the value of averting soil erosion on a field simply the value of changed productivity on that field over time? Or does it include costs and benefits on the fields and waterways to which the soil moved? How long is long enough to measure effects that cumulate over time?

The economic surplus framework for impact assessment aims to capture both consumer and producer net benefits from new technologies. New technologies change the total amount of economic surplus as well as its distribution between consumers and producers. In applying the economic surplus approach to NRM impact assessment, estimating the supply shifts due to new NRM technologies and determining how consumers will value those changes requires confronting the triple challenges of attribution, measurement and valuation.

This chapter outlines elements of methods for incorporating NRM indicators into the economic surplus approach to impact assessment. Along the way, it first summarises the economic surplus approach. Next, it discusses types of NRM research and associated impacts. Alternative methods for placing values on NRM impacts are briefly reviewed, emphasising cost-effective ways to address measurement and valuation. Finally, the chapter surveys and comments upon recent attempts to integrate sustainability indicators into the economic surplus framework, identifying needs for further development of both methods and applications.

Economic Surplus Approach to Impact Assessment

The economic surplus approach to impact assessment is rooted in the microeconomics of supply and demand. The basic idea is simple and is illustrated in Fig. 7.1. Consumer demand can be described by a downward sloping demand curve illustrating that some consumers are willing to pay more than others for a given commodity, such as sorghum grain. At a market-clearing equilibrium price, p*, those consumers who were willing to pay more than p* realise benefits by getting the product for less money than they were willing to pay. Across all consumers, the area beneath the demand curve, D, and above the equilibrium price, p*, measures the total value of *consumer surplus*. This area measures the aggregate difference between what consumers were willing to pay and what they did pay. Note that some consumers were willing to pay only prices lower than p*, so they did not buy.

Producer supply can be described by an upward sloping curve that illustrates that some producers can supply a product for a lower price than others. At the market-clearing equilibrium price, p*, those producers who could supply the product at a lower price obtain extra benefits. The aggregate benefits described by the area above the supply curve, S, and below the equilibrium price, p*, measure the total *producer surplus*. Together, consumer surplus and producer surplus sum to the *economic surplus*.

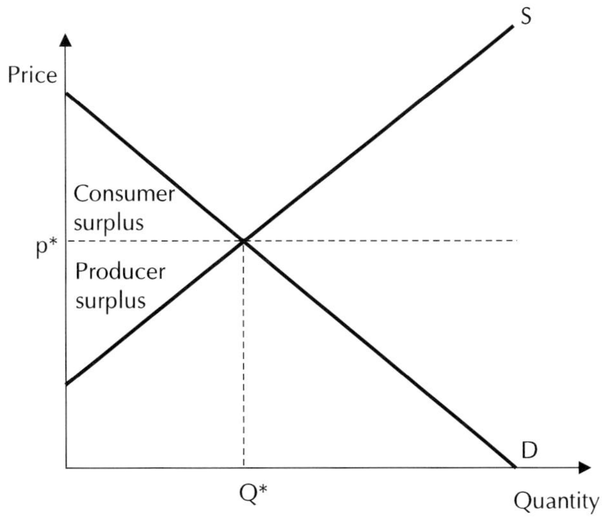

p*, equilibrium price; S, supply curve; D, demand curve

Fig. 7.1. Economic surplus divided between consumer and producer surplus.

The economic impact of a new production technology can be estimated as the changes in the economic surplus that result from a shift in the supply curve. For change in economic surplus to describe economic impact accurately, two assumptions must be met. First, supply and demand curves must accurately depict the values that consumer and producers assign to the product. Second, benefits (surplus) to all actors in the market must be valued equally (Alston *et al.*, 1998).

New production technologies typically reduce the cost of producing a unit of output. Both yield-enhancing and cost-reducing technologies have the net effect of reducing the average cost of production. The comparative static effects on product supply and economic surplus are illustrated in Fig. 7.2. The reduction in unit costs means that producers can now afford to supply the same amount of product at a lower price (or more of the product at the same price). The new, lower-cost supply curve, S′, shifts down (for cost reducing technological change) and/or to the right (for productivity increasing technological change), resulting in a new equilibrium price, p′. All consumers are better off, because the price is lower. Consumers who were buying the product before can now buy it for less, and some new consumers enter the market at the lower price, so that the quantity sold rises from Q* to Q′. Consumer surplus increases by the sum of areas **a** + **c**. The effect on producers is mixed. Producers receive a lower price for their product, so their producer surplus decreases by area **a**. But they are selling more at a lower cost of production, so producer surplus increases by area **b**.

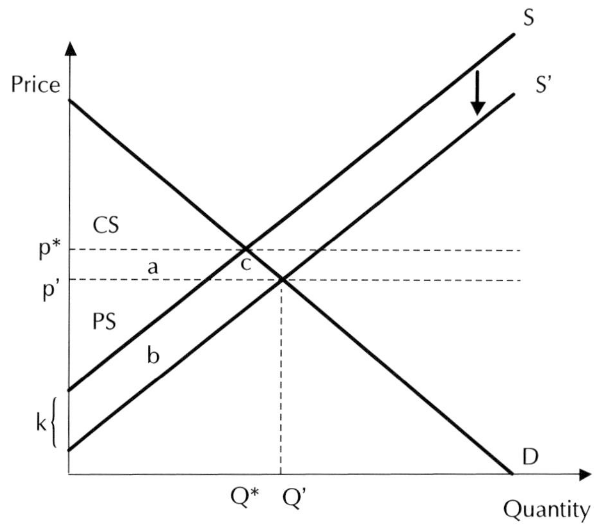

CS, consumer surplus; PS, producer surplus; p*, equilibrium price; p', new equilibrium price; Q*, equilibrium quantity; Q', new equilibrium quantity; S, supply curve; D, demand curve; k, magnitude of cost-reducing shift; S' new supply curve after shift k

Fig. 7.2. Change in economic surplus due to a cost-reducing shift in supply.

How the effects of a new technology are divided between producers and consumers depends upon the slopes of the supply and demand curves in the neighbourhood of equilibrium prices. The price elasticity of consumer demand is especially important. If consumers are willing to buy any quantity at a given price (the case of perfectly elastic demand where the demand curve is a horizontal line), then cost-reducing technological change creates no consumer surplus and all benefits go to the producers. By contrast, if consumer demand is very inelastic (nearly vertical demand curve), then technological change may lead to a large transfer of surplus from producers to consumers (meaning that areas **a** and **c** are large, potentially larger than area **b**, which depends on the new technology's supply effect alone).

For the class of technologies that reduce the unit production costs of agricultural commodities, the economic surplus approach to evaluating the impact of research and development has been thoroughly described by Alston *et al.* (1998). Indeed, those authors even offer a graphical analysis of how an environmental externality could be incorporated into the economic surplus model (Alston *et al.* pp. 294–296), assuming that it could be properly measured. However, most NRM technologies present special complications when it comes to measuring the quantity and value of environmental impact for which they are responsible.

Attribution and Measurement of NRM Research Impacts

As practised in crop research institutions, NRM research chiefly focuses on those natural resources that are most closely tied to crop production: soil, water, crop genetics, biodiversity of crop–pest complexes, and human health. A natural resource can usefully be conceived as a stock of natural capital that yields service flows over time that can be enhanced with supplemental investments (Pearce and Atkinson, 1995). Soil quality can be thought of as stock of soil fertility that will deteriorate if drawn down by crop production without fertility renewal. Soil quantity can likewise erode if soil loss occurs at a faster rate than replacement. Water quantity can diminish if used at rates exceeding recharge. Water quality can also diminish if the rate of contamination exceeds the rate of decontamination. Crop genetic resources are a stock that is valued both for current use values and for the option value of potential future productivity gains that they might yield (Evenson *et al.*, 1998). The biodiversity resources of pest–crop complexes include resources in a more abstract sense that includes the ways that species relate to one another, such as the genetic susceptibility of a pest to a given pest control mechanism. Finally, human health is obviously a resource that is fundamental to any system that humans manage. Yet nutrition and exposure to health risks in the production process may render human health another resource whose productivity is endogenous to the NRM system.

The attribution, measurement, and valuation of NRM technologies pose challenges in both time and space. All are complicated by the dynamics of how natural resource stocks evolve over time. Many NRM technologies also have effects that cut across multiple commodities. For example, reduced soil erosion and better water retention due to soil ridging technologies affect all crops on which they are used. But effects vary by geographical setting and the magnitude of an effect often changes over time.

Attribution of identified effects can be accomplished with controlled experiments or simulation models over time. Both can be used to identify and measure changes in crop productivity from soil conservation practices, for example. Gebremedhin *et al.* (1999) used randomly placed experimental plots on Ethiopian farm fields to monitor crop productivity effects from soil movement due to stone terraces of different ages. The experimental design permitted both attribution and the measurement of crop yield responses in relation to the distance of the plot from the nearest terrace (Gebremedhin *et al.*, 1999). It also established that crop yields were declining in the absence of terraces. Measurement of such a counterfactual for conservation investments is crucial to establishing the value of NRM impacts that may prevent productivity deterioration, rather than directly increase productivity.

Simulation models provide an apt environment for comparing scenarios 'with' vs. 'without' NRM technology over time. Simulation models can shorten the time it takes to observe slowly evolving NRM effects on resource stocks and related productivity outcomes. Likewise, they can permit a quasi-experimental setting that may be costly or difficult to maintain in the real world. Crop growth simulation models have been developed that are specifically designed to model changes in productivity in response to several types of NRM technologies, including soil erosion (Williams *et al.*, 1989; Sharpley and Williams, 1990; Pierce, 1991; Yoder and Lown, 1995), soil nutrient availability (Hanks and Ritchie, 1991; Shaffer *et al.*, 1991), and soil water availability (Skaggs *et al.*, 1986; Hill, 1991).

What to measure and how to do it are related challenges. For on-site productivity effects, controlled experiments and simulation models are very suitable. The consequences of such effects are felt chiefly on-site by the farm household. However, NRM technologies have two other kinds of effects. Some on-site effects are delayed, and may not be recognised at first by the manager. Examples are chronic effects of pesticide use that may not have been properly accounted in the farmer's decision making (Rola and Pingali, 1993; Crissman *et al.*, 1998). Other effects are not experienced by the farm household, but rather are experienced off-site as 'externalities' to the farmer's privately optimal management choices. For example, in some settings, soil erosion may reduce water quality or lead to sedimentation of waterways (Barbier, 1998). By the same token, NRM and yield-enhancing agricultural research may create positive externalities in the form of land-saving effects that protect amenities associated with forests and natural uses (Nelson and Maredia, 1999).

NRM technologies may potentially affect a wide variety of environmental and natural resource (ENR) services, so what to measure depends upon the NRM technology in question and the environmental setting where it is used. What to measure is linked also to those NRM impacts likely to have the greatest social value.

Valuation of Private vs. Public NRM Benefits

As noted above, the benefits of NRM practices can broadly be divided between those captured privately (by the NRM practitioner) and those external to the NRM practitioner that are captured publicly. Table 7.1 identifies illustrative cases of three NRM practices. Privately captured benefits are

Table 7.1. Common agricultural natural resource management (NRM) practices.

NRM practice	Main private benefits	Main public benefits
Soil fertility management	Reduced yield decline Reduced fertiliser costs	None
Soil conservation	Reduced yield decline Reduced fertiliser costs	Reduced erosion on neighbouring lands Reduced sedimentation of waterways
Pest-tolerant crop variety or integrated pest management (IPM) practice	Reduced pesticide costs; increased crop yields for farmer Reduced exposure risks to applicator Reduced residue risk for consumer	Reduced pesticide risks to drinking water supply and non-target species

the easiest to measure, especially when they are tied to marketed products. If the counterfactual scenario can be established to estimate the change in productivity with and without the NRM innovation, then the annual value of the innovation to adopters equals the net increase in income over the counterfactual alternative. The simplest case would be a NRM practice such as soil fertility management whose effects are wholly captured on-site (i.e. in a locale where off-site effects are negligible). The private value of soil fertility management is the value of yield loss averted plus the costs of any fertilisers that might have been applied to stem yield losses.

Within the realm of private benefits, the next level of benefit covers effects that are still privately experienced but hidden, due to lags or lack of obvious market valuation. Reduction in pesticide-related human health effects is a case in point (Crissman *et al.*, 1998; Maumbe and Swinton, 2003; Rola and Pingali, 1993). If the health damage from pesticide misuse were wholly limited to applicator effects, then all benefits would be privately realised from NRM practices that reduced applicator risk (e.g. pest-tolerant crop variety, safer pesticide, integrated pest management (IPM) practices that reduce pesticide use). However, these health benefits might be delayed, because they involve averting not just acute but also chronic health problems that are slow to develop.

Some NRM practices have public effects felt beyond the NRM practitioner. Such economic externalities are common among ENR services. In particular, production processes for marketed commodities sometimes generate by-products that are bad for the environment. Yet harmful by-products that have no market (e.g. nitrate or pesticide leaching) are prone to be ignored in the producer's benefit–cost calculus. Hence, the value of an NRM innovation that reduces the externality problem may need to be calculated indirectly.

Consider a hypothetical case where the conventional crop production practice requires a toxic pesticide that leaches into drinking water supplies. The dangers posed by pesticide leaching into drinking water represent an economic externality that is ignored by sorghum growers in deciding on input use, but it imposes social costs for pesticide poisoning and treatment. Figure 7.3 illustrates economy-wide marginal benefit and marginal cost curves that are analogous to demand and supply curves for the crop. The two supply curves differ in that the marginal private cost curve (MPC=S^p) represents the private production costs incurred by sorghum growers. By contrast, the marginal social cost curve (MSC=S^s) includes the MPC plus the externality cost for pesticide-induced suffering and medical treatment. Because the equilibrium market price, p^p, is based on the MPC curve, it results in higher demand for this crop than would result from the actual social costs reflected in the MSC curve (Tietenberg, 1984).

Release of a new crop variety with pest tolerance that does not require the leaching pesticide would generate two kinds of direct social benefits. First, the avoided cost of the pesticide would result in a downward shift of the MPC supply curve with effects similar to those illustrated in Fig. 7.2. Second, substitution of the new variety for the old one would remove the health externality cost that caused the MSC curve to lie above the MPC curve.

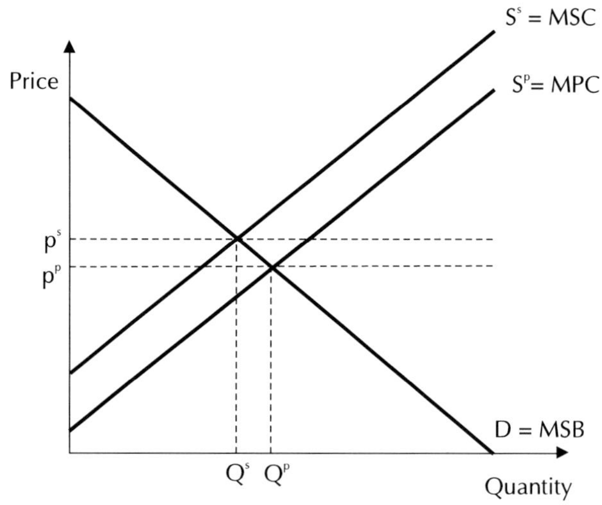

MPC, marginal private cost; MSC, marginal social cost; S^p, supply curve based on private costs; S^s, supply curve based on social costs; p^s, equilibrium price if all social costs factored in; p^p, equilibrium market price; MSB, marginal social benefit; Q^s, equilibrium quantity if all social costs factored in; Q^p, equilibrium quantity if only private costs factored in

Fig. 7.3. Difference between marginal private cost and marginal social cost when production involves a negative environmental externality.

Removing the health externality cost would create a pure gain in economic surplus. These two effects result in a double benefit from the new variety due to reduced direct production costs and reduced externality costs. Note that even if growers had to pay as much for the new seed as they had paid for the pesticides, society would still benefit by the reduced externality.[1] The major measurement challenge here lies in estimating the value of the externality, in this instance the value of pesticide-induced illness that could be averted with the new technology.

The thorniest NRM impact valuation challenge occurs when the NR impacts are publicly borne and associated with private use of a public good. A public good is defined as one whose consumption neither excludes nor directly reduces someone else's consumption. The classic problem with public goods is that they tend to be overexploited because individual actors do not face the full costs of their stock decline. Hence, NRM practices that benefit common property resources may not be adopted at socially optimal levels. For example, a productive forage crop may be little adopted because shared natural pastures can be exploited – despite the fact that natural pastures may be losing favoured forage species and diminishing in their carrying capacity. ENR public goods include common property resources, such as pastures, forests, water supplies and the atmosphere. We will not address further the special case of NRM impact valuation on common property resources.

Economic Valuation of ENR Services

While markets serve to place values on privately marketed products of NRM research, other methods are required for the economic valuation of human health and ENR services. Three classes of valuation methods dominate: direct market measures, revealed preferences inferred from market behaviour, and stated preferences for ENR services that are contingent on hypothetical market settings.

Direct market methods include:

- Cost of remediation
- Cost of illness (including work days lost and medical treatment)
- Cost of alternative production practices.

Revealed preference methods include:

- Hedonic valuation of ENR characteristics embedded in marketed commodities (e.g. real estate value differences due to air quality levels, or wage differentials explained by exposure level to toxic chemicals)
- Averting expenditures made to avoid exposure to some undesired ENR state
- Mitigating expenditures made to reduce emission of some undesired ENR service
- Travel costs incurred to gain access to some desired ENR services.

Stated preference methods are based on survey methods. They include:

- Contingent valuation to estimate willingness to pay (WTP) for ENR services or programmes
- Conjoint analysis for ranking ENR alternatives.

There is a large and growing literature on non-market valuation methods for human health (Viscusi, 1993; Kenkel, *et al.*, 1994) and ENR services (Braden and Kolstad, 1991; Freeman III, 1993; Haab and McConnell, 2002). The value of ENR services can be divided between value from direct use (e.g. clean water consumption, avoidance of illness) and from non-use (e.g. the value gained from existence of a resource that could be used in the future or bequeathed to the next generation). Broadly speaking, the direct market and revealed preference methods listed above fail to capture non-use values. Stated preference methods are theoretically the most complete measures of ENR value, but their use has been criticised based on practical difficulties with unbiased implementation (Diamond and Hausman, 1994; Hanemann, 1994).

Evaluating the many alternative non-market ENR valuation methods for use with diverse agricultural NRM technologies goes beyond the scope of this chapter. However, some indicative illustrations are worthwhile. For soil erosion, off-site values associated with sedimentation have been estimated using the cost of restoration approach associated with dredging navigable waterways (Ribaudo and Hellerstein, 1992; Barbier, 1998). For health risk reduction associated with reduced pesticide technologies, the cost of illness approach has been employed (Pimentel *et al.*, 1992; Rola and Pingali, 1993; Crissman *et al.*, 1994). For a broader set of benefits associated with adoption of IPM or avoidance of pesticide risks other than health alone, contingent valuation methods have been employed (Mullen *et al.*, 1997; Owens, 1997; Brethour and Weersink, 2001). Hedonic valuation methods have been used to estimate the value to US farmers of herbicide safety characteristics embodied in herbicide price differences (Beach and Carlson, 1993) and of soil conservation investments embodied in farmland prices (Palmquist and Danielson, 1989).

A key limitation of most health and ENR valuation methods is that they are costly to implement. A small but growing area of research into 'benefits transfer' examines the conditions under which environmental values reported in one study may be applied to a different setting. The simplest method of benefit transfer is to take a mean value from a reported study site and apply it to a new site. This method has been criticised because differences across sites in both socio-economic characteristics and biophysical setting may lead to different ENR valuation estimates. An alternative is to transfer a *benefit function*, typically an econometric forecasting equation into which typical values for explanatory variables from the new site may be inserted in order to tailor the predicted ENR benefit values to conditions at the new site. The benefit function approach is generally believed to be more accurate, and was found to be marginally so in a recent controlled study, although both approaches sometimes deviated substantially from on-site surveys (VandenBerg *et al.*, 2001). For economic surplus estimation purposes, the benefit function approach has the important advantage that it can be applied to simulate the variability in benefit valuation across a sample population at a new site, thereby capturing not just the average value of the benefit, but changes along the demand curve of marginal WTP for increasing levels of ENR services.

Implementing NRM Impact Assessment in the Economic Surplus Framework

How to accommodate the idiosyncrasies of NRM technology impacts in an economic surplus analysis? Although private and social costs are sometimes combined in theory (as in Fig. 7.3), for empirical work it is more practical to separate privately captured changes in economic surplus due to marketable goods and services from publicly captured externality effects due to non-marketed health and ENR services. Keeping private and public costs separate implies a parallel measurement and valuation process, such as the one illustrated for IPM impact assessment in Fig. 7.4 (Norton *et al.*, 2000).

Details for conducting an economic surplus analysis of returns to cost-reducing research into marketed agricultural commodities may be found in Alston *et al.* (1998). The key variables for measuring the *ex post* cumulative value of changes in economic surplus for some marketed commodity **j**, as illustrated in Fig. 7.2, are the downward shift in the supply curve (commonly denoted k_j and based upon proportionate changes in output supplied and cost of production), the equilibrium price elasticity of supply (ε_j), and the price elasticity of demand (η_j). Estimates of change in economic surplus are typically more sensitive to the estimate of k_j than to the elasticities (Alston *et al.*, 1998).

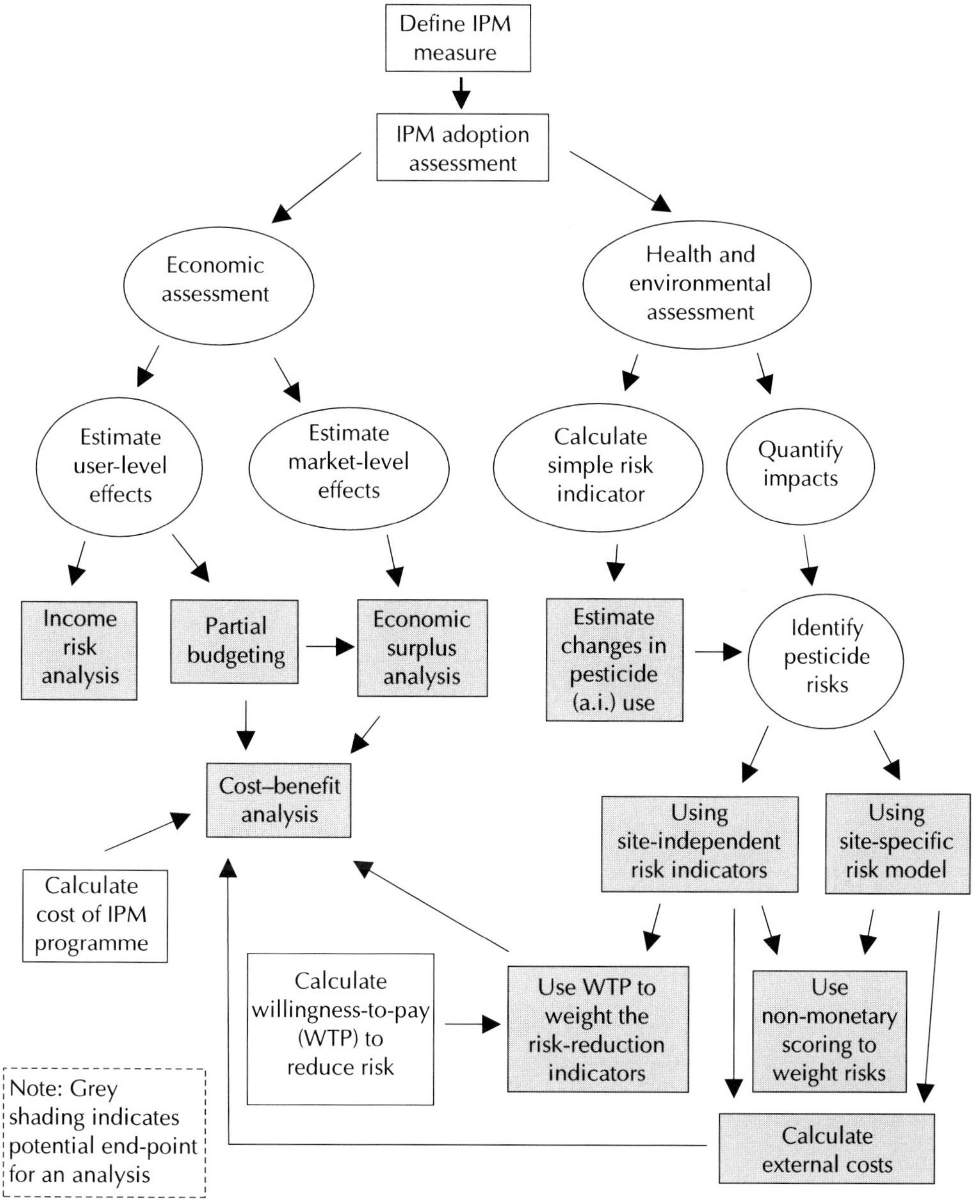

Fig. 7.4. Summary chart of the integrated pest management (IPM) impact assessment process (Norton *et al.*, 2000, Fig. 4).

Among NRM research impacts, the cost-reducing technology approach discussed applies to instances where NRM reduces costs of marketed products. The soil fertility management research in Table 7.1 is a case in point: The economic impact could be measured by the change in economic surplus, because the benefits accruing from reduced costs and increased yields are entirely captured in a single supply shift, k.

Two other kinds of NRM research impact require different measures of economic surplus. The first is the case of NRM technologies that cause changes in product qualities appreciated by consumers. Such technologies can induce a shift in consumer demand as well as one in producer supply. An example would be pest management that reduces pesticide residue risks to consumers. If consumers are wary of pesticide health risks, such a technological change should result in an upward demand shift, with willingness to pay higher prices. Such a research-induced demand shift requires measurement of the demand shift, as well as the supply shift (k). Note that the supply shift need not be negative. Figure 7.5 illustrates a case where rising production costs shift the supply curve upwards from S to S′, but the accompanying upward shift in consumer demand from D to D′ causes a net gain in producer surplus. Although Fig. 7.5 shows equilibrium quantity unchanged at Q′, quantity could increase or decrease, depending on the price elasticities of supply and demand.

The second class of NRM technology requiring a different approach to economic surplus estimation is the case of research affecting economic externalities not faced by the producer. In the special case where the externality

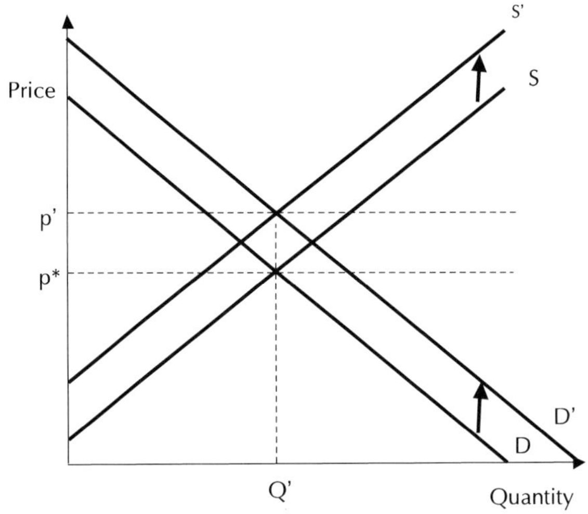

D, demand curve; D', new demand curve after shift; p*, equilibrium price; p', new equilibrium price; Q', new equilibrium quantity; S, supply curve; S', new supply curve after shift.

Fig. 7.5. Change in economic surplus due to joint upward shifts in supply from S to S' and demand from D to D'.

incurs a constant social cost per unit produced, illustrated in Fig. 7.3, one can adapt the Alston *et al.* (1998) approach to measure the parallel difference between marginal social cost embodied in the supply curve with and without a new NRM technology. For example, if the external cost of soil erosion were constant per sack of grain produced, such an approach would be valid.

However, the economic externalities associated with NRM technology adoption are typically not constant per unit of marketable product. First, they often exhibit increasing marginal costs, as when increasing output requires shifting production to more marginal settings (e.g. crop farming on more erodible lands). Second, economic externalities typically involve different goods and services than the one being produced for market. Consider the case of soil erosion; that causes sediment to deposit in a navigable waterway. Two markets are involved: 1. the market for the crop whose production entails soil erosion; and 2. the market for shipping services on the waterway. A soil conservation technology may cause a cost-reducing shift in supply of the crop. It will also cause a cost-reducing shift in the supply of shipping services. The latter is most accurately estimated directly, despite the common tendency to apply cost of restoration methods (Barbier, 1998). Why not simply apply a fixed cost of restoration per ton of soil eroded? There are two reasons. First, restoration is not necessarily feasible or desirable. Second, the true economic cost is the cost of switching to the next best alternative; that alternative may just as well be shipping by train as dredging the waterway to permit continued barge shipping (Bockstael *et al.*, 2000). Insights into the best alternative and the cost of switching to it are best obtained through direct observation. For cases where economic externalities are important, changes in economic surplus should be measured for a market related to the externality, as well as one related to the marketed product. Such measurements typically entail environmental goods and services that are not marketed, so they require inferences either from related indicator commodities for which markets exist or else from constructed markets, as discussed briefly above. Although demand elasticities have been estimated for agriculturally related ENR services (Owens, 1997), none have been incorporated into an economic surplus analysis of NRM impacts, to this author's knowledge.

Those few studies that have estimated the cumulative value of NRM impacts on non-marketed ENR services over time have lacked suitable elasticity estimates and so used a benefit–cost approach. All have assumed constant WTP ('price') for the ENR amenity, implying perfectly elastic demand. Most have likewise assumed that any increased costs associated with producing the ENR amenity were fully covered by privately captured benefits through marketed products, so production costs were excluded from the ENR benefit accounting. The net effect is to estimate the value of a shift in perfectly inelastic supply, like the one illustrated in Fig. 7.6. Ordinarily, this would imply that producers capture all surpluses. However, because the ENR amenity is not marketed, the normal distinction between producer and consumer surplus is meaningless; intuitively, it would seem that consumers chiefly capture the benefits of this shift.

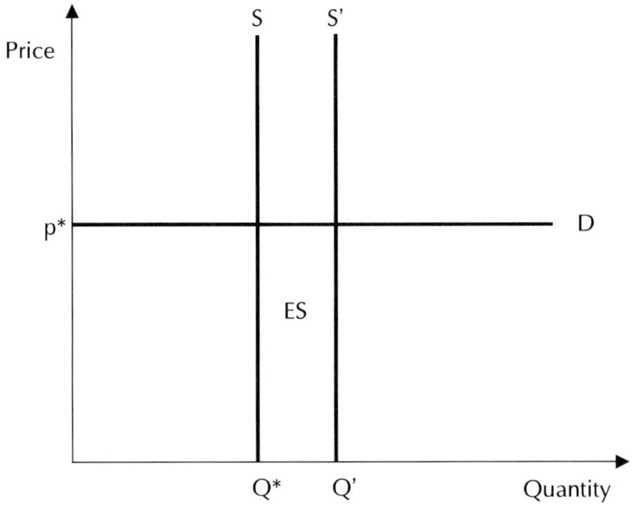

D, demand curve; ES, economic surplus (equal to producer surplus); p*, equilibrium price; Q*, equilibrium quantity; Q', new equilibrium quantity.

Fig. 7.6. Change in economic surplus due to an outward shift in inelastic supply from S to S', when demand is perfectly elastic at price p*.

Translation from consumer WTP units to producer NRM impact units

Even when monetary values can be estimated for non-marketed ENR benefits from NRM technologies, a secondary challenge is to associate consumer WTP for ENR amenities with producer measures of ENR amenities produced by adopting NRM practices. Two examples serve to illustrate.

Beddow supplemented his estimate of producer surplus associated with the adoption of IPM in sweet corn in Pennsylvania, USA, by estimating the mean value of ENR services gained (Beddow, 2000). He adjusted mean monthly WTP values for reducing eight types of pesticide risks from a contingent valuation survey of consumer households (Mullen, 1999), so that they corresponded with levels of IPM adoption by producers.

A limitation of Beddow's (2000) ENR valuation is that it used unchanging mean values rather than marginal WTP from a downward-sloping demand curve for ENR amenities. Labarta *et al.* (2002) recently drew upon a contingent valuation study that published marginal WTP for improved water quality (Poe and Bishop, 2001) in outlining a method to estimate the ENR value of soil fertility management to reduce groundwater contamination (Labarta *et al.*, 2002). In doing so, they illustrated a method for converting WTP denominated in consumer annual water consumption into values per unit of nitrate leached into drinking water.

A useful extension of the nascent efforts to incorporate NRM innovations into the economic surplus approach would be to apply empirical estimates of supply and demand elasticities for ENR amenities that arise from NRM practices. Supply elasticities would have to be estimated from survey data or multilocational experimental trials that reflect geographic and other differences in producer costs. For example, the marginal cost of pesticide reduction may be less where pest pressure is low. Demand elasticities would likely have to come from survey estimates such as the contingent valuation studies on which the Beddow (2000) and Labarta *et al.* (2002) efforts relied. Compared with the Beddow approach, which was based on average WTP values, the Labarta *et al.* effort uses marginal WTP values that should more accurately reflect consumer values for less-than-total elimination of risk. However, Labarta *et al.* did not build their analysis into an economic surplus model.

Care must be taken in transferring benefits between settings. This is especially true when the settings are very different in biophysical or socio-economic traits. Useful contingent valuation studies have been conducted of WTP to reduce pesticide-related risks for several crops in the USA, Canada and the Philippines (Higley and Wintersteen, 1992; Mullen *et al.*, 1997; Owens, 1997; Brethour and Weersink, 2001; Cuyno, *et al.*, 2001). However, not only do these apparently similar studies vary in production setting and income level of respondents, some are surveys of consumers, whereas others are surveys of producers. The inferences to be drawn from such different data sources are quite divergent, despite the common focus on valuation of pesticide risk reduction.

Measuring adoption

Estimating the discounted cumulative value of NRM impacts over time obviously depends not just upon the value of one individual's adoption, but also on how many adopt. As ably discussed by Alston, Norton and Pardey (Alston *et al.*, 1998), adoption rates may be projected based on expert opinion about key parameters (e.g. maximum adoption level, beginning date of diffusion, rate of diffusion, likely beginning of disadoption and corresponding rate). More accurate estimates of adoption rates may be had by surveys of adoption (Griliches, 1957; Byerlee and Hesse de Polanco, 1986; Fernandez-Cornejo and Castaldo, 1998; Fernandez-Cornejo *et al.*, 2002). Indeed, expert opinion may deviate substantially from actual adoption levels and may be affected by wishful thinking on the part of experts. A case in point is a 1999 survey of tart cherry growers in Michigan, USA, that found farmers were using no IPM methods on one-third of planted area, whereas IPM experts believed that virtually all farmers were using at least basic IPM practices (Norton *et al.*, 2000).

Conclusions

The nascent state of attempts to integrate sustainability indicators linked to NRM technologies into economic surplus analysis leaves ample room for innovation. The area ripest for new contributions is the incorporation of supply and demand elasticities for ENR services so that their valuation becomes more than a benefit–cost analysis exercise. A clear need exists for economic analyses of NRM impacts that incorporate welfare effects from both marketed products and non-marketed products with acknowledged welfare effects.

Additional research into benefits transfer will also be key to clarifying criteria and methods for adapting ENR amenity valuation estimates from one setting to other ones. Until now, most benefit transfer functions and meta analyses have been developed using mostly socio-economic data to capture differences in income influencing the budget constraint that affects consumer willingness to pay. But for agricultural NRM, spatial heterogeneity in the resource base makes integration of spatial biophysical determinants of WTP important as well. Such spatial integration has yet to be attempted.

Moving beyond the scope of the NRM technologies and economic surplus analysis discussed here, there are two areas worth exploring for ENR benefits linked to agricultural research. To the extent that plant-breeding innovations have intensified agricultural productivity per unit of land, they have likely saved land from agricultural use (Nelson and Maredia, 1999). More comprehensive efforts to place value on the ENR amenities so preserved could supplement impact assessments of ENR services due to direct NRM interventions.

The second potentially fruitful effort is to estimate the effect on ENR amenity valuation of rising incomes in developing countries. All the methods reviewed above have presupposed that the value of ENR services is static. However, it has been observed that as incomes rise in developing countries, levels of pollution at first begin to rise; then they decline with rising per-capita income. This bell-shaped relationship has been dubbed the 'environmental Kuznets curve'[2] (Dasgupta *et al.*, 2002; Yandle *et al.*, 2002). It is generally believed to result from two phenomena: 1. the replacement of old technologies with cleaner technologies that may also be more productive, and 2. rising demand for environmental quality as consumers become wealthier. While most research documenting the environmental Kuznets curve relationship has focused on urban air pollution, the dynamic effects could equally well apply to agriculturally related ENR amenities.

Finally, although NRM technologies can play an important role in reducing health and ENR risks linked to agricultural production processes, policy plays a crucial role for internalising the externalities that make these technologies worth adopting. Unlike products for which markets function, NRM technologies with increased costs will not be adopted for their ENR benefits alone. If producers perceive no incentive greater than a hypothetical consumer WTP number from a contingent valuation survey, then few will adopt environmentally beneficial technologies. If producers lack incentives to adopt sustainable NRM technologies, then researchers in turn will lack incentives to develop them (Swinton and Casey, 1999). Producer adoption is the *sine qua non* for impacts to occur. So another important role for *ex ante* assessments of NRM impacts is to reveal the value of ENR services that could be had if policy incentives for adoption of sustainable technologies were put in place.

Endnotes

[1]However, unless the new seed cost was less than the cost of pesticide use, farmers might not choose to adopt the new seed technology. Even when a new technology is available that creates net social benefits, public policy incentives may be necessary in order to induce its adoption (Casey *et al.*, 1999).

[2]The curve is named after Simon Kuznets who observed the bell-shaped pattern of correlation between income growth and inequality.

References

Alston, J.M., Norton, G.W. and Pardey, P.G. (1998) *Science Under Scarcity: Principles and Practice for Agricultural Research Evaluation and Priority Setting*. CAB International, Wallingford, UK, 585 pp.

Barbier, E. (1998) The economics of soil erosion: theory, methodology, and examples. In: Barbier, E. (ed.) *The Economics of Environment and Development: Selected Essays*. Edward Elgar, Cheltenham, UK, pp. 281–307.

Beach, E.D. and Carlson, G.A. (1993) Hedonic analysis of herbicides: do user safety and water quality matter? *American Journal of Agricultural Economics* 75, 612–623.

Beddow, J. (2000) Protocols for assessment of integrated pest management programs. MSc thesis. Department of Agricultural and Applied Economics, Virginia Tech University, Blacksburg, Virginia.

Bockstael, N.E., Freeman, A.M., Kopp, R.J., Portney, P.R. and Smith, V.K. (2000) On measuring economic values for nature. *Environmental Science and Technology* 34, 1384–1389.

Braden, J.B. and Kolstad, C.D. (1991) *Measuring the Demand for Environmental Quality*. North-Holland, Amsterdam, The Netherlands.

Brethour, C. and Weersink, A. (2001) An economic evaluation of the environmental benefits from pesticide reduction. *Agricultural Economics* 25, 219–226.

Byerlee, D. and Hesse de Polanco, E. (1986) Farmers' stepwise adoption of technological packages: evidence from the Mexican Altiplano. *American Journal of Agricultural Economics* 68, 519–527.

Casey, F., Schmitz, A., Swinton, S. and Zilberman, D. (1999) *Flexible Incentives for the Adoption of Environmental Technologies in Agriculture*. Kluwer Academic Publishers, Boston, Massachusetts, 370 pp.

Crissman, C.C., Antle, J.M. and Capalbo, S.M. (1998) *Economic, Environmental and Health Tradeoffs in Agriculture: Pesticides and the Sustainability of Andean Potato Production*. Kluwer Academic Publishers, Dordrecht, The Netherlands, 281 pp.

Crissman, C.C., Cole, D.C. and Carpio, F. (1994) Pesticide use and farm worker health in Ecuadorian potato production. *American Journal of Agricultural Economics* 76, 593–597.

Cuyno, L., Norton, G.W. and Rola, A. (2001) Economic analysis of environmental benefits of integrated pest management: a Philippine case study. *Agricultural Economics* 25, 227–233.
Dasgupta, S., Laplante, B., Wang, H. and Wheeler, D. (2002) Confronting the environmental Kuznets curve. *Journal of Economic Perspectives* 16, 147–168.
Diamond, P.A. and Hausman, J.A. (1994) Is some number better than no number? *Journal of Economic Perspectives* 8, 45–64.
Evenson, R.E., Gollin, D. and Santaniello, V. (1998) *Agricultural Values of Plant Genetic Resources.* CAB International, Wallingford, UK, 285 pp.
Fernandez-Cornejo, J. and Castaldo, C. (1998) The diffusion of IPM techniques among fruit growers in the USA. *Journal of Production Agriculture* 11, 497–506.
Fernandez-Cornejo, J., Alexander, C. and Goodhue, R.E. (2002) Dynamic diffusion with disadoption: the case of crop biotechnology in the USA. *Agricultural and Resource Economics Review* 31, 112–126.
Freeman III, A.M. (1993) *The Measurement of Environmental and Resource Values: Theory and Methods.* Resources for the Future, Washington, DC, 516 pp.
Gebremedhin, B., Swinton, S.M. and Tilahun, Y. (1999) Effects of stone terraces on crop yields and farm profitability: results of on-farm research in Tigray, northern Ethiopia. *Journal of Soil and Water Conservation* 54, 568–573.
Griliches, Z. (1957) Hybrid corn: an exploration in the economics of technological change. *Econometrica* 25, 501–522.
Haab, T.C. and McConnell, K.E. (2002) *Valuing Environmental and Natural Resources.* Edward Elgar, Cheltenham, UK, 326 pp.
Hanemann, W.M. (1994) Valuing the environment through contingent valuation. *Journal of Economic Perspectives* 8, 19–44.
Hanks, J. and Ritchie, J.T. (1991) *Modeling Plant and Soil Systems.* American Society of Agronomy, Crop Science Society of American and Soil Science Society of America, Madison, Wisconsin, 545 pp.
Higley, L.G. and Wintersteen, W.K. (1992) A novel approach to environmental risk assessment of pesticides as a basis for incorporating environmental costs into economic injury level. *American Entomologist* 38, 34–39.
Hill, R.W. (1991) Irrigation scheduling. In: Hanks, J. and Ritchie, J.T. (eds) *Modeling Plant and Soil Systems.* American Society of Agronomy, Crop Science Society of America and Soil Science Society of America, Madison, Wisconsin, pp. 491–509.
Kenkel, D., Berger, M.C. and Blomquist, G.C. (1994) Contingent valuation of health. In: Tolley, G.S., Kenkel, D. and Fabian, R. (eds) *Valuing Health for Policy: An Economic Approach.* University of Chicago Press, Chicago, Illinois.
Labarta, R., Swinton, S.M., Black, J.R., Snapp, S. and Leep, R. (2002) Economic analysis approaches to potato-based cropping systems: issues and methods. *Staff Paper 02–32.* Department of Agricultural Economics, Michigan State University, East Lansing, Michigan, 41 pp.
Maumbe, B.M. and Swinton, S.M. (2003) Hidden health costs of pesticide use in Zimbabwe's smallholder cotton growers. *Social Science and Medicine* 57, 1559–1571.
Mullen, J.D. (1999) Estimating environmental and human health benefits of reducing pesticide use through integrated pest management programs. MSc thesis. Department of Agricultural and Applied Economics, Virginia Polytechnic and State University, Blacksburg, Virginia, 170 pp.
Mullen, J.D., Norton, G.W. and Reaves, D.W. (1997) Economic analysis of environmental benefits of integrated pest management. *Journal of Agricultural and Applied Economics* 29, 243–253.

Nelson, M. and Maredia, M. (1999) *Environmental Impacts of the CGIAR: An Initial Assessment.* Consultative Group on International Agricultural Research (CGIAR), Technical Advisory Committee (TAC) Secretariat, World Bank, Washington, DC.

Norton, G.W., Swinton, S.M., Riha, S.J., Beddow, J., Williams, M., Preylowski, A., Levitan, L. and Caswell, M.F. (2000) *Impact Assessment of Integrated Pest Management Programs.* Economic Research Service (ERS), United States Department of Agriculture (USDA), Washington, DC (unpublished project report), 172 pp.

Owens, N.N. (1997) Farmer willingness to pay for herbicide safety characteristics. PhD dissertation. Department of Agricultural Economics, Michigan State University, East Lansing, Michigan, 139 pp.

Palmquist, R.B. and Danielson, L.E. (1989) A hedonic study of the effects of erosion control and drainage on farmland values. *American Journal of Agricultural Economics* 71(1), 55–62.

Pearce, D. and Atkinson, G. (1995) Measuring sustainable development. In: Bromley, D.W. (ed.) *The Handbook of Environmental Economics.* Blackwell, Cambridge, Massachussetts, pp. 166–181.

Pierce, F.J. (1991) Erosion productivity impact prediction. In: Lal, R. and Pierce, F.J. (eds) *Soil Management for Sustainability.* Soil and Water Conservation Society, Ankeny, Iowa, pp. 35–52.

Pimentel, D., Acquay, H., Biltonen, M., Rice, P., Silva, M., Nelson, J., Lipner, V., Giordano, S., Horowitz, A. and d'Amore, M. (1992) Environmental and economic costs of pesticide use. *BioScience* 42, 750–760.

Pingali, P.L. (2001) *Milestones in Impact Assessment Research in the CGIAR, 1970–1999.* Centro Internacional de Mejoramiento de Maíz y Trigo (CIMMYT), Mexico, D.F. Mexico. (http://www.cimmyt.org/Research/economics/map/impact_studies/milestones/pdfs/milestones.pdf).

Poe, G.L. and Bishop R.C. (2001) Information and the valuation of nitrates in ground water, Portage county, Wisconsin. In: Bergstrom, J.C., Boyle, K.J. and Poe, G.L. (eds) *The Economic Value of Water Quality.* Edward Elgar, Northhampton, Maryland, pp. 38–65.

Ribaudo, M.O. and Hellerstein, D. (1992) Estimating water quality benefits: theoretical and methodological issues. *Technical Bulletin no. 1808.* Economic Research Service (ERS), United States Department of Agriculture (USDA), Washington, DC, 28 pp.

Rola, A.C. and Pingali, P.L. (1993) *Pesticides, Rice Productivity, and Farmers' Health: An Economic Assessment.* International Rice Research Institute (IRRI), Los Banõs, The Philippines, 100 pp.

Shaffer, M.J., Halvorson, A.D. and Pierce, F.J. (1991) Nitrate leaching and economic analysis package (NLEAP): model description and application. In: Follett, R.F., Kenney, D.R. and Cruse, R.M. (eds) *Managing Nitrogen for Groundwater Quality and Farm Profitability.* Soil Science Society of America, Madison, Wisconsin, pp. 285–322.

Sharpley, A.N. and Williams, J.R. (1990) EPIC–Erosion/Productivity Impact Calculator: 1. Model Documentation. *Technical Bulletin no. 1768.* Agricultural Research Service (ARS), United States Department of Agriculture (USDA), Temple, Texas, 235 pp.

Skaggs, R.W., Hardjoamidjojo, S., Wiser, E.H. and Hiler, E.A. (1986) Simulation of crop response to surface and subsurface drainage systems. *Transactions of the American Society of Agricultural Engineers* 25, 1673–1678.

Swinton, S.M. and Casey, F. (1999) From adoption to innovation of environmental technologies. In: Casey, F., Schmitz, A., Swinton, S. and Zilberman, D. (eds) *Flexible Incentives for the Adoption of Environmental Technologies in Agriculture.* Kluwer Academic Publishers, Boston, Massachusetts, pp. 351–360.

Tietenberg, T. (1984) *Environmental and Natural Resource Economics.* Scott, Foresman and Co., Glenview, Illinois, 482 pp.

VandenBerg, T.P., Poe, G.L. and Powell J.R. (2001) Assessing the accuracy of benefits transfers: evidence from a multi-site contingent valuation study of ground water quality. In: Bergstrom, J.C., Boyle, K.J. and Poe, G.L. (eds) *The Economic Value of Water Quality.* Edward Elgar, Northhampton, Maryland, pp. 100–120.

Viscusi, W.K. (1993) The value of risks to life and health. *Journal of Economic Literature* 31, 1912–1946.

Williams, J.R., Jones, C.A., Kiniry, J.R. and Spanel, D.A. (1989) The EPIC crop growth model. *Transactions of the American Society of Agricultural Engineers* 32, 497–511.

Yandle, B., Vijayaraghavan, M. and Bhattarai, M. (2002) *The Environmental Kuznets Curve: A Primer.* The Property and Environment Research Center (PERC), Research Study 02–1, Center for Free Market Environmentalism, Bozeman, Montana. (http://www.perc.org/publications/research/kuznets.html).

Yoder, D. and Lown, J. (1995) The future of RUSLE: inside the new revised universal soil loss equation. *Journal of Soil and Water Conservation* 50, 484–489.

8 Bioeconomic Modelling for Natural Resource Management Impact Assessment

S.T. Holden

Agricultural University of Norway, Ås, Norway

Introduction

In this chapter various bioeconomic modelling approaches and methods which can be used to simultaneously evaluate the economic and environmental (sustainability) impacts of natural resource management (NRM) technologies and policies affecting natural resource use and management in the rural areas of developing countries are discussed. The importance of poverty to NRM, and poverty reduction as a millennium goal creates a need to carefully handle distributional and equity issues when NRM impact assessments are made.

Bioeconomic models link human behaviour to biophysical resource use and stock changes. Applied bioeconomic models are numerical programming models that may be based on theoretical dynamic models. The interdisciplinary and intertemporal nature of such models poses considerable challenges to modellers. The choice of modelling strategy is not obvious and requires careful consideration.

How can bioeconomic models be useful tools for NRM impact assessment? Because such assessment requires an interdisciplinary approach, bioeconomic models can be a useful tool for interdisciplinary analysis, since they allow integration of biophysical and socio-economic dimensions of the problem in a consistent manner. Bioeconomic models, with or without policy and technology experiments, can be useful to: predict adoption and impact of new (NRM) technologies; predict impacts of projects and policies targeting NRM; and make sensitivity analyses to assess the robustness of uncertain assumptions. They may also help to reveal knowledge gaps and provide guidelines for determining research priorities.

The operational question is how and which types of bioeconomic models best capture the efficiency, equity and sustainability issues in assessing NRM impacts. The choice of model is governed by a combination of theoretical

(eds B. Shiferaw, H.A. Freeman and S.M. Swinton)

pervasiveness of credit market imperfections, in combination with their poverty, also cause these households to have high discount rates (Holden *et al.*, 1998a) and this may affect their ability and willingness to invest in the conservation of natural resources. This may lead to a form of inter-temporal externality where government intervention may be necessary to prevent excessive resource degradation (Holden and Shiferaw, 2002).

Farm household models are not seen as representations of individual households, only as representations of the rational behaviour of groups of households. Econometric methods can be used to identify model parameters that are derived from the explained part of the model that represents a systematic response across a group of households. Typically there is insufficient information on individual households to analyse behaviour of individual households. Household models may therefore represent a group of households within a village, watershed, district, region, or country. Sometimes they may represent a whole village, watershed, district, or region. If this is so, socio-economic variation within the village, watershed, district or region is typically ignored. The following typology can help in handling socio-economic variation that is relevant to NRM impact assessment.

A typology of village economies[2]

Bioeconomic modelling can be simplified considerably if there is little socio-economic differentiation in the bioeconomy that is modelled. The economy can then be represented by one model of the average or typical farm household. There will be little or no trade inside such an economy. This is rather an exceptional case, however, and it may therefore be useful to start out with a more general village economy typology that allows for varying internal differentiation and varying degrees of isolation from the rest of the world (Holden and Binswanger, 1998; Holden *et al.*, 1998b). This then can be a useful basis for identifying a typology of sound bioeconomic village models.

Village economies are first characterised along two dimensions: a. Transaction costs related to the outside world (market access/market integration); b. Internal differentiation in access to resources and specialisation in activities.

This is illustrated in Fig. 8.1a that shows four extreme corner solutions while the real-world villages typically fall somewhere within the quadrant. Figure 8.1b shows which village model is more appropriate under specific conditions. The village may be represented by one separable farm household model if it is well integrated into markets (low transaction costs) and there is an egalitarian distribution of resources (lower left corner in Figs 8.1a and b). It may be represented by a single non-separable farm household model if the village is isolated from outside markets (e.g. no linkage to an external labour market) and distribution of resources within the village is egalitarian such that there is no need for local trade (upper left corner). With internal differentiation there will typically be more specialisation that may be captured by a number of separable household models when the economy is well integrated into markets such that prices are exogenous (lower right corner). With internal differentiation and isolation from the outside world, a model with several interacting non-separable household group models may be necessary if the internal transaction costs are such that production decisions of each group become non-separable from the consumption side (upper right corner). In this case there are some prices that are determined outside households but inside the village, giving local general equilibrium effects. Finally, there is a possible situation with internal differentiation and isolation, but low internal transaction costs causing production decisions to be separable from consumption decisions but leaving some prices to be exogenous to households that are endogenous to the village (middle right in Figs 8.1a and b).

a.

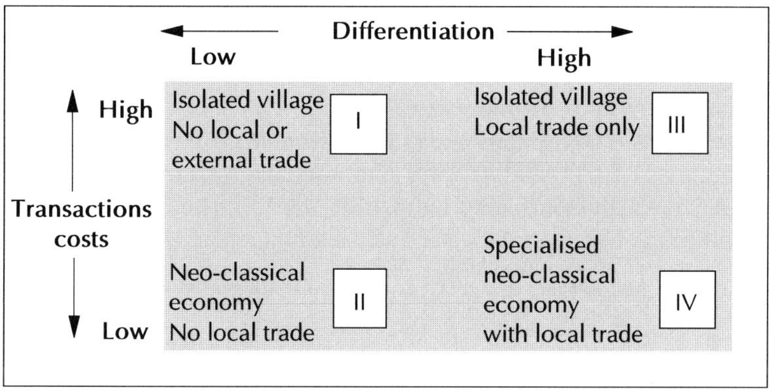

b.

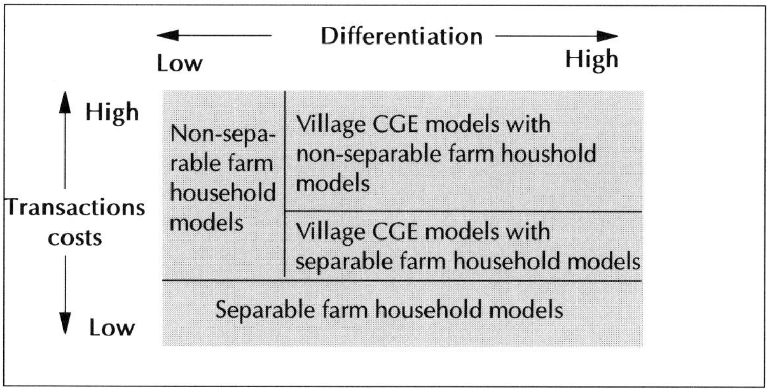

Fig. 8.1. a. Village economy typology and b. typology of village computable general equilibrium (CGE) economy models (Holden and Binswanger, 1998).

How can this typology of village economies and village economy models be related to the real rural world in developing countries? In reality there is a continuum of villages and the position of villages may also change over time or because of policy changes. Some further clarification of the types of rural economies that fall closer to the four corners (I, II, III and IV) in Fig. 8.1a may be helpful, and could clarify which type of model in Fig. 8.1b is most relevant to use for bioeconomic modelling.

Type I

In this category there are two main types of economies that can be represented through non-separable household models.

LAND ABUNDANT REMOTE RURAL ECONOMIES. There is likely to be little internal trade in such economies, causing land, labour and credit markets to be missing or highly imperfect, while outputs may be sold from the village. These are typical Chayanovian[3] economies (Chayanov, 1966). Holden *et al.* (1998b) provide an example of such a village model with two household groups that trade with the rest of the world but not with each other. The life cycle of households can be the main reason for some differentiation. The main environmental problems that could be of interest to incorporate in models of such land-abundant economies are deforestation and loss of biodiversity, because of the relative land abundance. But, preference for extensive land use may cause the carrying capacity of the system to be exceeded as seen, for example, in Amazonia, the outer islands of Indonesia, and central Africa.

LAND-SCARCE REMOTE RURAL ECONOMIES. Densely populated villages may be located far from external markets and, if there is an egalitarian distribution of local resources, there may be scant external or local trade, resulting in a stronger subsistence orientation. High population pressure can limit the ability to produce a surplus for sale. Such villages can also be modelled as non-separable farm households. A major difference from the typical Chayanovian situation, however, is that these villages face a land constraint. Under such conditions population growth should lead to agricultural intensification if it is technically and economically feasible. This should also increase land productivity, although labour productivity may decrease with increasing population pressure. Abundant labour could then turn out to be a less-important resource. Land degradation (causing loss of productivity over time) may be the main environmental problem in land-scarce economies due to the soil erosion and nutrient depletion it incurs. It becomes crucial to incorporate soil and nutrient stocks and stock changes and their productivity impacts over time into the models.

Type II

This rural economy has good market access for factors of production and commodities. The Singh *et al.* (1986) model is the well-known neoclassical farm household example. Production–environment analysis is simplified in such economies because the distribution of resources and poverty does not affect production and investment decisions. This also implies that profit-maximising behaviour is consistent with sustainability. The perfect markets ensure that resources are optimally combined. Biophysical process models may be used separately from the behavioural models in a recursive manner, where the behavioural models are used to assess the welfare impacts of NRM changes. The environmental impacts of changes in NRM technologies and policies can be captured by an integrated biophysical and production economy model.

Type III

Unequal distribution of land, livestock, labour and other resources in an isolated economy creates incentives for local trade unless the transaction costs are higher than the benefits from trade. For example, the distribution of oxen (bullocks), that may be crucial for land preparation, may be important in such economies since labour and draft power are often complements rather than substitutes. Unequal distribution of oxen may lead to rental markets for both oxen and land and to considerable productivity differentials if these markets do not function well (Bliss and Stern, 1982; Skoufias, 1995). Poverty and imperfections in inter-temporal markets can affect production and investment decisions through endogenous village prices and household group-specific shadow prices. This is the most difficult type of economy to model because of the simultaneous interactions between biophysical conditions, household group characteristics, and local market characteristics. An example of such a village model is provided by Holden and Lofgren (Chapter 13, this volume). Alternatively, if the local general equilibrium effects are ignored, such economies can also be modelled as a number of non-separable farm household group models with biophysical components where endogenous village prices and the general equilibrium feedbacks are considered less important. Holden and Shiferaw (2004) and Shiferaw and Holden (Chapter 12, this volume) are examples of such bioeconomic household group models.

Type IV

Economic development may lead to well-developed markets and more differentiation and specialisation, as found in developed countries. In such circumstances biophysical models linked to farm firm models in a time-recursive manner may be an adequate bioeconomic modelling approach. Commercial agriculture in developing countries could also possibly be modelled this way.

Most poor rural people in developing countries fall into Type I and Type III, and are facing significant market imperfections. This means that bioeconomic models that aim to capture poverty–environment linkages should be of the non-separable type, and that the significance of local general equilibrium effects should be assessed carefully. A major limitation of many bioeconomic models developed to date is that they have relied on perfect market assumptions and are of the separable type. Examples of models that relax these assumptions are emphasised later.

Overall, in choosing an empirical village (watershed) model type when the available budget is limited, the relative importance of the different concerns highlighted above must be assessed and a balance struck depending on the research objective, relative gains and costs, and the desired level of complexity and precision required.

Common property management in village models

Common property resource management cannot be captured easily or completely in farm household models because of the economy-wide feedback (externality) effects. Village models may be a more suitable way to capture these effects. But the challenge of formulating such models for impact assessment in a way that captures the incentives in terms of the dynamic cost and benefit streams for different agents in relation to the utilisation of the resource under common property remains. A game theoretic approach could be used. There are few strong predictions that can be made based on the theory of repeated games. However, it is more likely that cooperative solutions can be attained than can be drawn from a single-period prisoners' dilemma game (Balland and Platteau, 1996). This is an area that requires more research. Typical NRM issues include management of grazing lands, community woodlots, groundwater, and lake fisheries.

Time span and model integration

The pace of important processes and the frequency of decisions is crucial to the best construction of bioeconomic models for NRM impact assessment. Modelling issues in biophysical processes can be divided into: production decisions, consumption decisions, distribution, aggregation, and general equilibrium effects. The basic questions to ask when preparing for NRM impact assessment include:

1. Which NRM technology or policy is of interest?
2. What types of impacts should be assessed?
3. Which processes and decisions (listed above) must be included in the analysis?
4. Has the NRM technology or policy been introduced or how quickly will it be introduced (because the analysis could be *ex post* or *ex ante*)?
5. What is the time period to be included in the analysis?
6. How quickly and frequently do changes in processes and decisions need to be updated?
7. How frequently are important resource use, investment and consumption decisions made?
8. How frequently do the interactions between the different processes and decisions that are made need to be captured?
9. How should the inter-temporal decisions and trade-offs be captured?
10. What data are available for model construction and validation?

This can be illustrated with an example. If it is assumed that the area of study is a watershed where resources are distributed unequally among land users and poverty affects ability to invest in (adopt) a new NRM technology, this implies that inter-temporal markets and possibly other markets do not function well. The NRM technology reduces land degradation and increases future land productivity. The objective is to assess the potential of the new NRM technology to improve land productivity and sustainability on different types of land for different types of households and to determine how it can improve household welfare over time, especially for poor households. This information indicates that the only issue that could be eliminated from the list of modelling issues above may be general equilibrium effects, although these could also become significant over time. There will be a clear need for models for different household types as these differ in investment behaviour. This means that non-separable dynamic household models will be needed. The model will have to capture the distribution of different types of land across different homogeneous household categories. Let us assume (Question 4) that the new NRM technology was made available 3 years ago and that some households have started to adopt it. Technology adoption is labour-demanding and adoption is gradual (a limited area can be converted each year). Productivity impacts increase gradually from year to year on land where the technology is adopted. In this case, an annual update of area with adoption for each household group may be sufficient. Depending on technology used, annual changes in land productivity on different types of land may also be sufficient. It can also be assumed that adoption decisions are made seasonally along with cropping decisions. Imperfections in markets and unequal distribution of resources cause shadow prices to be specific to each land type for each household group, and that, taking broader constraints into account, this is the basis for technology adoption choice.

There may also be a need to capture constraints to adoption through seasonal labour constraints if the labour market works poorly and the family is the main source of labour. This implies that the shadow wage varies through the year and across household groups. The model then needs to capture annual and seasonal production, investment and consumption decisions (leisure demand) simultaneously for each household group. Expected future benefits (weighted benefit stream) have to be assessed in relation to the costs of the NRM technology. Changes in marginal utilities and the discount rate are typically used as a basis for inter-temporal decisions. The discount rate may vary across household groups, as may marginal utilities, and cause adoption and welfare effects to vary across household groups. In other words, this involves a number of dynamic bioeconomic non-separable farm household group models that together represent the watershed model for NRM technology impact assessment. To assess the overall impacts the impacts must be aggregated across the dynamic household group models. This part of the analysis may incorporate specific policy or project objectives and external effects about which household groups are ignorant, or to which they give less weight, since it is not assured that this aggregation involves direct optimisation as part of the model.

Static vs. dynamic models

The previous section highlighted the time dimension in relation to NRM impact assessment. It may, however, under certain conditions be preferable to use static models for NRM impact assessment. Firstly, static models are much simpler and cheaper to construct. Secondly, important inter-temporal constraints or preferences can be incorporated into static models. Thirdly, the total potential impact of the NRM technology or policy after some time may be of interest, not the gradual adoption process.[4] The before and after situation may then be represented through two static models, one without and one with access to the new technology, or with the new NRM policy.

Dynamic models can be derived from optimal control theory (Conrad and Clark, 1987; Clark, 1990; Chiang, 1992). 'Dynamic programming' has, however, a more narrow meaning than the dynamic optimisation models considered here. It refers to a situation where the final (end period) state is known and it is therefore possible, through backward induction, to arrive at an optimal pathway. The types of models under consideration are different in that the terminal conditions are not known. Optimal decisions at any point in time are based on current knowledge and expectations about the future. Different types of models must be used to analyse such situations. Examples of dynamic models are presented in the following section.

Scale of analysis

For NRM, impact assessment data from plot or land type, household, village/ watershed, district, regional or country levels can be used. Biophysical processes mostly take place at plot or land type level but shocks (like droughts or floods) can also hit at watershed or higher levels. NRM technologies may be introduced at local level but adoption may have impacts on markets and prices at higher levels. Policies may influence prices at higher levels and have impact on incentives down at the village, household and land type levels. Processes and decisions on different scales may be integrated within a single model or through a combination of models that are used sequentially. As shown in the next section, bioeconomic models can therefore range from bioeconomic (micro level) household models to bioeconomic macro-computable general equilibrium (CGE) models. The choice of scale for a bioeconomic model therefore depends on the type of NRM technology or policy and the types of impacts that are to be assessed.

Equity, distribution and aggregation

It has already been explained that market imperfections cause non-separability of production, investment and consumption decisions. It is worth repeating, however, as many of the bioeconomic models developed to date have resorted to perfect market assumptions that lead to separable models because they are much simpler to make. However, such models fail to capture how the distribution of resources and preferences affect NRM technology adoption and impacts. The use of bioeconomic models that more explicitly capture poverty–environment linkages and distributional effects of NRM technologies and policies is advocated. Aggregation errors can be reduced by: first identifying homogeneous household groups, and then aggregating across these groups afterwards.

Examples of Bioeconomic Models for NRM Impact Assessment

Building on the typology and guidelines described, examples of bioeconomic models for NRM impact assessment follow. This is not meant to provide an exhaustive or even a thorough review of such models, but rather to give examples of alternative approaches and how they may be used for NRM impact assessment. Rational behaviour can be captured through explicit or implicit optimisation. Household models typically have explicit optimisation, while CGE models are implicit optimisation models since they do not have an objective function that is maximised or minimised. Rather they are based on implicit optimisation as they incorporate the first-order conditions for optimal behaviour of the economic agents (institutions).

Biophysical process or system analysis models do not incorporate optimal behaviour but may include simple rules of thumb for behaviour. The most famous of these types may be the 'Limits to Growth' models (Meadows *et al.*, 1972). Such models are not regarded as true bioeconomic models as they do not capture behaviour and price signals well. However, they may have a better representation of biophysical conditions and changes than bioeconomic models. They may also form one part of a system of models that are used jointly to analyse the interaction between biophysical and socio-economic conditions.

Bioeconomic optimisation models

Optimisation models have an explicit objective function that is maximised or minimised. For rational agents this objective may be to maximise utility, maximise profit, minimise drudgery, or minimise risk, each of them subject to a set of resource availability, crop rotations and seasonality constraints. Basic needs requirements of agents may also be handled through a set of constraints that need to be satisfied. The theory of separable farm household models is well known and has been used as a basis for many bioeconomic household models.

Static, non-separable, bioeconomic farm household models

There are many ways of incorporating biophysical conditions into static models. One is to introduce sustainability constraints that could be imposed as hard or soft constraints. A hard constraint is one that is not allowed to violate the sustainability constraint. But, such a requirement may be too strict, and not match the real behaviour of farm households in many situations. It may even make it impossible to solve the model in many cases (e.g. sustainable land use may be infeasible) where people are poor and crowded into a fragile environment without exit options.

There are two approaches that may be useful for capturing important sustainability conditions in static models. These approaches involve introducing carrying capacity constraints and user cost constraints into the static model. The first may be most useful in more land-abundant areas (fallowing systems), while the latter may be more useful in land-scarce economies with permanent cultivation.

Carrying capacity approach

Carrying capacity (defined as 'the maximum population a particular environment can support indefinitely without leading to degradation' (Ellen, 1982)) constraints can be derived from specific crop-rotation and fallowing sequences that are practised and required to regain the initial land productivity. Based on carrying capacity it is possible to introduce soft constraints into the static model that account for the minimum land requirement per capita, or maximum population density per unit of land that can be sustained into perpetuity. If access to land resources is lower than this, i.e. when population density exceeds the carrying capacity, current land use practices are not considered to be sustainable. This may imply that over time farm households must shift from more land-extensive to more land-intensive technologies. However, returns to other factors may be such that this intensification is delayed and this may cause more rapid and accelerating deforestation. When the aim is to model real behaviour and incentives, it is wrong to impose this sustainability constraint as a hard constraint because the development path in the form of non-sustainable land use would be lost in the model. Similarly, it is also possible to incorporate production of perennial NRM technology in a static model by imposing an assumption of a stable (and possibly sustainable) rotation.

Example: Holden (1993a, b) uses static non-separable farm household models to analyse the evolution of farming systems in the *Miombo* woodlands of northern Zambia during the 20th century. The models were used to assess the impacts that the adoption of NRM technologies [including use of cassava, maize/fertiliser and agroforestry (alley cropping and planted fallows)] had on the carrying capacity of the land and on the incentives to deforest (choice of extensive vs. intensive cropping systems), together with their effects on labour demand and human welfare. Households were assumed to first prioritise their basic needs and beyond that to be drudgery-averse and to strike a balance between an additional income and a leisure goal. Dynamic relations in land use are captured by incorporating crop rotation and fallow requirements. These allow an assessment of the carrying capacity under different conditions. By modelling labour-rich and labour-poor, male- and female-headed households, with different access to land, markets and technologies, important socio-economic variations, responses and impacts were identified. Poverty and missing credit markets cause households to have a short planning horizon and to ignore the more long-term impacts of their activities on the natural resource base. Static models with crop-rotation requirements are therefore considered to adequately represent the real behaviour of farm households in this area.

User cost approach

User costs are the perpetual productivity losses resulting from soil-degrading production practices. The concept of social user cost may also include other off-site and on-site effects of resource degradation. User costs can be derived theoretically from an optimal control model with land degradation and can reflect the shadow value of land assets. Such user costs can be operationalised in a static programming model. They may be included as accounting constraints that do not affect the solution if they represent a true externality about which the land user is ignorant. Alternatively, a Pareto-relevant externality, i.e. an externality that can be internalised such that the benefits to society of doing so are higher than the associated costs, may be internalised by including the discounted user costs as a penalty function in the objective function of the land user, who then may modify his/her behaviour to reduce the land degradation problem. The level of response will depend on the productivity impact of resource degradation and the discount rates of the resource user. Changes in NRM practices will then affect user costs, which can be considered as environmental impacts of improved management. If the social discount rate is much lower than the private discount rate of the land user and this causes too low investment in environmental conservation, this implies the existence of an environmental externality. A Pigouvian tax-subsidy policy may then be implemented to internalise the externality and this should lead to a reduction or elimination of the (Pareto-relevant) externality. Such a tax may be imposed on erosive crops (output tax) or on land-degrading inputs, while the subsidy may be used to stimulate land conservation. If land degradation is irreversible, and its rate is estimated to cause a 1% loss of land productivity per year, and the social rate of discount is 10% per year, the annual on-site cost of the irreversible land-degrading activity would be 10% of the output value from that activity. If the private discount rate is much higher than 10%, something that is not unusual among poor people, they are willing to pay only a small share of the social cost of their land-degrading activity. If private costs are higher than private benefits, this may cause them to not even internalise a small share of the Pareto-relevant externality. For example, they may not adopt a soil-conservation technology although it would be socially optimal to do so. The share that poor people are willing and able to pay can be increased by allowing them to pay over a longer period of time (Holden and Shiferaw, 2002). If they are allowed to pay in kind (e.g. with their labour during periods when the opportunity cost of labour is low), their contribution can be further increased.

Example: Shiferaw and Holden (1999, 2000) have developed static bioeconomic non-separable farm household models for an area with good agricultural potential and relatively good market access in the Ethiopian highlands. Significant market imperfections caused the need to develop non-separable household models. The models jointly determine production, consumption and soil conservation decisions. They include (estimate) the on-site user costs of soil erosion. Soil loss was estimated based on the Universal Soil Loss Equation (USLE) adapted to Ethiopia. Experimental data were used to estimate a translog production function to find the impact of loss of soil depth on crop productivity. The loss in land productivity due to soil loss was assumed to be irreversible. The user cost was calculated as the net present value of the permanently lost land productivity for the dominant crop, teff (*Eragrostis tef*). Models were run without incorporating the user costs and with user costs included based on the estimated average (private) discount rates of farm households in the area and with social discount rates at 10% and 5%. Different assumptions were included with respect to the impact of conservation technologies on short-term crop yields: 1. 20% less than traditional farming; 2. equal to traditional farming; and 3. 20% more than traditional farming. The models were used to assess the resource-use patterns, sustainability impacts and economic benefits to farm households derived from adoption of soil conservation technologies (soil bunds and stone bunds). The models were also used to assess the social efficiency and economic and environmental impacts of such alternative NRM policy measures as input subsidies, output taxes, and the interlinkage of input subsidies with a conservation requirement (cross-compliance policies) (Shiferaw and Holden, 2000).

Dynamic optimisation models

Time-recursive optimisation models

Time-recursive models are formulated as a sequence of static models that are updated from period to period. Such models are typically run for 1 year at a time. Every year the resource stocks are updated and depend on the situation in the previous year. Weather conditions and market conditions may also change over time and can affect the development pathway. Such models may have a planning horizon of more than 1 year based on expectations about the future.

Example: Barbier and Bergeron (2001) have developed a time-recursive bioeconomic micro-watershed model with a 5-year planning horizon for single-year decisions. The model maximises the additive discounted utility of two household groups (ranchers and small-scale farmers) split into 18 farm sub-models based on spatial locations of land types in the hillsides of Honduras. The model contains a local labour market with an endogenous wage rate that is linked to an external labour market. The model is run for as much as 100 years (1975–2075) and is updated every year. Resources carried over from one period to the next include population, livestock, tree volume of different-aged trees, soil depth, soil conservation structures, and ploughs. The advantage of the recursive method is that it can make simulations for much longer periods than the 5-year planning horizon. It also offers the possibility of periodically shocking the model for exogenous changes in prices.

Soil erosion is modelled as a function of crop choice by area and the presence or absence of conservation technologies. Erosion affects yields through both loss of nutrients and soil depth. The biophysical model Erosion Productivity Impact Calculator (EPIC) is used to generate the crop productivity data, based on land-use practices and soil quality while the Modified Universal Soil Loss Equation (MUSLE) is used to estimate erosion levels used in the bioeconomic model.

Barbier and Bergeron's model has been used to assess the impact of population growth, new technologies (improved varieties and sprinkler irrigation), market liberalisation, road construction, and land reform on NRM, soil erosion, input use, crop yields and income. This type of model could also be useful for other NRM technology and policy impact assessments. The maximisation of aggregate utility across a number of household groups depends on the assumption that there are no externality problems cutting across household subgroups. Such externalities may sometimes be very important collective action problems in watersheds. For example, upstream farmers may encroach on forests and cause negative external effects on households living downstream. A model that maximises aggregate utility would automatically internalise such an externality while there may not be a mechanism that takes care of this situation in reality.

Non-stochastic dynamic bioeconomic models

Non-stochastic models may or may not incorporate risk. They can be formulated without knowing the exact end-point levels of stocks (free terminal value problems) but have a limited time horizon. The objective can be formulated as the net present value of the (discounted) welfare (utility) from all the time periods.

Such models will typically have multiple constraints in each period. Resource stocks must also be updated or corrected from period to period. At the same time, final period depletion of resource stocks cannot be accepted unless it is realistic. One method of avoiding such a problem is to include the return to the last period stock in the last period element of the objective function, assuming it is sustained forever at that level. Alternatively, restrictions can be imposed such that stock depletion is avoided. Market imperfections will typically cause these models to become non-separable.

Example: Holden and Shiferaw (2004) have developed a non-linear non-stochastic dynamic non-separable farm household model with risk. The model is applied to a less-favoured area with high population density and rugged topography in the Ethiopian highlands. Imperfections in markets, such as missing markets, price bands, credit rationing, and share tenancy, are included in the model. Risk is related to unreliable rainfall (drought), frost and hailstorms. Drought also affects prices of crops and livestock. Land quality classes are highly disaggregated (based on soil type, slope, soil depth and use of conservation technologies). A constant partial risk-aversion utility function is used. Discounted utility of certainty equivalent full income is aggregated and maximised for a specific time period (5 or 10 years).

The model and its recent extensions (Holden *et al.*, 2003; Holden and Shiferaw, 2004) are used to assess the impacts of NRM technologies (soil and stone bunds, tree planting) and NRM policies (food-for-work for soil conservation) on NRM, soil erosion, crop yields, food self-sufficiency, and household welfare over time.

Stochastic dynamic bioeconomic models

Discrete stochastic programming can be used to construct dynamic bioeconomic models. The discrete stochastic programming with recourse approach was developed by Cocks (1968) and Rae (1970, 1971). This is a type of time-recursive model where some of the decisions are based on expected probabilities about future events, while other decisions can be delayed till the outcome of the random event is known. This implies that the models have a decision-tree structure with the nodes of the tree as decision points and the branches as different states of nature. Such models can be relevant for NRM impact assessment in areas with frequent environmental hazards (droughts, floods, serious pest attacks) and when the NRM technology or policy affects the outcome of, and responses to these hazards. Such models are more useful when the sequence of events affects the outcomes and when decisions depend on the state of nature at a point in time.

Example: Barbier and Hazell (2000) developed a discrete stochastic programming with reports (DSPR) bioeconomic model for an agro-pastoral area in a semi-arid area in Niger. The model is used to simulate the longer-term consequences of changes in population growth and reduced access rights to transhumance grazing areas with special emphasis on the role of drought risk in conditioning the model's results. Furthermore, the model is used to assess how improved methods of managing drought-risk affect the development pathway followed by the community. Such a model could potentially be used to assess the impact of an NRM technology like water harvesting, on crop choice, land productivity and variability, and income or welfare where risk aversion may be included.

The model has two states of nature, normal years and drought years with a 10% probability of drought. With a planning horizon of 4 years, the model yields $2^4=16$ possible states of nature. Rainfall outcomes are assumed to be independent over time. In order to obtain outcomes for a longer period of time the model is solved recursively each year to provide a series of moving 4-year plans. This recursive framework allows adjustments of expected versus actual outcomes each year. The total production and closing stocks of livestock and grain are recalibrated and entered as initial stocks for the next period. The model is a linear programming (LP) model applied at village (community) level and includes use of common property resources.

The bioeconomic components of the model include crop and livestock interactions with uncertain weather. Crop production depends on soil conditions (phosphorus balance), use of manure from livestock and purchased fertiliser. Forage is produced by the village pastures, transhumance pastures or is purchased. Crop residues may be used as forage or for manure production.

Bioeconomic economy-wide models

Macro CGE models

Macro CGE models have typically modelled the agricultural sector as a pure profit-maximising producer. Such bioeconomic elements as pollution, deforestation and land productivity impacts can be integrated into such models, and the impact on human welfare and the environment (using indicators) of alternative NRM policies and technologies can be assessed. If these effects are externalities that do not have consequences for behaviour or productivity in the short term, they do not have an impact on the solution unless policies are implemented to internalise the externalities. If the environmental externality, e.g. land degradation, has productivity impacts such as reduced crop yields this should be captured through an impact on parameters in the production function in the model.

Macro CGE models can be useful for assessing NRM policies and NRM technologies that are already in use if the level of disaggregation is sufficient to distinguish the key resources and activities. It is a bit tricky (but possible) to use them to assess new NRM technologies (*ex ante* analysis) because these will lack the initial starting values in the social accounting matrix on which such models typically build.

Example: Glomsrød (2001) gives an overview of two approaches used to build environmental linkages within macro CGE models. The first incorporates soil productivity with nitrogen as the limiting factor in a CGE model for Tanzania. The agricultural sector is divided into 11 production activities for different crops. Use of fertiliser and recycling of crop residues determine the rate of soil productivity loss. Soil productivity is treated as a factor-neutral technical change in the Cobb-Douglas production function that is adjusted annually according to the previous year's farm practice. The soil module builds on the Tropical Soil Productivity Calculator developed by Aune and Lal (1995) and incorporates the nitrogen cycle, while soil loss is estimated based on the Universal Soil Loss Equation (USLE). The model was used to assess the impact of adjustment policies (devaluation and removal of fertiliser subsidies) on economic growth and on the land degradation externality. One of this model's limitations is that the structure (pure producers of single crops) does not accurately reflect the structure of rural economies in Tanzania (agro-ecosystems where households typically produce a mixed portfolio of crops for subsistence and markets in a setting with imperfect markets).

The second model for Nicaragua (Glomsrød *et al.*, 1999) incorporates deforestation by smallholder agriculture. Deforestation is caused by migration to the agricultural frontier (new colonisers) and others who have to clear new land because of rapid land degradation. The process is driven by the need for subsistence production and relates to alternative sources of income. The model is used to assess the relationship between income distribution, economic growth and deforestation, and the impact of economic policy reforms on deforestation.

Village CGE models

Village CGE models are only needed when there are significant local general equilibrium effects causing the existence of endogenous prices in the village while these prices are exogenous to households. This implies that there is no trade with the external world for these commodities or factors, i.e. these commodities and factors are traded only within the village. Internal transaction costs within the village may lead to internal, possibly household-specific, price bands between purchase and selling prices for household tradables. When such local markets exist, it is important to study carefully how they function. This also has implications for whether households can be modelled as separable or non-separable. Is there competition in the markets? What is the bargaining power of buyers and sellers? Is there some form of price regulation and rationing? Are there inter-linked transactions (e.g. share-tenancy)? How are inter-linked transactions for otherwise non-traded goods valued? It is important to get an overview of the relative importance of these transactions and how much prices and quantities fluctuate. If they represent a very small share of the total factor or commodity use, they may perhaps be ignored without committing too large an error. If they are significant, their internal logic should be revealed and be represented in the model. A village social accounting matrix (SAM) can shed light on the presence and relative importance of within-village transactions.

Example: Holden and Lofgren (Chapter 13, this volume) have developed a bioeconomic village CGE-model with market imperfections for a village economy with high agricultural potential and good market access in the Ethiopian highlands. Transaction costs are explicitly modelled through price bands. Complex crop–livestock interactions are modelled through multiple input and output production functions. This was done by using nested Constant Elasticity of Substitution (CES) functions that allow for different substitution elasticities at different levels. Land, traction power (oxen), labour, and seeds are combined with a low elasticity of substitution at the lower level to construct planted land, while planted land and fertiliser are combined with a high elasticity of substitution at the higher level (reflecting the much higher flexibility in fertiliser application). Production activities are made household-group specific because of the imperfect markets since buyers and sellers in village markets for factors and outputs face different farm-gate prices.

The major environmental problem in the area is land degradation. This is included through a user cost approach incorporating short- and long-term productivity effects of soil erosion. It is due to the high personal discount rates and short planning horizons of people in the area that this externality is not internalised through individual optimisation behaviour (Holden *et al.*, 1998a; Holden and Shiferaw, 2002).

The model is used to assess the impact of NRM policies (Pigouvian output taxes and input subsidies) on household welfare and on the land degradation externality. The question of whether a Pigouvian subsidy on fertiliser can be defended on environmental grounds is thoroughly assessed through a sensitivity analysis with different subsidy levels and different input substitution elasticities. It is possible to use such a model to evaluate new technologies, but this may be more demanding than it would be in household optimisation models.[5] Another limitation of such models is that it is relatively costly (time and skill-demanding) to add such constraints as seasonality and crop rotation to the model.

Conclusions

Some of the main advantages of using bioeconomic models for NRM technology and policy impact assessment are:

- They allow a *consistent treatment* of complex biophysical and socio-economic variables, providing a suitable tool for *interdisciplinary analysis*
- They allow *sequential and simultaneous interactions* between biophysical and socio-economic variables
- They can be used to assess the *potential impacts* of new NRM technologies and policies (*ex ante* impact assessment)
- They allow disturbing variation to be controlled (*ceteris paribus* conditions) for evaluation of impacts of certain interactions by isolating effects from other influences
- They can capture both *direct and indirect effects* (i.e. the total effect of technology or policy change can be estimated)
- They can be used to carry out *sensitivity analyses* in relation to various types of uncertainties.

These advantages versus the costs of bioeconomic modelling need to be judged against the advantages and costs of alternative approaches to NRM impact assessment.

What type of bioeconomic models should be used for NRM impact assessment? A theoretical and empirical basis for the choice of models used to assess technology and policy impacts in developing countries has been presented. Model choice depends on the type of technology or policy being assessed, local market characteristics, environmental issues, resource distribution and scale of analysis. A number of the bioeconomic models developed to date have relied on perfect market assumptions and have therefore assumed production decisions to be separable. Most rural economies in developing countries do not fit this assumption. The aim of this chapter has been to set a new standard for bioeconomic models for poor rural economies in relation to their relevance for evaluating the impacts of technologies and policies that affect NRM.

Market imperfections cause land use and poverty to be non-separable and create a need for models that simultaneously link production, investment and consumption decisions through shadow prices. Non-separable, bioeconomic farm household (group) models can do this in an adequate way. Such models can also be used to assess the distributional implications (poverty effects) of NRM technologies and policies.

The scale of the analysis needs to be determined by the type of impacts to be studied and the expected reach of these impacts in the bioeconomic system. Aggregation through identification of homogeneous household groups may be a cost-effective way to reduce the aggregation bias and link poverty/welfare more systematically to land use when markets do not work well. Non-separable, bioeconomic household models should be used when poverty affects land use due to imperfect markets for commodities that are produced (or used as inputs in production) and consumed by households. Imperfections in inter-temporal markets (credit, insurance) can also have strong impacts on the investment behaviour of poor people.

Market imperfections and exogenous shocks make it relevant to use stochastic time-recursive bioeconomic household group models. Stochastic weather also creates price fluctuations. When NRM technologies and policies directly affect stochastic risk and the sequence of events matters for the outcome, discrete stochastic models may be useful.

Economy-wide models should be used when there are important general equilibrium effects related to NRM technologies and policies. A new generation of bioeconomic CGE models is required to adequately capture the micro-foundations and environmental linkages in rural economies in developing countries. This indicates nesting non-separable, bioeconomic household models into CGE models. CGE models could also be useful at micro- and meso-levels when there are important local general equilibrium effects that need to be captured in the impact assessment.

Endnotes

[1] Constrained Pareto-efficiency refers to a world with transaction costs while unconstrained Pareto-efficiency refers to a world without transaction costs.

[2] What is called a village here may also be a watershed or a larger unit if preferable.

[3] Chayanov studied the Russian peasantry in the early 20th century. This was a land abundant economy with an output market but without a labour market. Chayanov developed his theory of the subjective equilibrium of peasant households where they work up to the point where the drudgery of work equals the marginal benefit of the work.

[4] Bioeconomic models can also be used to jointly assess the adoption process and welfare and environmental impacts of NRM technologies but this requires a dynamic model.

[5] These models, unlike household optimisation models, need exact starting values for calibration of the new technologies.

References

Aune, J. and Lal, R. (1995) The tropical soil productivity calculator – a model for assessing the effects of soil management on productivity. In: Lal, R. and Stewart, B.A. (eds) *Soil Management: Experimental Basis for Sustainability and Environmental Quality III: Advances in Soil Science.* CRC Press, Boca Raton, Florida.

Balland, J.M. and Platteau, J.P. (1996) *Halting Degradation of Natural Resources. Is there a Role for Rural Communities?* Clarendon Press, Oxford, UK.

Barbier, B. and Bergeron, G. (2001) Natural resource management in the hillsides of Honduras. Bioeconomic modeling at the microwatershed level. *IFPRI Research Report no. 123.* International Food Policy Research Institute (IFPRI), Washington, DC.

Barbier, B. and Hazell, P. (2000) Implications of population growth and declining access to transhumant grazing areas for the sustainability of agropastoral systems in the semi-arid areas of Niger. In: McCarthy, N., Swallow, B., Kirk, M. and Hazell, P. (eds) *Property Rights, Risk, and Livestock Development in Africa.* International Livestock Research Institute (ILRI), Nairobi, Kenya, and International Food Policy Research Institute (IFPRI), Washington, DC.

Binswanger, H. and McIntire, J. (1987) Behavioral and material determinants of production relations in land-abundant tropical agriculture. *Economic Development and Cultural Change* 36, 73–99.

Binswanger, H. and Rosenzweig, M.R. (1986) Behavioral and material determinants of production relations in agriculture. *Journal of Development Studies* 22, 503–539.

Binswanger, H., McIntire, J. and Udry, C. (1989) Production relations in semi-arid African agriculture. In: Bardhan, P. (ed.) *The Economic Theory of Agrarian Institutions.* Clarendon Press, Oxford, UK.

Bliss, C.J. and Stern, N.H. (1982) *Palanapur: The Economy of an Indian Village.* Oxford University Press, New Delhi, India, and New York.

Boserup, E. (1965) *The Conditions of Agricultural Growth. The Economics of Agrarian Change under Population Pressure.* Earthscan Publications, London, UK.

Chayanov, A.V. (1966) *The Theory of Peasant Economy.* Thorner, D., Kerblay, B. and Smith, R.E.F. (translators and eds). Irwin, Homewood, Illinois.

Chiang, A.C. (1992) *Elements of Dynamic Optimization.* McGraw-Hill, New York.

Clark, C.W. (1990) *Mathematical Bioeconomics. The Optimal Management of Renewable Resources,* 2nd edn. John Wiley and Sons, New York.

Cocks, K.D. (1968) Discrete stochastic programming. *Management Science* 15, 72–79.

Conrad, J.M. and Clark, C.W. (1987) *Natural Resource Economics. Notes and Problems.* Cambridge University Press, Cambridge, UK.

Ellen, R. (1982) *Environment, Subsistence and System. The Ecology of Small-scale Social Formations.* Cambridge University Press, Cambridge, UK.

Glomsrød, S. (2001) Soil degradation in macro CGE models. In: Heerink, N., van Keulen, H. and Kuiper, M. (eds) *Economic Policy Reforms and Sustainable Land Use – Recent Advances in Quantitative Analysis for Developing Countries.* Springer-Verlag, Berlin, Germany.

Glomsrød, S., Monge, M.D. and Vennemo, H. (1999) Structural adjustment and deforestation in Nicaragua. *Environment and Development Economics* 4(1), 19–43.

Greenwald, B. and Stiglitz, J.E. (1986) Externalities in economies with imperfect information and incomplete markets. *Quarterly Journal of Economics* 101, 229–264.

Hazell, P. and Norton, R.D. (1986) *Mathematical Programming for Economic Analysis in Agriculture.* Macmillian Publishing Company, New York.

Holden, S.T. (1993a) Peasant household modelling: farming systems evolution and sustainability in northern Zambia. *Agricultural Economics* 9, 241–267.

Holden, S.T. (1993b) The potential of agroforestry in the high rainfall areas of Zambia: a peasant programming model approach. *Agroforestry Systems* 24, 39–55.

Holden, S.T. and Binswanger, H.P. (1998) Small-scale farmers, market imperfections, and natural resource management. In: Lutz, E., Binswanger, H.P., Hazell, P. and McCalla, A. (eds) *Agriculture and the Environment. Perspectives on Sustainable Rural Development.* The World Bank, Washington, DC.

Holden, S.T. and Shiferaw, B. (2002) Poverty and land degradation: peasants' willingness to pay to sustain land productivity. In: Barrett, C.B., Place, F.M. and Aboud, A.A. (eds) *Natural Resource Management in African Agriculture: Understanding and Improving Current Practices.* CAB International, Wallingford, UK, in association with the World Agroforestry Centre (ICRAF), Nairobi, Kenya.

Holden, S.T. and Shiferaw, B. (2004) Land degradation, drought and food security in a less-favoured area in the Ethiopian highlands: a bioeconomic model with market imperfections. *Agricultural Economics* 30(1), 31–49.

Holden, S.T., Shiferaw, B. and Wik, M. (1998a) Poverty, market imperfections, and time preferences: of relevance for environmental policy? *Environment and Development Economics* 3, 105–130.

Holden, S.T., Taylor, J.E. and Hampton, S. (1998b) Structural adjustment and market imperfections: a stylized village economy-wide model with non-separable farm households. *Environment and Development Economics* 4, 69–87.

Holden, S.T., Benin, S., Shiferaw, B. and Pender, J. (2003) Tree planting for poverty reduction in less-favoured areas of the Ethiopian Highlands. *Small-scale Forest Economics, Management and Policy* 2(1), 63–80.

de Janvry, A., Fafchamps, M. and Sadoulet, E. (1991) Peasant household behaviour with missing markets: some paradoxes explained. *The Economic Journal* 101, 1400–1417.

Meadows, D.H., Meadows, D.L., Randers, J. and Behrens III, W.W. (1972) *Hvor går grensen?* J.W. Cappelens Forlag Ås, Oslo, Norway.

Rae, A.N. (1970) Capital budgeting, intertemporal programming models with particular reference to agriculture. *Australian Journal of Agricultural Economics* 14, 39–52.

Rae, A.N. (1971) Stochastic programming, utility and sequential decision problems in farm management. *American Journal of Agricultural Economics* 53(3), 448–460.

Ruthenberg, H. (1980) *Farming Systems in the Tropics,* 3rd edn. Clarendon Press, Oxford, UK.

Shiferaw, B. and Holden, S.T. (1999) Soil erosion and smallholders' conservation decisions in the highlands of Ethiopia. *World Development* 27(4), 739–752.

Shiferaw, B. and Holden, S.T. (2000) Policy instruments for sustainable land management: the case of highland smallholders in Ethiopia. *Agricultural Economics* 22, 217–232.

Singh, I., Squire, L. and Stauss, J. (1986) *Agricultural Household Models. Extensions, Applications, and Policy.* The World Bank, Washington, DC, and Johns Hopkins University Press, Baltimore, Maryland.

Skoufias, E. (1995) Household resources, transactions costs, and adjustment through land tenancy. *Land Economics* 71(1), 42–56.

Stiglitz, J.E. (1986) The new development economics. *World Development* 14(2), 257–265.

Part IV.

NRM Impact Assessment in Practice

9 Valuing Soil Fertility Change: Selected Methods and Case Studies

P. Drechsel[1], M. Giordano[2] and T. Enters[3]

[1]International Water Management Institute (IWMI), Accra, Ghana
[2]IWMI, Colombo, Sri Lanka
[3]Food and Agriculture Organization of the United Nations (FAO), Bangkok, Thailand

Introduction

Quantitative assessments of the extent and impact of soil degradation remained in their infancy until the early 1980s, largely due to a dearth of relevant data. Over the last 20 years, knowledge gaps have been filled and our understanding of changes in soil quality has increased. However, confusion over terminology remains (Box 9.1) and the relationships between soil fertility decline, soil erosion and soil degradation are often not well articulated. As an example of the implications of such confusion, economic analyses frequently treat nutrients lost through differing processes (e.g. erosion, leaching or crop harvest) equally, although the actual impacts on crop production and economic output can vary substantially (Bishop and Allen, 1989; Drechsel and Gyiele, 1999). For example, in contrast to nutrient loss through crop removal, only a small percentage of the nutrients lost through erosion are typically of relevance for plant growth and thus the value of their loss in terms of actual impact on crop production is relatively low (for discussions see Bishop and Allen, 1989; Bojö, 1996).

The economic impacts of nutrient and soil fertility depletion, as opposed to those of soil erosion, continue to be under-researched. To some extent, the reason for this neglect is that general fertility decline is far less obvious or visible than soil erosion and so draws less attention outside the field of soil science. As a result of this neglect, the true costs of nutrient depletion and the benefits of improved soil fertility management remain largely unknown. The example of sub-Saharan Africa (SSA) illustrates the possible consequences. Although erosion in SSA is a significant cause of soil nutrient loss and receives considerable attention, it is less well known that nitrogen (N), phosphorus (P) and potassium (K) losses via crop harvest and residue removal are at least as important (Drechsel and Gyiele, 1999). Failure to value nutrient losses

Box 9.1. Processes of soil degradation affected by natural resource management (NRM).

Soil degradation is a broad term for the decline in the capacity of the soil to produce goods of value to humans. The term encompasses the deterioration in physical, chemical and biological attributes of soils. Soil degradation is a long-term process that can result from erosion, soil nutrient depletion, soil pollution, salinisation and/or breakdown in soil structure. Soil degradation typically results in the loss of soil fertility.

Soil nutrient depletion describes the net loss of plant nutrients from the soil or production system due to a negative balance between nutrient inputs and outputs. Typical processes contributing to nutrient depletion are harvest, leaching, denitrification, fire, erosion, and runoff.

Erosion is one process of soil degradation that contributes to nutrient and fertility depletion but also to other (physical) soil-degrading processes. Erosion reduces soil productivity in general through removal of topsoil, reduction in rooting depth, removal of plant nutrients, and loss of water.

Source: Adapted from Blaikie and Brookfield (1987), Lal (1990), and Eaton (1996)

through these less-obvious channels has direct implications for the targeting and evaluation of natural resource management (NRM) interventions.

In this chapter case studies are used to examine the significant challenges in assessing, i.e. quantifying and valuing, the impacts of soil fertility and nutrient depletion. After briefly reviewing the main economic approaches that are used to value soil nutrient change, four case studies are presented. These illustrate the use and limitations of the two valuation approaches most commonly applied in developing country analysis, the replacement cost approach (RCA) and the change in productivity approach (PCA). The cases provide insights into the costs of nutrient decline – as well as enrichment – at scales ranging from the field to the continent and illustrate the disparity between different stakeholder perspectives. The first case study from Ghana assesses the private costs, experienced by farmers, of soil nutrient depletion in two different farming systems. Focusing on wastewater irrigation in Mexico, the second case demonstrates the divergence of perspectives between local farmers and regional communities. The third case highlights the economic costs of nutrient depletion in SSA. While the first three cases use the RCA to assess soil fertility change, the final case, from India, focuses on erosion as a nutrient-depleting process. It illustrates the use of the PCA and the differences between the PCA and RCA. The final section highlights issues in applying the methods described and discusses the contribution that soil nutrient valuation can provide to decision making, whilst at the same time highlighting the limitations of economic valuation in general.

Valuing Soil Fertility Change: Different Methods and Divergent Perspectives

As Stocking (2003) recently pointed out, interventions in soil management designed to reverse declining trends in food security must recognise the resilience and sensitivity of major tropical soil types. While this is no doubt true, to decision makers – from the farm to the national level – the economic implications of action or inaction on nutrient depletion can be as important as the physico-chemical change itself. To assess those implications, economic evaluation methods are necessary.

There are various methods that can be used to value natural resources and the services they provide in general and with respect to soil nutrient depletion in particular. By far, the most common are the RCA and PCA. Other approaches, such as hedonic pricing (property valuation) or contingent valuation, have rarely been applied successfully (see Box 9.2).

Box 9.2. Hedonic pricing and contingent valuation.

With respect to agricultural production, hedonic pricing (property valuation) uses land prices to estimate the economic value of soil quality. The sale or rental prices of land with different soil qualities can then be assessed using regression analysis (Pearce and Turner, 1990; Clark, 1996). The basic assumption is that higher-quality soils and investments in soil conservation translate into higher land values, i.e. higher future benefits to a producer (Barbier, 1998). However, in the developing-country context, the property valuation approach has seen little use, perhaps because land markets, and the institutional arrangements (such as property rights) that foster the development of markets and meaningful prices, are often poorly developed, especially for farming (Grohs, 1994; Nunan *et al.*, 2000). The utility of hedonic pricing is probably further reduced in cases where traditional rules, rather than strict market economics, steer land allocation, and of course where there is still sufficient land for shifting cultivation. The approach might find an appropriate context where there are clearly defined ownership rights to land, significant soil quality differences, farming is more commercialised, land is a limiting factor, and land acquisition/distribution is based on payment, i.e. not on share-cropping or family/clan distribution or inheritance pattern. Where these conditions do not apply, an alternative valuation approach could be contingent valuation, i.e. asking farmers to directly state their willingness to pay for maintaining soil fertility, based on a hypothetical scenario (Pearce and Turner, 1990).

In general, the most common approach to the economic assessment of nutrient depletion is the RCA, probably to a large extent because it is relatively simple to apply when nutrient loss data are available (Bojö, 1996; Predo *et al.*, 1997; Drechsel and Gyiele, 1999). The RCA estimates economic values based on costs of replacing or restoring damaged assets to their original productivity and uses this cost as measure of the benefit of restoration. With regard to soil fertility decline, the RCA calculates the costs that are or would be incurred to replace lost nutrients (Grohs, 1994). Other factors influencing soil fertility, such as the soil organic matter status, are seldom considered. Thus total replacement cost is usually based on the annual costs of fertiliser application needed to compensate for the loss of soil nutrients (see the first case study). Most studies that apply the RCA rely on the cost of inorganic fertiliser. The RCA suffers from a number of potential problems (Barbier, 1998). For example, soil nutrients may not be the factor limiting production and so their loss is of no direct economic consequence to farmers, farmers may have more cost-effective means to counter nutrient decline than using costly inorganic fertilisers, or nutrient loss may be accompanied by a degradation in soil structure that would not be addressed through fertiliser application (Enters, 1998).

In contrast, the most common method for assessing the economic impacts of soil degradation in general and soil erosion in particular is the PCA. Its main advantage is that it is logical, straightforward to apply – as long as relevant data such as crop yield changes over time are available – and relatively easy to comprehend for non-economists. Still, most analyses to date have used PCA only to assess the overall effects of soil erosion and not changes in soil nutrients *per se* (Enters, 1998). Using PCA, the cost of nutrient depletion is assumed to be the difference in crop yields with and without soil fertility decline, multiplied by the unit price of the crop and potentially considering differences in production costs. As a physical measurement, the PCA relies on crop yield changes independent of their cause(s) and can be used to measure actual change or, when coupled with yield simulations, to assess the likely impacts of possible interventions. Like the RCA, the PCA also suffers from a number of inherent problems. For example, crop production is highly variable and yield decline is not easily ascribable to a single cause such as erosion (Nye and Greenland, 1960; Lindgren, 1988; Theng, 1991; Enters, 1992; Prasad and Goswami, 1992). Further, the technique has to ensure that technological progress and change in farming practices, and their effects on yields, are isolated from the analysis. This is difficult since farmers might adapt their farming systems in the face of soil fertility loss and other changes. Finally, the possible existence of irreversibility – that it may be impossible to return soil productivity to its pre-degradation state – suggests that a higher cost should be given to nutrient depletion than just the actual loss in income so as to account for the irreversible reduction in the (soil) capital stock (Sanders *et al.*, 1995). The irreversibility problem also applies to the practical application of the RCA.

Application of both approaches in assessing NRM interventions are based on comparisons between sets of scenarios such as before/after nutrient depletion or among various NRM options. However, it is obvious that the exact definition of the 'control' can significantly influence conclusions on the costs and benefits of intervention. For example, Barbier (1998) argued that in fact no feasible technology exists (on sloping land) to produce crops without some degree of erosion. Similarly, all crop harvests necessarily also remove nutrients. Thus if the question is one of farming system options, it is meaningless to compare farming and no-farming scenarios, since any level of production will necessarily have soil nutrient implications.

Having the advantages and limitations of each of the assessment methods in mind, it is then necessary to choose one technically appropriate to a given situation and understandable to the target audience. The vantage point of the assessment also needs to be clear. This means that, for example, it has to be made explicit whether the focus is on society at large or an individual farm household. This difference is important, because 'society' is typically interested in economic (i.e. social) cost–benefit analysis while the household is typically concerned primarily with financial (i.e. private) cost–benefit analysis.

Economic analysis differs from financial analysis in three ways. Firstly, economic analysis considers 'social' costs and benefits whereas financial analysis considers actual costs as faced by an individual. Distortions induced by regulations, subsidies, overvalued currencies and market imperfections all give rise to differences in social and financial costs and reduce the applicability of market prices for valuing inputs and outputs for economic analysis. For example, with respect to soil fertility analysis, tariffs and subsidies can cause considerable differences between domestic and international fertiliser and crop prices. Secondly, social and private discount rates usually differ, with social rates typically assumed lower than private. Thirdly, externalities or off-site costs (as well as benefits) are typically ignored in financial analysis while they are an integral part of economic analysis (Barbier, 1998; Enters, 1998). NRM impact assessment usually evaluates the multidimensional social impacts of various interventions. This requires a social economic analysis, which accounts for externalities and market imperfections. The case studies presented in this chapter are mainly analysed from the farmer's perspective. Thus the values estimated do not reflect social values for changes in soil fertility.

The following four case studies have been selected to illustrate how the PCA and RCA are used to value changes in soil fertility and provide useful information for NRM impact assessment. While the studies presented in essence analyse the costs of soil fertility decline (or gain) and not the impact of a particular intervention, the same techniques could easily be used to estimate – as suggested by Barbier (1998) for soil erosion – the difference between the (present value) net return of an agricultural practice with soil fertility management and the (present value) net return with soil fertility decline, in other words the net benefits from adoption of improved soil fertility management practices.

The Costs of Nutrient Depletion in Kumasi, Ghana: An Example at the Farm Scale

The International Water Management Institute (IWMI) used the RCA to analyse the costs of soil nutrient depletion in farming systems along an urban–rural gradient in and around Kumasi, Ghana in West Africa's tuber belt. At one end of the gradient, in an urban agricultural system, vegetables are grown on scarce open spaces with access to irrigation water, and soil fertility decline can only practically be countered through fertiliser applications, as possibilities for shifting location do not exist. At the other end of the gradient, a 'traditional' maize–cassava system, there is no significant land shortage, giving peri-urban and rural farmers the flexibility to shift production to alternative fields, as crop yields decline. The goal of the study was to estimate the costs of soil nutrient depletion from the farmers' (private) perspective. The study demonstrates, among other things, that the results are significantly influenced by the specific farming conditions, measures that farmers apply to maintain production levels and input and output prices, especially the cost of fertiliser.

The costs of nutrient depletion in mixed-vegetable farming systems

In the Gynease suburb of Kumasi, land availability is severely constrained and farmers respond to soil fertility decline by applying fertilisers. In these farming systems, vegetables are grown continuously in the same beds resulting in three cabbage and at least 8–9 spring onion or lettuce harvests per year. Nutrient losses are high due to frequent harvests, the few remaining crop residues, and leaching on the sandy soils.

To compensate for nutrient losses, farmers apply substantial amounts of organic fertiliser, mainly poultry manure, which is available locally, at a rate of 20–50 t/ha for cabbage and 50–100 t/ha for lettuce and spring onions. Inorganic fertilisers are also applied. Irrigation with highly polluted stream water provides additional nutrients. Such water typically contains significant amounts of N when extracted downstream from the city. Irrigation is done with watering cans, mostly in the dry season and during dry spells in the rainy season. Annual application rates are high and range from 640 to 1600 l/m^2, with an average of about 1000 l/m^2. Table 9.1 summarises nutrient application rates at the study site.

Table 9.1. Annual nutrient application (kg/ha) rates for vegetable (cabbage/lettuce/spring onion) production, Kumasi, Ghana environs.

	Annual application rate by source (kg/ha)			
			Irrigation water	
Soil nutrient	NPK fertiliser (only used on cabbage)	Manure	Upstream of Kumasi	Downstream of Kumasi
N	75–180	770–1650	10	50
P_2O_5	75–180	420– 900	7	11
K_2O	75–180	350– 750	50	80

To analyse the costs of nutrient depletion, the RCA was used based on local nutrient input prices. To conduct the assessment, a simplified version of the standard NUTMON model that is used to calculate soil nutrient balances was applied (www.nutmon.org). This model considered the major nutrient in- and outflows including crop harvest, plant residues, manure, fertiliser, irrigation water, and precipitation. Leaching and losses through erosion were calculated via transfer functions. Data were verified through soil analyses comparing fields with and without nutrient application. The local price of poultry manure was used to calculate replacement costs as it reflected the farmers' actual choice. The analysis showed that, despite significant N and K losses of 180 kg N and 50 kg K_2O/ha, the annual costs of nutrient depletion were only about US$45/ha, consisting of US$10/ha for the manure and US$35/ha for handling and application. As average farm sizes are about 0.1–0.2 ha, annual costs per farm are about US$5–9. If the inexpensive poultry manure were not available in Kumasi, the use of mineral fertiliser would have increased the replacement costs four times, assuming constant costs for handling and application (Table 9.2).

Table 9.2. Costs of soil nutrient depletion in two contrasting farming systems, Kumasi, Ghana.

	Urban mixed-vegetable system		Peri-urban maize–cassava system	
	(0.1-ha farm)	(US$)	(0.7-ha farm)	(US$)
Annual costs (actual)	Nutrient replacement	1[a]	Land rent	21
	Manure handling/ application	4.5	Land clearing	14
	Total	3.5	Total	35
Annual costs (per ha)	Nutrient replacement	10	Land rent	30
	Manure handling/ application	35	Land preparation	20
	Total	45	Total	50
Net annual income (actual farm size)		400–800		200–450
Cost of actual nutrient depletion as a % of net income	About 1%		About 10%	

[a]US$14 on average for inorganic industrial fertiliser if poultry manure is not available.

The costs of nutrient depletion in traditional maize–cassava systems

The costs of nutrient depletion were also assessed using the RCA for a 'traditional' rainfed farming system located in the Atwima district of peri-urban Kumasi. In this system, nutrient losses occur mainly through the removal of harvested cassava and maize and their residues. NUTMON analysis showed that nutrient losses were mainly centred on N (58 kg/ha) as the other nutrients are largely replenished through fallow burning. In general, farmers do not attempt to compensate for N losses by fertiliser applications. Instead, they utilise new N pools by opening 'new' plots (shifting cultivation) allocated by local chiefs. In fact, they may never return to their old fields. In conducting the analysis, depletion or replacement costs were assessed by calculating the costs of acquiring and preparing a new location for cropping.

Although this represents a departure from the conventional RCA, moving to a new location is, in essence, the farmers' method of nutrient replacement. In addition to movement costs, a one-time rental payment, which is demanded by some local chiefs, was included in the cost calculations. As farmers actually shift production, this version of the RCA is more consistent with the actions of the target audience (i.e. the farmers). Therefore it made the cost assessment more applicable to them.

In the study area, different tenure arrangements are common, including share-cropping and land rental. The annual rent for a new plot ranges from about US$10–50/ha, paid for the whole term in advance, with lower rates often indicating a higher risk of eviction for land development (Nunan *et al.*, 2000). A correlation between land rental prices and soil quality could not be confirmed, although Nunan *et al.* (2000) mentioned such an influence. In addition to the costs of land acquisition, the farmers also incur land-clearing costs in the order of US$40/ha. Maize productivity decline is generally severe enough to induce farmers to shift production after 2 years. Thus, the financial cost of nutrient depletion per hectare can be calculated as follows:

2 years × US$30 (average land rent) +
1 × US$40 (land clearing) = US$100 over 2 years (1)

If N could be replenished annually by applying fertilisers it would be possible to crop on the same field for at least 4 years. The amount of inorganic fertiliser needed would cost US$116 over this period. Thus the calculation is:

4 years × US$30 (average land rent) + 1 × US$40 (land clearing) +
US$116 (fertiliser) = US$276 over 4 years (2)

If the average land rent of US$30/ha is a one time payment to village chiefs and farmers cannot rent out this land, the actual cost to the farmer reduces to US$140/ha over 4 years with shifting cultivation alone or US$186/ha with fertiliser use after the first land clearing. As other cost factors do not differ much, the example shows that under both assumptions about land rents, fertiliser application appears less profitable than shifting cultivation. The situation would look quite different if the farmers were able to obtain poultry manure at the same low cost at which it is now available to the mixed-vegetable farmers, an option not currently available. Quansah *et al.* (1998) explain that farmers not only shift between fields because of the declining effect of ash fertilisation, but also to avoid increasing weeding costs under longer cultivation. In this study example, extra weeding costs over 4 years were largely balanced by the costs of the second land clearing in the 2-year system.

Comparing the two systems

Table 9.2 provides a comparison of the costs of soil nutrient depletion in the two systems. On a per hectare basis, farmers in both locations are faced with similar annual costs for nutrient decline (around US$50). However, the exact techniques used in the two RCA applications to arrive at this figure were varied so as to reflect the actual practices used by farmers to counter crop yield decline. This approach – following what farmers actually do – avoids what Barbier (1998) called an estimation that 'can only be an accurate reflection of on-site costs by chance'. Where land availability is constrained (the first case), the costs of nutrients and their replenishment with manure are most relevant. Where farmers can easily find new land or open 'new' N pools (the second case), land acquisition and preparation are the relevant and determining variables affecting cost calculations.

The results highlight that cost assessments can be highly dependent on factors not related to soils and nutrients. Large-scale poultry production in Kumasi provides a ready supply of inexpensive nutrients. Without this local poultry production, the costs of soil nutrient depletion to the farmers would be significantly higher. It is also interesting to note that labour costs make up a significant share of total replacement costs. This aspect is frequently omitted in RCA studies and explains much of the cost underestimation of which some studies have been accused (Enters, 1998). In fact, in urban Kumasi, labour costs (actual or opportunity costs) are higher than in rural Ghana, and most farmers use their own labour (especially for manual watering). This keeps the urban plots smaller (0.1–0.2 ha) than the average 0.7 ha of those in the maize–cassava system.

Valuing Nutrient Gains from Wastewater Irrigation in Mexico: Contrasting Perspectives and Divergent Results

While most research on the valuation of soil nutrients has concentrated on the costs of nutrient depletion, Scott *et al.*'s (2000) study of Guanajuato, Mexico focused on nutrient enrichment or gains. Wastewater is usually considered as a negative externality, but it can also have positive aspects if its nutrients, when applied through irrigation, reduce the need to apply inorganic or other fertilisers (Box 9.3). In this study, the impact of wastewater treatment on decreasing nutrient availability was calculated.

As in the previous example from Ghana, the RCA was used. However, while in the Ghana case the costs of lost nutrients were assessed, in the Mexican case the benefits of additional nutrient supply were calculated. In carrying out the analysis, the amounts of N and P delivered to fields through untreated wastewater irrigation under current practices were first estimated and compared with a scenario using less nutrient-rich treated wastewater (Table 9.3). Nutrient costs were then calculated based on prices provided by local commercial fertiliser suppliers (Table 9.4) (see the next case study for additional details on how such calculations are made).

From this information, the costs of replacing the reduced amounts of N and P in the water incurred by the construction of a wastewater treatment plant were calculated. These costs include the fertiliser itself as well as the labour needed to apply it. Based on survey data, the researchers concluded

Box 9.3. Diverse impacts of wastewater use.

The composition of municipal wastewater must be taken into account to calculate its true benefit. Industrial waste can introduce pathogens and chemical pollutants that can be harmful to humans and plants. Further, if the total nutrient content in wastewater exceeds crop needs or if certain nutrients are over-represented soil nutritional imbalances can occur. These imbalances can affect the availability and uptake of under-represented nutrients. For example, if wastewater irrigation exceeds the recommended nitrogen dosage for optimal yields, it may stimulate vegetative growth, but delay ripening and maturity, cause micronutrient deficiencies and, in extreme circumstances, crop failure. Likewise, a predominance of domestic wastewater may affect the yield of salt-sensitive crops in addition to soil structure and groundwater quality. Some effects might not be obvious immediately, but represent hidden or long-term costs for the environment, farmers and society. In short, the economic impact of wastewater use is much more complex than discussed in this chapter. A variety of valuation techniques can be used to quantify the different socio-economic, health, and environmental impacts of wastewater use (Hussain *et al.*, 2002).

Table 9.3. Simulated total nitrogen (N) and phosphorus (P) deliveries (kg/ha) from actual measurement of both untreated and treated wastewater, Guanajuato, Mexico.

	Untreated (kg/ha)		Treated (kg/ha)		Change through treatment (%)	
Study site	N	P	N	P	N	P
San Jose de Cervera	455	76	36	7	–92.2	–90.6
Santa Catarina	1,597	258	285	42	–82.2	–83.5
Comparison with lucerne requirements	88	115	88	115		

Table 9.4. Unit costs of nitrogen (N) and phosphorus (P) fertilisers in Mexico, 1999.

Source of nutrient	Content	Cost (US$/kg)
N	N (%)	N
Urea	46.0	0.40
Ammonium nitrate	33.5	0.52
Ammonium sulphate	20.5	0.37
Average		0.43
P	P (%)	P
Triple superphosphate	46.0	0.51
Single superphosphate	18.0	0.56
Di-ammonium phosphate (DAP)	46.0	0.63
Mono-ammonium phosphate (MAP)	52.0	0.57
Average		0.57
Application cost (combined N+P) in US$/ha		31.58

that the forgone annual value of the reduced nutrient delivery would be about US$906/ha. However, this value constitutes an overestimate as the nutrient requirements for lucerne, the principal crop grown, are greatly exceeded when untreated wastewater is applied (Table 9.3). A more realistic estimate excludes the value of the difference between crop nutrient demand and nutrient supply from the untreated wastewater, because any excess does not represent a true economic on-site benefit to farmers. Accordingly, the annual value of the 'lost' nutrients is reduced to around US$135/ha, which constitutes the tangible benefit of the wastewater to the farmers.

Multiplying the crop-required share of the total water treatment capacity of the plant with the value of nutrients lost to farmers, the operating plant 'costs' farmers some US$18,900 per year in forgone nutrients. The result demonstrates that any economic impact assessment must be comprehensive enough to capture unintended side effects and unexpected benefits or costs. In addition, it illustrates the importance of reference point. From the farmer's perspective, the construction of the treatment plant has a negative impact in that it reduces the provision of free nutrients – a positive externality – and results in additional costs if soil fertility is to be maintained. From the perspective of the plant's intended beneficiaries, local and regional communities who expect a cleaner environment, safe drinking water and improved sanitary and health conditions, this is irrelevant (assuming they are not also farmers). It is thus left to a social cost–benefit analysis to determine whether the plant should be constructed and, if so, how the costs and benefits of operating the plant should be distributed.

Nutrient Depletion in Sub-Saharan Africa

The on-site impacts of nutrient depletion can be conducted at various scales. The previous two cases focused predominantly on the field or farming system scale, taking the farm household as the decision-making unit. Such analyses are useful in improving understanding of the economic impact of soil nutrient change, and they serve as a valuable input in decision making about interventions in the agricultural sector. However, at this scale the magnitude of the problem for a region, nation, or continent does not become apparent. The importance of such an assessment should not be underestimated as it provides valuable insights to policy makers at national and international levels. In particular, it can be a useful instrument for identifying 'hot spots' or priority areas for soil-conservation interventions and areas with a high potential risk of food insecurity in addition to raising awareness of the problem.

With this in mind, the International Board for Soil Research and Management (IBSRAM) conducted a continental-scale assessment of the costs of soil nutrient depletion in SSA (Drechsel *et al.*, 2001a, b). The research goal was to inform policy makers of the 'hidden' costs of soil nutrient mining so as to highlight the potential impact and benefit of soil-conservation investments.

To undertake an analysis on such a scale requires an approach that can be applied across a large number of states and very diverse environmental and socio-economic conditions. The approach has to: 1. be undemanding with respect to data requirements; and 2. produce results that can be compared across countries together with outputs that are understandable by and acceptable to policy makers. To meet these prerequisites, it was decided to employ the RCA using nutrient balance predictions (N, P, and K deficits) for the year 2000 provided by Stoorvogel and Smaling (1990) and data obtained through a fertiliser retail price survey in 15 countries.

Since the types of fertilisers available vary among countries, calculations were based on the costs of nutrient units rather than the costs of specific marketed fertilisers. This required developing a cost or price ratio between the main macronutrients (i.e. N, P and K). Based on world market prices for products and applying knowledge of product content, macro-unit prices were calculated (Table 9.5) along with standardised nutrient ratios. Based on these price ratios, average nutrient costs in K_2O equivalents were determined. The results for Nigeria are shown in Table 9.6. The last column in the table

Table 9.5. World market prices (US$) of fertiliser raw materials (FERTECON, 1998).

	Fertiliser raw material		
Costs	Ammonia (NH_3)	Phosphate (H_3PO_4) salts	Potassium chloride (KCl)
Raw material (US$/t)	140.0[a]	276.8	94.7
Nutrient in raw material	N	P_2O_5 [b]	K_2O[b]
(%)	ca. 77	ca. 53	ca. 60
Cost/t (US$)	182	522	158
Cost/kg (US$)	0.182[a]	0.522	0.158
Price ratio/K_2O unit	1.15	3.3	1.0

[a]Calculation example: 1 t ammonia costs US$140, about 77% is N. Thus, 1 t pure N would cost US$182.
[b]By standard convention, the oxidised forms of P (P_2O_5) and K (K_2O) are used.

Table 9.6. Costs (US$) per unit of nutrient in K_2O price equivalents in Nigeria.

Fertiliser product	N	P_2O_5	K_2O	All three nutrients	Price survey (US$/100 kg)	Cost/K_2O equivalent (US$)
	K_2O equivalents					
15:15:15	17.3	49.5	15.0	81.9	31.0	0.340
20:10:10	23.0	33.0	10.0	66.0	28.9	0.44
20:10:10+10 Ca[a]	23.0	33.0	10.0	66.0	27.4	0.42
25:10:10	28.8	33.0	10.0	71.8	31.7	0.44
Urea (46% N)	53.0	0.0	0.0	53.0	30.1	0.57
SSP (18% P_2O_5)[b]	0.0	59.4	0.0	59.4	28.1	0.47
Mean						0.45

[a]Ca, calcium.
[b]SSP, single superphosphate.

provides the average cost per K_2O unit (US$0.45/kg). Based on the price ratio of the raw materials for nitrogen and phosphorus per K_2O unit (Table 9.5), the nutrient costs could then also be calculated for N (US$0.52/kg) and P_2O_5 (US$1.49/kg).

The procedure was repeated for the 14 other SSA countries included in the survey where the required data was supplied by the country offices of the fertiliser industries. Average unit prices and their standard deviations were then multiplied by the corresponding quantities of depleted nutrients. A correction factor of 5% was used for nutrients lost through erosion (i.e only 5% of the nutrients were valued), as only a small percentage of the total loss is actually plant-available. This was discussed by Drechsel and Gyiele (1999) and suggested by Bishop and Allen (1989).

The average values of all calculations were:

- N: 0.50 ± 0.10 US$/kg
- P_2O_5: 1.22 ± 0.20 US$/kg
- K_2O: 0.43 ± 0.06 US$/kg

These figures were used as estimates of nutrient costs for countries not covered in the survey after taking into account variations in fertiliser type and transport distance (seaport vs. land-locked).

The results of the analysis indicate that soil nutrient depletion is a significant on-site cost for the agricultural sector in Africa. For SSA as a whole, nutrient depletion accounts for about 7% of the agricultural gross domestic product (GDP) (both crop and livestock production). This amounts to an annual cost of approximately US$32 per farm household, or about US$20 for each hectare of arable land (currently cultivated and fallow land) In some African nations, particularly those in the East African Highlands (Burundi, Malawi, Rwanda, Uganda), nutrient depletion per hectare is especially severe, even after adjusting for nutrients lost through erosion (Fig. 9.1, left). The primary reason for this is the high land-use intensity and resultant higher nutrient exports through crop removal combined with the low percentage of arable land under fallow (Drechsel *et al.*, 2001b). In terms of share of the overall agricultural economy, nutrient depletion is largest in Mozambique, Niger, Rwanda, Ethiopia and Tanzania (Fig. 9.1, right). In general, the estimates from the study can be considered conservative, since they include only the fertiliser cost of nutrients already lost and not the additional fertilisers required because of limited fertiliser efficiency after replacement. Neither do the estimates consider labour costs, which as the previous examples have shown, can significantly affect the results of the calculations. While comparisons of these results to other regions of the developing world would be useful, sufficient data are currently not available. It is also important to note that any aggregate data mask the tremendous variation that may exist within each country. If nutrient depletion data are available at sufficiently fine scales, a similar approach to that outlined here could be used to estimate more localised nutrient depletion costs (Drechsel and Gyiele, 1999).

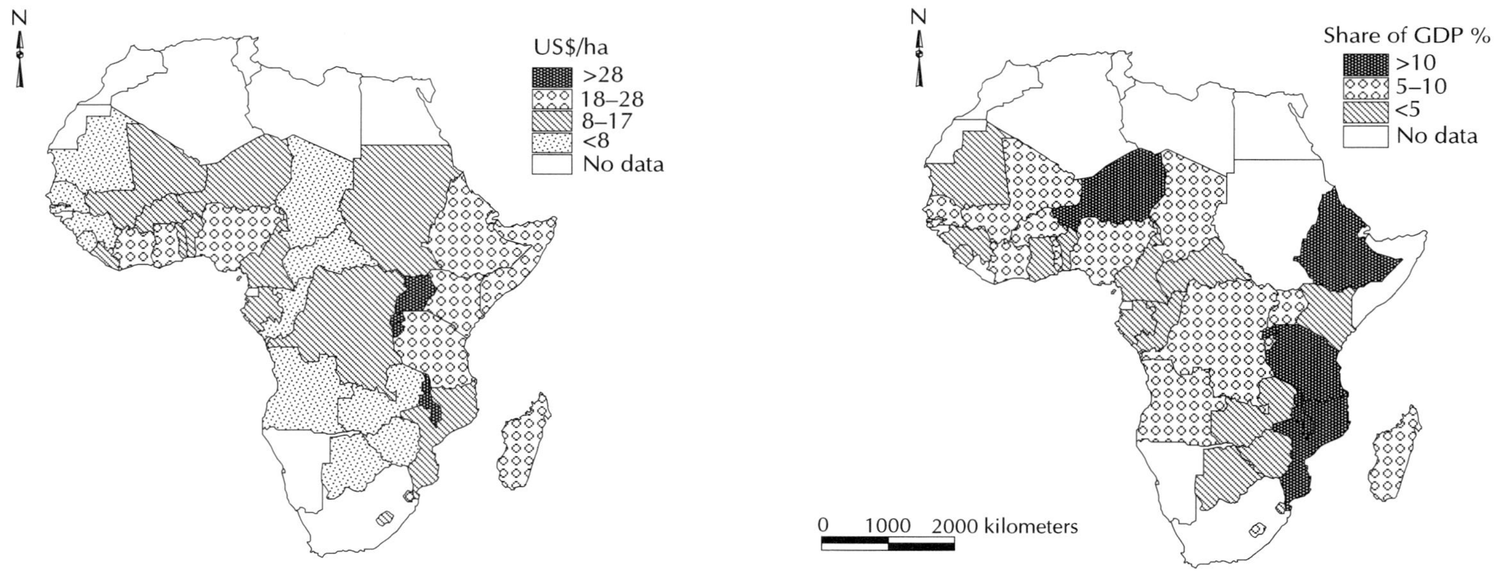

Fig. 9.1. Costs of nutrient depletion (US$/ha) on total arable land including fallow but excluding pasture (left) and as share of the agricultural gross domestic product (GDP) (right) in sub-Saharan Africa. Only countries covered by the survey of Stoorvogel and Smaling (1990) and with information available on their GDP in 2000 are considered.

Erosion in India, the Productivity Change Approach and a Comparison with the RCA

The RCA applied in the previous case studies attempted to value the on-site costs of nutrient depletion. In contrast, the PCA usually values the total change in soil productivity as expressed through changes in crop yields, multiplied by the unit price of the crop, preferably less the differential in production costs. The approach is especially useful when addressing soil erosion, which not only translocates nutrients but also reduces rooting depth, organic matter content and soil water retention capacity.

For the case of erosion, Enters (1998) describes the different steps in applying the PCA. As a first step soil erosion is quantified. To do this, most researchers make use of empirical–mathematical models that express soil erosion rates in t/ha. Although their limitations are widely recognised, the most commonly used models are the Universal Soil Loss Equation (USLE) developed by Wischmeier and Smith (1965), modified versions of the USLE, or more process-based models such as the Water Erosion Prediction Project (WEPP) Model. In the second step, the impact of soil erosion on crop yields is estimated. There are numerous approaches to determine soil erosion–productivity relationships. In some studies, known estimates from comparable areas have been used (Wiggins and Palma, 1980) or results derived through linear regression (Pagiola, 1993). Models developed in the early 1990s have found more widespread acceptance only recently. Grohs (1994) used the Erosion Productivity Impact Calculator (EPIC) and Nelson *et al.* (1996) used the Agricultural Production System Simulator (APSIM). Once the environmental effects on yield have been quantified, in the third step, the estimated crop production losses are valued in market prices, i.e. they are translated into monetary terms. Although the approach appears to be straightforward, the comparison of exploitative (i.e. soil-eroding) practices with soil-conservation requires an analysis of cost and benefit streams over time, so discounting becomes necessary. In addition, labour inputs and costs should enter the equation, as they usually vary among practices.

Kumar (2004) used the PCA to assess the on-site costs of soil erosion in the Doon Valley of Uttar Pradesh, India, an area experiencing deforestation and increased population pressure. To carry out the study, satellite data from 1972 and 1994 were used to map potential soil-erosion hazard classes that were then used to estimate erosion rates caused by natural and anthropogenic factors. Calculations were made using the USLE. By combining estimated erosion rates with experimental data on the impact of soil erosion on crop yield under various soil types (Table 9.7 gives a general example) and information on local producer prices (from 1992–1995), the costs of erosion were calculated by subtracting the expected value of output with erosion from the value expected in a 'no-erosion' scenario, which, as pointed out by Barbier (1998, p. 288) is flawed, as it is impossible 'to eliminate soil erosion altogether'.

Unlike in the previous cases, Kumar (2004) attempted to value not just the cost of nutrient loss for 1 year, but rather the present value of the annual

Table 9.7. General relationship between erosion and crop yield losses (%) by soil type as summarised by the Indian National Bureau of Soil Survey and Land Use Planning, Nagpur (source: Singh *et al.*, 1992; Kumar, 2004).

Degree of erosion	Annual erosion (t/ha)	Loss (%) in yield[a] Alluvial soil	Black soil	Red soil
Slight	0–10	0–5	0–10	0–25
Moderate	11–20	5–10	11–25	26–50
Strong	21–40	11–25	26–50	>50
Severe	>40	26–50	>50	nd[b]

[a]The annual productivity losses are high, but are based on empirical results on highly erodible soils on steep slopes (Kumar, 2004).
[b]nd, not determined.

productivity losses occurring over a longer time horizon. Erosion rates and crop yields were projected into the future based on exponential functions with the cost of the expected productivity losses valued at average local prices for 1993 to 1995 period. As presented here, two discount rates, 5% and 10%, were applied to future losses over 5, 10, and 20 years time horizons. For simplicity, results for three selected assessments are presented in which it is assumed that the entire valley (209,000 ha) was planted only to wheat, rice or maize (Table 9.8).

Table 9.8. Productivity change approach (PCA)-estimated discounted costs of soil erosion for three crops in the Doon Valley, Uttar Pradesh, India over a time horizon of 20 years.[a]

		Planning horizon (years)							
		1		5		10		20	
		Discount rates (%)							
Crops	Costs	5	10	5	10	5	10	5	10
Wheat	US$ (millions)	16.7	15.9	134.6	115.8	279.6	216.4	486.7	319.9
	US$/ha	80	76	647	557	1344	1040	2340	1538
Rice	US$ (millions)	22.4	21.4	193.8	166.4	418.0	321.7	744.6	484.9
	US$/ha	108	103	932	800	2009	1547	3580	2331
Maize	US$ (millions)	11.5	11.0	97.5	83.8	207.5	160.0	366.5	239.4
	US$/ha	55	53	469	403	997	769	1762	1151

[a]The following formula with an exponential yield loss factor derived in Kumar (2004) was used to compute the discounted values:

$$DV_T = \sum_{t=1}^{T} (P_c[Y_0 - Y_0(1-d)^t]/(1+r)^t$$

where DV_T is the total discounted value for planning horizon T, Y_0 is the yield in base year, d is the annual percentage decline in yield, P_c is the price of the crops, t is specific year, and r is the discount rate.

Table 9.8 highlights a number of important issues in understanding the valuation of soil productivity loss. Firstly, a higher discount rate reduces the present value of the costs of soil erosion. While 5 and 10% rates were used here, in Kumar's (2004) original study rates of only 3 and 4% were applied.

In fact, the determination of the 'appropriate' discount rate is complex and has been debated at great length. Most financial analysts use discount rates between 10 and 20% while a discount rate of below 10% is more common in economic analysis (Enters, 1998). Secondly, the magnitude of the costs depends to a substantial extent on the crops produced and their market prices. In the examples shown, the estimated costs of soil erosion are roughly twice as high for farmers growing rice than maize due largely to differences in prices. This is similar to the study from Ghana described above in which the RCA valued nutrient depletion at a relatively low level because poultry manure was readily available.

Kumar (2004) also applied the RCA to compare estimates with the results obtained through the PCA. The calculations of the replacement costs were based on estimated erosion losses of three nutrients and organic carbon, which could be considered as a proxy for other soil services not explicitly considered (estimated annual losses were: N 310 kg/ha, P 6 kg/ha, and K 1157 kg/ha, while estimated organic carbon loss was 2437 kg/ha). Prices were based on local fertiliser and manure prices and calculations are again presented here using 5% and 10% (3 and 4% in the original study) discount rates over 5-year, 10-year and 20-year time horizons (Table 9.9).

Table 9.9. Estimated discounted replacement costs of soil erosion in the Doon Valley, Uttar Pradesh, India over a time horizon of 20 years (modified from Kumar, 2004).

Planning horizon (years)	Discount rate (%)	Total study area (US$, million)	Average cost/ha (US$)
1	5	139	664
	10	132	663
5	5	631	3,017
	10	552	2,641
10	5	1,125	5,381
	10	895	4,282
20	5	1,815	8,684
	10	1,240	5,932

A comparison of Tables 9.8 and 9.9 indicates that the replacements costs are considerably higher (up to ten times) than the costs of crop production losses. This echoes the findings by Clark *et al.* (1998) from Sri Lanka, Predo *et al.* (1997) from The Philippines and Grohs (1994) from Zimbabwe. Besides the reasons discussed above, another possible explanation for the discrepancy is that the RCA as used by Kumar (2004) assumes that all lost (and valued) nutrients were relevant for crop growth. The PCA, on the other hand, only considers actual changes in crop yields. Such changes mean that nutrient demand is indirectly valued, but total supply is not valued. On the other hand, the PCA considers an array of other factors affecting soil productivity such as changes in rooting depth. By measuring all such factors together, the PCA produces results that are often considered to be more directly relevant to and comprehendible by farmers than those calculated using the RCA.

Discussion

This paper examines common methods for valuing the costs (and benefits) of soil nutrient decline (or enrichment) in developing countries which can also be used to measure or compare the impacts of NRM interventions. Progress in resource and environmental economics in the last two decades has provided a suite of methods for such assessments. However, not every method fits the context of developing countries and many theoretically justified approaches have had little or no application. Instead, the Replacement Cost Approach and the Productivity Change Approach remain the most commonly used methodologies.

The RCA and PCA are fundamentally different in nature. The RCA attempts to place a value on actual nutrient loss or gain while the PCA attempts to value the change in production caused by that change. Naturally two methodologies with such different approaches are likely to assign different economic values to soil nutrients or their change. Replacements cost (RCA) estimates are usually considerably higher (often up to ten times) than estimates of corresponding crop production changes (PCA) [see also descriptions by Grohs (1994), Bojö (1996), Clark *et al.* (1998), Predo *et al.* (1997), and Kumar (2004)]. The divergence would be even larger if RCA studies were expanded beyond their typical focus on only the best known and most easily analysed macronutrients (i.e. N, P, K). The consideration of soil carbon by Kumar (2004) described above is still unusual and resulted in a doubling of replacement cost estimates. Existing spreads between RCA and PCA estimates would further increase if the economic value of other relevant nutrients (e.g. Mg, S, Ca, Zn, Cu, etc.) were added.

One reason for the divergence between estimates is that the RCA typically values the total volume of studied nutrients, including that quantity not of relevance for current crop growth (though this problem could be overcome through relatively simple adjustments). In contrast, the PCA only considers those nutrients and soil services directly impacting yield. Put another way, the RCA tends to implicitly focus on long-term impacts (by valuing a permanent change in nutrient stocks) while the PCA tends to focus on shorter time horizons (by focusing on changes in actual output over discrete periods). Grohs (1994) and Barbier (1998) provide additional theoretic explanations for differences in RCA and PCA outcomes.

In deciding between replacement cost and productivity change approaches for use in assessing NRM interventions, it is critical to explicitly consider the questions to be answered and the use to which results will be put. For analyses of nutrients specifically, as opposed to broader soil services, the RCA has an obvious advantage in that it is tied directly to the nutrients themselves. When the focus is on soil fertility change or soil degradation (or improvement) in general, the PCA becomes increasingly attractive, as it implicitly considers all biological, chemical and physical soil properties affecting soil productivity.

Another consideration in choosing between the two approaches is data requirements and availability. The RCA has the clear advantage that it is simple to apply once net nutrient losses or gains are known, since market prices for key nutrients are usually available, as the examples from Mexico and sub-Saharan Africa have shown. However, this advantage can also skew results. The ability to incorporate easily available commercial fertiliser prices may encourage analysts to ignore more cost effective, but more difficult to quantify, options actually available to farmers such as manure application or shifting cultivation. Again, this problem can be addressed, for example as was done in the case study from Ghana. The data demands of the PCA can also be partially overcome by farmers themselves, and the approach's ability to cost alternative crop production practices can easily be built into cost–benefit analyses with direct applicability for farmers. Involving farmers in participatory research also has the decided advantage that it helps to ensure that critical socio-economic components are not ignored and that results are farmer relevant.

A further point, as the case from Ghana made clear, is that the costs of nutrient decline or the benefits of nutrient increase are related as much to socio-economic factors as to biophysical ones. While many analysts do not consider labour costs, especially in the case of subsistence economies [in fact, the increased use of labour has even been counted as a benefit (Wiggins and Palma, 1980)] the actual or opportunity costs of labour and land may drive the behaviour of farmers more than the physical costs of fertility change or its amelioration. In the case of the mixed-vegetable farming system in Kumasi, Ghana, labour costs for applying poultry manure were more than three times higher than the cost of manure itself. While the picture at greater distances from markets and off-farm income-generating opportunities is somewhat different, the role of labour costs does help to explain much of the rationale for farmers to engage in nutrient mining or shifting cultivation.

Conclusions

The above discussion leads us to the following conclusions. Although there is undoubtedly a need for further research on the biophysical aspects of soil fertility change and its effects on crop yields, more attention needs to be directed at the socio-economic characteristics of farmers, tenure arrangements, and the economic conditions under which they produce their crops so as to increase our understanding of the cost of nutrient decline and the value of NRM interventions. As long as farmers' livelihood strategies remain external to the valuation of soil fertility change, the results of any analysis will remain largely irrelevant to farmers' decision making. As Rigg (1997) pointed out for Southeast Asia, to farmers and rural life there is 'more than the soil'. More attention must be paid to the numerous explanatory variables of the economic system that determine farmer decision making and the costs that farm households face in soil fertility management. A clear understanding of the nature of factor markets, particularly land, labour, and capital is necessary for understanding farmers' choices (Pender and Kerr, 1998) and in selecting the valuation method to be used in the analysis. Only an integrated understanding of all these factors will allow insights into farmer's decision making about whether to invest in conservation or not.

The second conclusion is that it is not the cost of soil fertility decline *per se* that is important to decision making but rather whether the long-term benefits of soil conservation or land husbandry make the current cost of abatement worth bearing. The use of technologically unattainable standards as the basis for valuing soil fertility decline should be avoided. There is neither a no-erosion scenario on sloping lands (Barbier, 1998), nor is there crop production without alteration or exploitation of soil fertility. In the same vein, it is essentially pointless to make comparisons between farming and no-farming options. The issues are essentially fertility management vs. non-management, or one management practice vs. another. However, as the Mexican example shows, any management – in this case the replacement of nutrients via wastewater – has to enter the analysis as an investment cost. Many studies have been rightly criticised for failing to recognise that income derived from crop production is frequently decreasing in the near term because of soil-conservation interventions (Barrett, 1997). Soil fertility decline has a cost, so does soil conservation. In fact, as described by Fox and Dickson (1988) 'farmers have been reluctant to adopt conservation tillage, not because of a lack of information, not because of "perceptions" and not because "old habits die hard", but because they would lose money, both in the short and in the long run'. Hence, assessment of the impacts of NRM interventions need to include the investment cost incurred by farmers along with the estimated benefits.

Clearly the use of economic methods in developing countries for assessing soil nutrients and soil nutrient change as well as the impact of NRM strategies on both is filled with challenges. To overcome them, research aimed at finding ways to better apply theoretically valid methods in the contexts of smallholder agriculture, where rural land and other markets tend to function poorly, would be especially useful. This would allow the improvement of assessment of the economic impacts of NRM interventions and the identification of cost-effective strategies for the mitigation of soil degradation.

References

Barbier, E.B. (1998) The economics of soil erosion: theory, methodology and examples. In: Barbier, E.B. (ed.) *The Economics of Environment and Development. Selected Essays.* Edward Elgar, Cheltenham, UK, pp. 281–307.

Barrett, S. (1997) Microeconomic responses to macroeconomic reforms: the optimal control of soil erosion. In: Dasgupta, P. and Mäler, K.-G. (eds) *The Environment and Emerging Development Issues,* Volume 2, Clarendon Press, Oxford, UK, pp. 482–501.

Bishop, J. and Allen, J. (1989) The on-site costs of soil erosion in Mali. *Environment Working Paper no. 21.* Environment Department, The World Bank, Washington, DC, 71 pp.

Blaikie, P. and Brookfield, H. (1987) *Land Degradation and Society*. Routledge, London, UK, and New York, 296 pp.

Bojö, J. (1996) The costs of land degradation in Sub-Saharan Africa. *Ecological Economics* 16, 161–173.

Breman, H. (1998) Soil fertility improvement in Africa: a tool for or a by-product of sustainable production? *African Fertilizer Market* 11(5), 2–10.

Clark, R. (1996) Methodologies for the economic analysis of soil erosion and conservation. *CSERGE (Centre for Social and Economic Research on the Global Environment) Working Paper GEC 96-13*. University of East Anglia, Norwich, UK, 71 pp.

Clark, R., Manthrithilaka, H., White, R. and Stocking, M. (1998) Economic valuation of soil erosion and conservation: a case study of Perawella, Sri Lanka. *Advances in Geo Ecology* 31, 879–888.

Drechsel, P. and Gyiele, L. (1999) The economic assessment of soil nutrient depletion – Analytical issues for framework development. *Issues in Sustainable Land Management no. 7*. IBSRAM (International Board for Soil Research and Management), Bangkok, Thailand, 80 pp.

Drechsel, P., Gyiele, L., Kunze, D. and Cofie, F. (2001a) Population density, soil nutrient depletion, and economic growth in sub-Saharan Africa. *Ecological Economics* 38(2), 251–258.

Drechsel, P., Kunze, D. and Penning de Vries, F. (2001b) Soil nutrient depletion and population growth in sub-Saharan Africa: A Malthusian nexus? *Population and Environment* 22(4), 411–423.

Eaton, D. (1996) The economics of soil erosion: a model of farm decision-making. *Discussion Paper no. 96-01*. Environmental Economics Programme, International Institute for Environment and Development (IIED), London, UK, 46 pp.

Enters, T. (1992) Land degradation and resource conservation in the highlands of Northern Thailand: the limits to economic valuation. PhD Dissertation. The Australian National University, Canberra, Australia (unpublished), 249 pp.

Enters, T. (1998) Methods for the economic assessment of the on- and off-site impact of soil erosion. *IBSRAM Issues in Sustainable Land Management no. 2*. IBSRAM (International Board for Soil Research and Management), Bangkok, Thailand, 60 pp.

FERTECON (1998) www.fertecon.com

Fox, G. and Dickson, E.J. (1998) What's economic about the costs of soil erosion to Canadian farmers? *Discussion Paper DP88/3*. Department of Agricultural Economics and Business, University of Guelph, Ontario, Canada.

Grohs, F. (1994) *Economics of Soil Degradation, Erosion and Conservation: A Case Study of Zimbabwe*. Arbeiten zur Agrarwirtschaft in Entwicklungsländern. Wissenschaftsverlag Vauk Kiel KG, Kiel, Germany, 199 pp.

Hussain, I., Raschid, L., Hanjra, M.A., Marikar, F. and van der Hoek, W. (2002) Wastewater use in agriculture. Review of impacts and methodological issues in valuing impacts. *IWMI Working Paper no. 37*. International Water Management Institute (IWMI), Colombo, Sri Lanka, 61 pp.

Kumar, P. (2004) *Economics of Soil Erosion. Issues and Imperatives from India*. Concept Publishing Company, New Delhi, India, 177 pp.

Lal, R. (1990) *Soil Erosion in the Tropics: Principles and Management*. McGraw-Hill, New York, 579 pp.

Lindgren, B.-M. (1988) Economic evaluation of a soil conservation project in Machakos District, Kenya. *International Rural Development Centre Working Paper no. 95*. Swedish University of Agricultural Sciences, Uppsala, Sweden, 32 pp.

Nelson, R.A., Dimes, J.P., Silburn, D.M. and Carberry, P.S. (1996) Erosion/productivity modelling of maize farming in the Philippine uplands. Part 1: a simple description of the agricultural production systems simulator. *Working Paper no. 12.* SEARCA (Southeast Asian Ministers of Education Regional Center for Graduate Study and Research in Agriculture) Uplands Research Project, Los Baños, The Philippines, 21 pp.

Nunan, F., Bird, K. and Bishop, J. (2000) *Valuing Peri-urban Natural Resources: A Guide for Natural Resource Managers.* Natural Resources Systems Programme, Department for International Development (DFID), University of Birmingham, UK, 101 pp.

Nye, P.H. and Greenland, D.J. (1960) The soil under shifting cultivation. *Technical Communication no. 51.* Commonwealth Bureau of Soils, Harpenden, UK.

Pagiola, S. (1993) Soil conservation and the sustainability of agricultural production. PhD Dissertation, Stanford University, California, 183 pp.

Pagiola, S. (1994) Soil conservation in a semi-arid region of Kenya: rates of return and adoption by farmers. In: Napier, T.L., Cambini, S.M. and El-Swaify, S.A. (eds) *Adopting Conservation on the Farm.* Soil and Water Conservation Society, Ankeny, Iowa, pp. 171–188.

Pearce, D.W. and Turner, R.K. (1990) *Economics of Natural Resources and the Environment.* Harvester-Wheatsheaf, London, UK, 378 pp.

Pender, J.L. and Kerr, J.M. (1998) Determinants of farmers' indigenous soil and water conservation practices in semi-arid India. *Agricultural Economics* 19, 113–125.

Prasad, R. and Goswami, N.N. (1992) Soil fertility restoration and management for sustainable agriculture in South Asia. In: Lal, R. and Stewart, B.A. (eds) *Soil Restoration, Advances in Soil Science,* Volume 17. Springer-Verlag, New York, pp. 37–77.

Predo, C., Grist, P., Menz, K. and Rañola, Jr, R.F. (1997) Two approaches for estimating the on-site costs of soil erosion in the Philippines: 2. The replacement cost approach. *Imperata Project Paper 1997/8.* CRES (Center for Resource and Environmental Studies), The Australian National University, Canberra, Australia, pp. 25–36.

Quansah, C., Asare, E., Safo, E.Y., Ampontuah, E.O., Kyei-Baffour, N. and Bakang, J.A. (1998) The effect of poultry manure and mineral fertilizer on maize–cassava intercropping in peri-urban Kumasi, Ghana. In: Drechsel, P. and Gyiele, L. (eds) *On-farm Research on Sustainable Land Management in Sub-Saharan Africa: Approaches, Experiences, and Lessons. IBSRAM Proceedings 19.* IBSRAM (International Board for Soil Research and Management), Bangkok, Thailand, pp. 73–90.

Rigg, J. (1997) *Southeast Asia – The Human Landscape of Modernisation and Development.* Routledge, London, UK, and New York, 488 pp.

Sanders, J.H., Southgate, D.D. and Lee, J.G. (1995) The economics of soil degradation: technological change and policy alternatives. Soil Management Support Services *(SMSS) Technical Monograph no. 22.* USDA/Natural Resources Conservation Service, World Soil Resources, Washington, DC, 82 pp.

Scott, C.A., Zarazua, J.A. and Levine, G. (2000) Urban wastewater reuse for crop production in the water-short Guanajuato river basin, Mexico. *IWMI Research Report no. 41.* International Water Management Institute (IWMI), Colombo, Sri Lanka, 34 pp.

Singh, G., Babu, R., Narain, P., Bhushan, L.S. and Abrol, I.P. 1992. Soil erosion rates in India. *Journal of Soil and Water Conservation* 47(1), 97–99.

Stocking, M.A. (2003) Tropical soils and food security: the next 50 years. *Science* 302, 1356–1359.

Stoorvogel, J.J. and Smaling, E.M.A. (1990) Assessment of soil nutrient depletion in Sub-Saharan Africa: 1983–2000. *Winand Staring Centre Report no. 28.* Wageningen, The Netherlands, 137 pp.

Theng, B.K.G. (1991) Soil science in the tropics – the next 75 years. *Soil Science* 151, 76–89.

Wiggins, S.L. and Palma, O.G. (1980) Cost-benefit analysis of soil conservation. *Project Report no. 104.* Acelhuate River Catchment Management Project, El Salvador, Land Resources Development Centre. Overseas Development Administration (ODA), Surbiton, Surrey, UK.

Wischmeier, W.H. and Smith, D.D. (1965) *Predicting Rainfall Erosion Losses from Cropland East of the Rocky Mountains, AH-282.* United States Department of Agriculture (USDA), Washington, DC, 47 pp.

10 Evaluating the Impacts of Watershed Management Projects: A Practical Econometric Approach

J.M. Kerr and K.R. Chung

Michigan State University, East Lansing, Michigan, USA

Introduction

Watershed management is widely used as an approach for managing natural resources and raising agricultural productivity in sloped, rainfed areas of many developing countries, especially India (Farrington *et al.*, 1999; Lal, 2000). Despite the growing importance of watershed projects, there is still relatively poor information on their impact and the factors that determine it.

This chapter discusses various econometric approaches to project evaluation and presents a case study of an econometric evaluation of Indian watershed projects. Like many project evaluations, this one was commissioned long after the projects had been initiated. No prior steps were taken to establishing the projects and collecting baseline data in a way that would facilitate subsequent evaluation. The pragmatic evaluation approach that was taken may be of use to evaluators working under similar conditions.

The Government of India and the World Bank commissioned the evaluation presented here as part of a larger effort to identify priorities for investing in predominantly rainfed agricultural areas. It was conducted in collaboration between the International Food Policy Research Institute (IFPRI) and the National Centre for Agricultural Economics and Policy Research (NCAP), New Delhi. The study covered dryland watershed projects operated by government agencies and non-governmental organisations (NGOs) in Andhra Pradesh and Maharashtra, two states in India's semi-arid tropical region. This chapter presents findings from Maharashtra.

The chapter is divided into five sections. Following this introduction, some distinctive characteristics of watershed development that have implications for impact assessment are reviewed. Then approaches for econometric evaluation of watershed projects are presented followed by the Indian case study with its econometric findings, and the chapter concludes with some reflections on future evaluation of watershed projects.

Watersheds and Watershed Projects

A watershed or catchment is commonly defined as an area in which all water drains to a common point. From a hydrological perspective a watershed is a useful unit of operation and analysis because it facilitates a systems approach to land and water use in interconnected upstream and downstream areas. Watersheds mean different things in different contexts. In temperate areas, for example, watershed projects often focus on movement of water pollution through runoff and stream flow, and in catchment areas of hydroelectric dams they focus on reducing siltation. In dryland areas such as the Indian semi-arid tropics – the focus of this chapter – watershed projects aim to maximise the quantity of water available for crops, livestock and human consumption through on-site soil and moisture conservation, infiltration into aquifers, and safe runoff into surface ponds.

Watersheds in hilly or gently sloping areas of India are often densely populated and typically contain a variety of land uses, including forests, pastures, rainfed agriculture on sloping lands, and both irrigated and rainfed agriculture in the lowlands. Off-site concerns are typically limited to the local, intra- or inter-village level due to low chemical use and the near absence of perennial streams and large water bodies.

Watershed projects have numerous distinguishing features that have important implications for both project implementation and impact assessment. Among these are the preponderance of spatial inter-linkages and externalities, the multitude of project objectives and dimensions, and the long gestation of project benefits.

Hydrological connections between upper and lower catchment areas create spatial interlinkages. Optimising land and water management in a watershed often requires coordination or collective action between people living in upper and lower areas. Uneven distribution of benefits and costs often makes this difficult. Evaluation may require examining separate project impacts on subgroups of people within the watershed.

Projects typically have multiple objectives and dimensions such as increasing water quantity, improving water quality, reducing sedimentation, or increasing the supply of certain types of biomass, among other things. Some may focus more on organising people to manage externalities. Project approaches and budgets vary with local conditions and objectives.

Many watershed project interventions have long-term impacts, some of which may be difficult to perceive. Conserving soil, stabilising hillsides through vegetative cover, recharging groundwater, and increasing soil moisture and organic matter all take time. Perceiving benefits is particularly difficult where interventions do not raise productivity but merely prevent gradual degradation. As a result, it is difficult to know what conditions would have prevailed in the absence of project interventions.

Other factors besides a watershed project's activities may determine whether or not it achieves its objectives. Such factors include local agroclimatic conditions, land tenure arrangements, people's willingness and ability to work together to devise arrangements to share benefits and costs, and infrastructure and market conditions that help shape farmers' incentives to manage their land. As a result, identifying the specific contribution of a watershed project and comparing across projects can be difficult.

Econometric Approaches to Project Evaluation

Econometric evaluation begins with the premise that the analyst fully understands the nature and determinants of a programme's success and can obtain the data needed to measure and relate them statistically. Econometric evaluation attempts to attribute changes in various outcome variables to a project intervention (or 'treatment') and to determine whether such effects are statistically significant.

Ideally, quantitative evaluation involves an *ex ante* experimental design in which project beneficiaries (e.g. individuals, villages, or project sites) are randomised across treatment and control groups. When sample sizes are large enough this methodology is powerful. The randomisation process creates groups that may be considered equal in all observed and unobserved attributes. The analysis requires a sufficient sample size, generated through the randomisation process, rather than a 'convenience sample' of a few sites. Randomisation removes the possibility of sample selection bias, an analytical problem that arises when systematic, pre-existing differences between programme and non-programme locations are correlated with project participation and the outcome variable of interest (Greene, 2000). With no sample selection bias, the analyst is confident that the differences in outcome truly result from the differences in treatment; the programme's impact is estimated by calculating the difference between the mean outcome of each treatment group and the control.

An experimental approach is often considered the gold standard of quantitative evaluation. However, there are many situations in which an experimental approach may not be possible. Firstly, it may be politically or administratively infeasible to randomly assign project sites to treatment groups. Secondly, many watershed projects do not deal with sample sizes that make randomisation a feasible strategy for study design.

Even if an experimental approach to quantitative evaluation is possible, the results from a project's initial evaluation stage may not be achieved when the project expands to a wider sample (Manski, 1995). Firstly, the conditions of the experimental project site are unlikely to be replicated exactly at other sites. Differences in social, economic and physical factors may lead to changes in programme outcomes. Secondly, an experimental programme is likely to be conducted differently from the actual programme established subsequently. This might occur due to issues of scale. For example, a small experimental programme may not strain the supply of competent programme administrators or affect the market wage, which would influence the programme's effectiveness. Scaling up the programme, however, might introduce such constraints and limit performance.

As a result, many evaluations have proceeded with non-randomly determined treatment and control groups. Various approaches have been used, each with their own strengths and weaknesses. One approach is a 'before/after' study. The evaluator measures the levels of outcome indicators in a watershed area before and after an intervention. With this design, the 'before' scenario is used as a control against which the effects of the intervention can be compared. This is a weak but feasible design (Campbell and Russo, 1999) that involves the unlikely assumption that there have been no other significant changes during the study period. As a result, this approach often gives biased results as it assumes that without the project, the pre-intervention values of the outcome indicator would have remained the same over time. For example, this approach would not identify any benefit from a project that arrested degradation of a resource that would have taken place had there been no project. It poses a serious threat to the validity of the findings.

A second approach, a 'with/without' design, is useful when no baseline data are available. This is often the case when an evaluation is commissioned after a project has been implemented. Randomisation is impossible and sample selection bias is likely in this situation. To reduce this threat, the evaluator must find a control site that is similar to the treatment sites in as many factors as are hypothesised to affect the outcome. However, in practice, sites are likely to vary in many ways, and evaluators try to match sites only in those factors that suggest likely threats to validity.

To improve their ability to create comparable treatment and control groups in a with/without setting, Jyotsna Jalan and Martin Ravallion (2003) used a statistical technique called 'propensity matching' to match on the basis of multiple factors. This involves modelling the probability that each site participates in a project as a function of all observable variables known to affect participation, and then matching pairs of participating and non-participating sites that have an equal probability of having been selected for the project. Project impact is estimated as the mean of the differences between all matched pairs in the outcome variable.

Such approaches to with/without analysis may succeed in creating treatment and control groups that are equivalent in terms of observable characteristics, but they cannot control the effects of unobservable characteristics. To the extent that some factors that determine programme placement are unknown, selection bias may persist (Baker, 2000). Given this problem, it is not surprising that evaluators often suggest a combination of the before/after and with/without approaches. This 'difference of differences' or 'double difference' approach calculates the difference between control and treatment groups at baseline and post-intervention to assess whether this difference changed over time. For example, to examine the effects on crop yields or employment resulting from a project, the evaluator would take the difference in yields or employment between project and non-project areas before and after the project period. Statistically significant differences in these differences would constitute evidence of impact. Simple differences in yield or employment with and without the project would not be sufficient since baseline levels might have been unequal in project and non-project areas. This approach has the advantage of 'differencing out' any time-invariant unobservable factors that might cause sample selection bias (Baker, 2000). But it also requires the assumption that these unobservable factors have not changed during the study period. In addition, the evaluation must be commissioned *ex ante* because data on participants and non-participants are required before and after the intervention.

All of the above approaches have been modelled after the scientific tradition of experimental design and are thus termed 'quasi-experimental'. Social scientists have developed another approach to deal with the inherent problems of sample selection bias when quasi-experimental designs are infeasible or insufficient. Rather than comparing treatment and control groups, a statistical technique known as instrumental variables is used to remove the bias introduced by sample selection bias (Greene, 2000). Typically, a two-stage model is used; one equation models the probability that a given observation is selected (or self-selects) for a given programme. A second estimates the outcome in question, replacing the endogenous treatment variable with its predicted value. This process adjusts for selection bias if: 1. exogenous 'instruments' can be found that are significant determinants of project participation but do not directly affect the outcome of interest conditional on participation, and 2. the participation model is valid.

The instrumental variables procedure carries the advantage that impact evaluations may be conducted *ex post*, so long as appropriate data exist for the non-participating sites. Its disadvantages are: 1. the estimated effect is highly dependent on the validity of the chosen instruments, and 2. appropriate instruments are often difficult to find. In cases where inappropriate instruments are used, the bias introduced by the two-step procedure can be worse than the bias it aimed to correct (Bound *et al.*, 1995).

Case Study: IFPRI–NCAP Evaluation of Indian Watershed Projects

As introduced above, watershed management in semi-arid areas of India focuses on augmenting water quantity through on-site moisture conservation, groundwater recharge, and surface water harvesting. The essential component of watershed development that distinguishes it from other rural development approaches is treatment of the upper watershed, which reduces soil erosion and leads to increased infiltration of rainwater, reduced siltation of downstream ponds, and groundwater recharge. These improvements in water management lead in turn to increased irrigation, higher cropping intensity, higher yields, changes in cropping patterns, development of the local dairy industry, and higher employment, which are the ultimate objectives of watershed development.

The IFPRI–NCAP evaluation studied all these objectives and they are discussed in Kerr *et al.* (2002). Due to space limitations, this chapter focuses on the efforts to reduce erosion and increase water infiltration in the upper catchment, since these are the project interventions most purely related to watershed development.

Despite the large budgets for watershed development in India, reliable evaluation studies were scarce when the IFPRI–NCAP watershed evaluation study was initiated in 1996. Some early studies indicated high adoption rates of soil and water conservation practices and favourable benefit–cost ratios (see, for example, the special section on watershed management in the *Indian Journal of Agricultural Economics*, 1991). However, these studies focused on heavily supervised projects with subsidies of 90–100% for adoption of prescribed packages. As such, the estimates of adoption rates were not meaningful. Also, the benefit–cost studies were conducted before actual outcomes could be known. They estimated net project benefits using yield impacts based on experimental data and assuming adoption and maintenance rates by farmers (e.g. Singh *et al.*, 1989). *Ex post*, however, some evidence suggested that many farmers abandoned watershed measures once the project subsidies ended (Kerr and Sanghi, 1992). Taken together, these factors suggested that many of the early, favourable evaluations were overly optimistic.

On the other hand, there was detailed documentation of a small number of highly successful projects that highlighted innovative social organisation arrangements or the influence of exceptional leadership in addition to technical interventions (e.g. Chopra *et al.*, 1990). Many NGOs gave reports of their own successful watershed development initiatives, and while there were undoubtedly many favourable projects, it is also likely that these reports focused mainly on the best cases and gave less attention to the problems they faced.

Study design

The village was selected as the unit of analysis since most Indian watershed projects operate at the village level and the people affected by the projects are organised in villages. The quantitative component was conducted as a 'with and without' design, covering four project categories (treatments) together with villages without projects. These included the following:

- Ministry of Agriculture (MoA): Projects under the National Watershed Development Project for Rainfed Areas (NWDPRA) that focused primarily on technical aspects of developing rainfed agriculture
- Jal Sandharan (JS): Engineering-oriented projects sponsored by the Government of Maharashtra, with funds from the Ministry of Rural Development (MoRD),[1] that focused on water harvesting through construction of percolation tanks, contour bunds, and other structures
- Non-governmental organisations (NGOs): Projects that typically placed greater emphasis on social organisation and less on technology relative to the government programmes. Those specific projects covered in this research were operated by BAIF, Social Centre, Don Bosco, Gramayan, and a few others

- Non-governmental Organisations–Government Organisations (NGO–GO): Collaborative projects between government and NGOs that sought to combine the technical approach of government projects with the NGOs' orientation toward social organisation. Specific projects covered in this research are the *Adarsh Gaon Yojana* (AGY) and Indo-German Watershed Development Programme (IGWDP)
- Control: Villages that had never had a watershed project.

All of these project categories are discussed in detail in Kerr *et al.* (2002).

To avoid choosing only conveniently located sites or success stories, researchers generated a stratified random sample from a census of villages where watershed projects were concentrated. Ultimately 70 villages, stratified by the five project categories, were sampled from a frame of over 600 villages in the rainfed areas of Pune and Ahmednagar districts in Maharashtra.[2] While it was important to randomly sample the sites to be studied, generating the census of watershed projects was particularly time-consuming because such information was not available from official records. The sample includes the following number of villages in each category: 10 MoA, 17 Jal Sandharan, 12 NGO, 14 NGO–GO collaboration, and 17 with no project. No village in the study area had more than one type of project operating in it at the same time. All but two of the project villages were previously treated under the earlier Comprehensive Watershed Development Project (COWDEP), sponsored by the Government of Maharashtra in the 1980s and early 1990s. On average, villages in each category had been treated under COWDEP for 7–9 years, and under the new generation of projects covered in this study for 5–6 years.

This study encountered many of the challenges cited earlier in this chapter, and its design reflects the constraints imposed upon the research team. Firstly, there was no baseline data on the performance criteria that were of interest to the evaluation team. As such, multiple indicators were used to assess project performance, some of which were based on respondents' perceptions. Respondents' recall was used for indicators that could be defined in terms of an easily observed, discrete change between one period and the next, such as adoption of new varieties, changes in infrastructure, and ownership of assets. Table 10.1 shows how performance criteria of interest were operationalised into indicators.[3]

Secondly, a lack of secondary data on the sites from the initial census precluded the use of propensity matching to construct control and treatment groups. Rather, the groups were stratified by project type and topography of the project site (hilly vs. flat). Stratifying by project type was necessary because some project categories are much more common than others. Villages in all project categories had to have a similar topographic range because topography strongly influences water harvesting potential and thus the likelihood of project success.

Thirdly, the project sites were not originally assigned through a random process, so sample selection bias was an issue. Site-selection criteria differed by project type and this is discussed further below. Given these constraints an instrumental variables approach was selected.

Table 10.1 Ideal and operationalised indicators of performance.

Performance criteria	Ideal indicators[a]	Operational indicators used in this study[b]
Soil erosion	Measurement of erosion and associated yield loss	Visual assessment of rill and gully erosion (current only)
Measures taken to arrest erosion	Inventory, adoption and effectiveness of SWC[c] practices	Visual assessment of SWC investments and apparent effectiveness (current only) Adoption of conservation-oriented agronomic practices Expenditure on SWC investments
Groundwater recharge	Measurement of groundwater levels, controlling for aquifer characteristics, climate variation and pumping volume	Approximate change in number of wells Approximate number of wells recharged or defunct Change in irrigated area Change in number of seasons irrigated for a sample of plots Change in village-level drinking water adequacy
Soil moisture retention	Times series, intra-year and inter-year variations in soil moisture, controlling for climate variation	Change in cropping patterns Change in cropping intensity on rainfed plots Relative change in yields (higher, same or lower)
Agricultural profits	Net returns at the plot level	Net returns at the plot level, current year only
Productivity of non-arable lands	Change in production from revenue and forest lands (actual quantities) Wildlife habitat	Relative change in production from revenue and forest lands (more, same or less than pre-project) Extent of erosion and SWC on non-arable lands Change in wildlife and migratory bird populations
Household welfare	Change in household income and wealth Nutritional status	Perceived effects of the project on the household Perceived change in living standard (better, same, worse) Change in housing quality Change in percentage of families migrating Perceived changes in real wage and availability of casual employment opportunities (higher, same, lower)

[a] All ideal indicators would be collected both before and after the project.
[b] Only a few of the indicators shown here are presented in this chapter due to lack of space. This chapter focuses only on village-level indicators related to management of the upper catchment area.
[c] SWC = soil and water conservation.

Evaluation: an instrumental variables approach

To evaluate the performance of the watershed projects, the following instrumental variables model (Greene, 2000; Baker, 2000) represents the analytical framework for this study:

$$S = a + bV + cZ + e_1 \quad (1)$$

$$Y = f + g\hat{S} + hV + e_2 \quad (2)$$

where:
S = a categorical variable indicating one of five project categories
$\hat{S}$ = the predicted probability that the village falls in each project category
Y = a vector of performance indicators (project outcomes)
V = a vector of village-level explanatory variables affecting both S and Y
Z = a vector of variables affecting S but not Y
e_1 and e_2 = error terms

The instrumental variables approach is used because S – participation in a particular type of project – is endogenous and data limitations prevent the use of a double difference design. Experimental and quasi-experimental design approaches to overcome endogeneity were not possible due to the lack of baseline data for the watershed project. In short, if each project category has different criteria for selecting villages, then it is possible that differences in performance can result more from differences in initial, pre-project conditions than from the work undertaken by the watershed project. Differences in selection criteria for each project may be based on both observed and unobserved village characteristics. Unobserved differences cause standard econometric approaches to yield biased coefficients (Baker, 2000).

Variables in the analysis

As indicated above, the model has two stages. The first is a multinomial logit that estimates the likelihood that a village falls into a particular project category. This analysis yields predicted probabilities that replace project category in the second stage, in which performance indicators are regressed on factors hypothesised to influence them.

DEPENDENT VARIABLES. In the multinomial logit (first stage) the dependent variable represents the five project categories. Second stage equations analyse the determinants of performance indicators including: drainage line conditions, erosion on uncultivated land, and changes in access to products from common lands. Due to space limitations, these variables represent only a sample of the performance indicators used in the study and are presented in Table 10.1; others are analysed in Kerr *et al.* (2002).

The erosion status of permanently uncultivated land in each village's upper catchment area provides a good indicator of its condition, and thus of watershed project performance. Most of this land is under common property.

Watershed projects invest in soil and water conservation and afforestation in these areas to regulate the flow of runoff water, increase infiltation, and prevent siltation of water-harvesting ponds further down the slope. Trained soil surveyors transacted the upper catchment and estimated erosion scores. Each segment of the transect was assigned a score from 1 to 3, where 1 = low erosion, 2 = medium erosion, and 3 = high erosion. Resources were not available for more sophisticated measurement of erosion. A 'segment' is an area in which there is no change in slope, soil type, erosion status, or tenure status. Any time one of these factors changes, a new segment begins. Each observation is weighted by length of the segment; obviously there are multiple observations per village. An ordered probit regression is used for the analysis.

Condition of a village's drainage line is another good indicator of watershed project performance as it measures the extent of erosion in the upper catchment. If uncontrolled runoff causes erosion in the upper catchment, the drainage line will itself become eroded and uncontrolled and this will cause siltation of water-harvesting structures. The dependent variable in this regression is a score between 1 and 3, where 3 = good condition and 1 = poor, based on visual assessments by trained surveyors. Investigators transected the entire drainage line and gave a score for every 100-m segment. The score for the entire drainage line is the mean value of all the segments. The score takes a continuous value because it is the average of multiple observations of the drainage line within each village; each village has only one observation. Details of the scoring system and its strengths and weaknesses are provided in Kerr *et al.* (2002). Multiple regressions are used for this model because the village-level scores take continuous values.

One step in a watershed project is to impose restrictions on access to common lands in the upper catchment to allow the vegetation to regenerate. Such restrictions can have distributive impacts on people living in a watershed. As such, changes in access to products from common lands between 1987 and 1997 are modelled econometrically as the third performance indicator in this study. Grass fodder, tree fodder and fuel were the most common products collected from the commons; only grass fodder is discussed here to save space. An ordered probit model is used for econometric analysis of the determinants of whether people in a village have: 1. less, 2. the same amount, or 3. more access to grass fodder from the government revenue lands.

For all the models, the survey regression commands in Stata Version 7 are used to account for stratification, sampling weights and clustering and render all estimates robust to heteroskedasticity.

EXPLANATORY VARIABLES. For the analysis of determinants of project category, the factors determining a village's inclusion in a given project represent conditions prevailing in 1987, before the projects began. They can be categorised into agroecological factors, infrastructure conditions, and socio-economic characteristics. Beginning with agroecological factors, altitude range (the difference between the highest and lowest points, in metres) is important since many projects focus on areas with high potential for water harvesting. Hillier areas have greater water-harvesting potential. Area of the village is also included.

Percentage of cultivated area that was irrigated before the project has agroecological implications but is a form of infrastructure. Other infrastructure variables (measured in the pre-project period) include the distance to the nearest bus stop, whether or not the village is connected by a paved road, the distance to the *taluka* (sub-district) headquarters, adequacy of drinking water availability, distance to the nearest public health centre, distance to the market for agricultural inputs, and the population density in 1990, which is positively correlated with many indicators of infrastructure development. Other infrastructure variables are omitted due to their high correlation with those that are included. One such variable is previous watershed investment under COWDEP (discussed above), since all but two villages with current watershed projects were previously under COWDEP. Multinomial logit estimation is infeasible using this variable since it almost perfectly predicts the presence of a current project.

Explanatory variables representing social conditions and social institutions include whether the village practised voluntary community labour (*shramdan*), the number of communal groups, the percentage of inhabitants from scheduled castes and tribes, the approximate percentage of households with at least one seasonal labour migrant, and whether the village contained government revenue land (common land).

The regressions on performance indicators contain some of the same variables used in the project placement regression along with some others. For condition of uncultivated land, private land tends to be better managed than common land in India and this is accounted for by including a dummy variable for tenure status. Villages with a higher percentage of shepherds and with more communal diversity are expected to face greater challenges in protecting common lands. Villages with a higher percentage of people with off-farm income are expected to have less enthusiasm for managing common lands since it is less important to their livelihoods.

Project categories are represented by their predicted values from the multinomial logit in the instrumental variables approach used in this analysis. For each village, the probabilities for the five project categories (including no project) sum to one. The amount that the project (including the earlier COWDEP project) spent per hectare in the village represents the extent of project effort. This information incorporates the total number of years the watershed projects have worked in the village since it is based on annual budgets and extent of area covered. This expenditure is interacted with the predicted probability for each project category.[4]

Results of the econometric model

This section presents the econometric findings and relates them to respondents' perceptions of the distribution of project benefits.

Determinants of project category

Results of the analysis are shown in Table 10.2. Non-project villages are the base category in the multinomial logit, so estimated parameters are in relation to the non-project category. A positive, significant coefficient for a given project category indicates that the villages selected for that project have a significantly greater value for that variable than the non-project category. For example, all projects have a greater range in altitude between the highest and lowest point in the village, than non-project villages, and this difference is significant for all except the NGO–GO collaborative projects. This is to be expected since hilly areas are most suited to water harvesting. It should be noted that while the altitude range appears to be greater for the MoA category than the NGO and NGO–GO categories, the analysis does not make clear whether this difference is significant. To determine whether project categories differ significantly from each other (as opposed to those from the non-project category), the analysis would have to be done repeatedly, each time with a different project category as the base against which others are compared.

NGO project villages are the only category with a higher percentage of irrigated area than non-project villages in the pre-project period; the reason for this difference is not known. NGO–GO villages are significantly larger in area; the reason for this difference is not clear. MoA villages were likely to be more densely populated and other villages less densely populated than non-project villages, but this difference is significant only for the MoA villages. This is consistent with the published guidelines of the MoA's National Watershed Development Project for Rainfed Areas (NWDPRA), which calls for working in more accessible, visible villages (Government of India, 1992). It probably reflects a non-random selection process. MoA villages are also closer to public health clinics and markets, though only the former is significant. NGO villages, on the other hand, were significantly likely to be located further from markets and *taluka* headquarters, and NGO–GO villages were significantly further from the nearest public health office. Only JS villages were less likely to have an adequate supply of drinking water, consistent with the JS project's mandate, but the difference was not statistically significant.

The villages under NGO–GO collaborative projects were significantly more likely to practise *shramdan* in 1987. MoA villages were actually significantly less likely to practise *shramdan* than non-project villages; the reasons for this finding are not known. MoA, JS, and NGO villages all had more communal diversity and more people of scheduled castes and tribes and backward classes than non-project villages, and the latter is consistent with published guidelines. NGO–GO collaborative project villages, on the other hand, had no significant differences from non-project villages in communal diversity and scheduled castes and tribes. If the analysis is conducted using NGO–GO projects as the base category (not shown) the communal diversity and population of scheduled castes and tribes are significantly lower than other project categories. The Indo-German Watershed Project, an NGO–GO collaborative project, requires consensus-based decision making, which may be easier with communal homogeneity, and the two projects require a ban on

Table 10.2 Determinants of project category in Maharashtra,[a] multinomial logit regressions (standard errors in parentheses).

	Project category			
Variable	Ministry of Agriculture (MoA)	Jal Sandharan (JS)	NGO	NGO–GO collaboration
Altitude range ('00 metres)	3.34 (1.02)***	1.93 (1.00)*	2.44 (1.06)**	2.16 (1.34)
Area of the village ('00 ha)	0.17 (0.13)	1.29 (1.34)	0.09 (0.13)	0.30 (0.13)**
Area irrigated in 1987 (%)	2.90 (3.28)	–2.39 (5.76)	8.29 (3.55)**	1.94 (4.52)
Population density in 1990 ('00 persons/km^2)	3.71 (0.82)***	0.88 (1.76)	–1.81(–1.43)	–0.59 (0.88)
Distance to nearest public health centre, 1987 (km)	–0.38 (0.15)**	0.17 (0.15)	0.18 (0.15)	0.33 (0.14)**
Distance to market for agricultural inputs in 1987 (km)	–0.15 (0.11)	0.23 (0.15)	0.34 (0.16)**	0.10 (0.13)
Distance to *taluka* headquarters (km)	0.21 (0.05)***	0.01 (0.05)	0.35 (0.43)	–0.03 (0.04)
Distance to nearest bus stop in 1987 (km)	0.83 (0.34)**	–0.16 (0.27)	0.16 (0.32)	–0.34 (0.29)
Paved road in 1987 (dummy)	0.29 (1.27)	–1.58 (1.63)	0.41 (1.11)	–2.49 (1.53)
Whether the village had sufficient drinking water in 1987 (dummy)	3.31 (1.38)**	–1.35 (1.27)	0.26 (1.54)	0.93 (1.49)
Village practised community voluntary labour (*shramdan*) in 1987 (dummy)	–2.01 (1.10)*	–1.31 (1.51)	1.57 (1.57)	8.42 (2.35)***
Number of communal groups in the village	1.18 (0.25)***	0.76 (0.29)**	0.85 (0.30)***	0.13 (0.35)
Inhabitants of SC, ST, BC (%)	0.047(0.025)*	0.08 (0.03)***	0.12 (0.03)***	–0.03 (0.06)
Households with at least one seasonal migrant, 1987 (approximate %)	–0.10 (0.03)***	–0.06 (0.04)	–0.10 (0.06)	0.09 (0.03)***
Whether the village contained government revenue land, 1987	–0.32 (1.16)	–2.10 (1.22)*	–4.96 (1.17)***	–1.16 (0.88)

[a] Reference category is no project; variables reflect values in the pre-project period. 70 observations. Model is not corrected for choice-based sampling, i.e. that the sample is stratified on the dependent variable. Coefficients and standard errors are adjusted to account for sampling weights, stratification and finite population size. *, **, and *** indicate statistical significance at the 10%, 5% and 1% level, respectively. F (46,15) = 41.3.

open grazing and tree cutting, which may be more difficult for poor, landless or near-landless people to accept because they rely on products from the commons for their livelihoods. NGO and JS villages were significantly less likely to contain government revenue land, possibly suggesting that these projects sought to reduce the potential for equity trade-offs. MoA villages were likely to have fewer households with at least one seasonal labour migrant, whereas NGO–GO villages were likely to have more households with at least one seasonal migrant. This may indicate better economic conditions in the MoA villages and worse in the NGO–GO villages, but on the other hand it could just indicate differences in propensity to migrate. MoA and NGO–GO villages were also less likely to contain government revenue land, but the difference is not statistically significant.

Erosion of uncultivated land

The first column of results in Table 10.3 shows the determinants of erosion status of uncultivated lands in the villages' upper catchments. The NGO–GO collaborative projects appear to have had the greatest, most highly significant impact on reducing erosion, followed closely by the NGO projects. For every thousand rupees spent per ha by the NGO–GO projects, the erosion status score fell by 0.45 on a scale between 1 and 3 (a negative number indicates less erosion); for NGOs the fall was 0.35. The JS projects had a smaller but statistically significant effect, with the erosion status score falling by 0.20. The MoA projects had an equally small but statistically insignificant effect. In short, three of the project categories appear to have had an effect on erosion in the uncultivated upper catchment, and it is by far the strongest in those projects that devoted greater attention to social organisation.

Private land is much less likely to be eroded than common land, as expected. Land in villages with a lower population density also has less erosion. The reason for this is unclear; it could be because there are not so many people to overuse the land, or it could be that areas with lower population density also have less-diversified economies and thus more people with an interest in taking care of the uncultivated land.

Condition of the drainage line

The second column of Table 10.3 shows the regression results for the determinants of condition of the drainage line. The results of this model are very similar to those for erosion status of uncultivated land. The model is highly significant, but with an R^2 value of 0.30 it does not explain a large extent of total variation. All the project expenditure variables are positive, and all are statistically significant except the JS, which is nearly significant. This suggests that at least three of the projects were successful in improving the condition of the drainage line. The NGO–GO and NGO categories have much higher coefficients than the other categories as well as higher statistical significance, so these projects appear to have performed the best. For every thousand rupees spent per ha by the NGO–GO and NGO projects, the drainage line score rose by 0.23 on a scale between 1 and 3; for MoA and JS projects it was less than 0.10.

Table 10.3 Regression results (performance indicators).

Variable	Coefficients (standard errors in parentheses)[a]		
Dependent variable	Erosion on uncultivated lands[b]	Drainage line condition[c]	Access to grass fodder[d]
Mean expenditure per ha in MoA village ('000 Rs)	–0.20 (0.14)	0.08 (0.04)**	0.06 (0.60)
Mean expenditure per ha in JS village ('000 Rs)	–0.20 (0.07)***	0.07 (0.05)	–0.89 (0.31)***
Mean expenditure per ha in NGO village ('000 Rs)	–0.35 (0.17)**	0.23 (0.08)***	1.35 (2.29)
Mean expenditure per ha in NGO–GO village ('000 Rs)	–0.45 (0.13)***	0.23 (0.05)***	–2.04 (0.38)***
Availability of grass fodder in 1987	N.A.[e]	N.A.	2.09 (0.61)***
Whether the village contains common land (dummy)	N.A.	0.38 (0.13)***	N.A.
Altitude range ('000 m)	0.33 (1.40)	–5.89 (9.19)	3.71 (0.96)***
Distance to nearest bus stop in 1987 (km)	–0.02 (0.05)	0.03 (0.04)	0.53 (0.19)***
Paved road in 1987 (dummy)	0.31 (0.33)	0.16 (0.12)	0.92 (0.66)
Population density in 1990 (1000s/100 persons/km^2)	–0.66 (0.21)***	0.05 (0.14)	–1.06 (0.51)**
Distance to *taluka* headquarters (km)	0.04 (0.09)	–0.06 (0.06)	0.33 (0.33)
Inhabitants working primarily in non-agricultural sector (%)	0.008 (0.017)	–0.06 (0.06)	0.10 (0.04)**
Inhabitants working primarily as shepherds (%)	0.04 (0.05)	–0.06 (0.04)	0.62 (0.17)***
Whether land is operated privately (dummy)	–0.57 (0.36)*	n.a.	n.a.

[a] Coefficients and standard errors are adjusted to account for sampling weights, stratification and finite population size. ***, **, and * indicate statistical significance at the 10%, 5% and 1% level, respectively. Predicted values based on the multinomial logit regression in Table 10.2 are used for the project category variables. Standard errors are not adjusted for use of predicted values in complex, two-stage regressions; the author is not aware of analytical formulas to do so. Bootstrapping (a method of checking the reliability of data by repeatedly analysing sub-samples of the data) was not viable given the small number of observations per stratum.

[b] OLS regression; possible transect scores range from 1 to 3, including fractional values. 64 observations (6 villages have no main drainage line). $F(12,43)=4.81$ ($P>0.0001$); $R^2 = 0.30$.

[c] Ordered probit regression; possible transect scores range from 1 to 3 where 1=less, 2=same, 3=more, with no fractional values. 174 observations from 64 villages (6 villages have no uncultivated land). $F(13,42)=3.45$, $P>0.002$.

[d] Ordered probit regression. Possible scores range from 1 to 3, where 1=less, 2=same, 3=more. 40 observations (30 villages have no common land). $F(13,19) = 6.88$, $P> 0.01$.

[e] Not available.

The dummy variable indicating the presence of common land is positive and statistically significant, which was unexpected. Other variables have the expected sign but are insignificant.

Institutions for restricting access to common lands

NGO and NGO–GO projects aim to create conditions of controlled, equitable access to common property resources. Government-sponsored projects, as discussed above, paid relatively little attention to social institutions under the pre-1995 guidelines. For all of these projects, there was some risk that protection of common lands would come at the expense of the poorest people who depended on them the most. In any case, managing common lands was challenging because most villages lacked good institutional arrangements for doing so.

Examination of where projects chose to operate suggests that some of them aimed to avoid equity trade-offs by working in villages that had no common land. Among the study villages, only 40 out of 70 (57%) contained common land, including only 33% of those under NGO projects, 60% under MoA and JS projects, and 57% under the NGO–GO projects. By contrast, 71% of control villages contained common land.

The most common institutions for restricting access were bans on grazing and cutting trees. A traditional penalty against illicit grazing is to impound the grazing animals in the *panchayat* (village government) office and release them only upon payment of a fine.

Investigators collected data on access restrictions and their enforcement. In both 1987 and 1997, banning grazing on the commons was the exception, not the rule. Only 5 out of 40 villages with common land (12.5%) had banned grazing before the projects, rising to 35% afterwards. The numbers of people who received imposed punishments for illicit grazing were even lower, with 5% in 1987 and 22% in 1997. Two findings are particularly interesting. Firstly, even some of the non-project villages imposed grazing bans, showing that this step does not necessarily require a watershed project. Secondly, while none of the NGO–GO villages had imposed bans or penalties in 1987, by 1997 four out of eight (50%) of them had done so compared to no more than 25% for other project categories. Only in the NGO–GO category did a significantly higher percentage of villages impose access restrictions than those under the non-project category. No regression analysis was performed on the determinants of banning grazing and tree cutting, because so few villages actually imposed these restrictions.

Change in access to fuel and fodder from the common lands

Table 10.3 (last column) suggests that the projects have led to a reduction in access to grass fodder from common lands compared to non-project villages. The variables for expenditure per hectare in the NGO–GO and JS project categories have negative, statistically significant signs; the NGO–GO coefficient also has a much higher magnitude. Other categories are insignificant, including the large positive coefficient for NGOs that is based on a limited number of villages as mentioned above.

This finding is consistent with those presented above showing that the NGO–GO projects were particularly successful in restricting access to common lands and reducing erosion in the drainage line and pasture lands. Improving the condition of these lands requires restricting access to them, and Table 10.3 suggests that access was in fact restricted. Several other variables are also significant. Population density has a negative sign, while the variables with positive signs include availability of grass fodder in 1987, altitude range, distance to the nearest bus stop in 1987, percentage of households working primarily outside of agriculture, and percentage of households working primarily as shepherds. The highly significant, strongly positive coefficient for shepherds is consistent with the finding that it was more difficult to manage the drainage line in the villages with the most shepherds, presumably because access restrictions in the upper catchment were difficult to enforce. The positive sign for altitude range may reflect high rainfall, which is omitted because it is highly correlated with altitude range. High rainfall stimulates rapid growth of natural vegetation, so it may be that access restrictions can be less strict in these villages. The negative sign for population density either means that availability of fodder declined due to population pressure, or that more-densely populated villages were more likely to impose access restrictions. The positive sign for the percentage of households working outside of agriculture means either that this caused less competition for fodder, or that there was less pressure to impose restrictions.

Discussion: Productivity, Conservation and Equity Impacts

As mentioned above, this chapter presents only a small set of the overall findings from the entire watershed evaluation study. Although data related to increases in irrigation, cropping intensity, cropping patterns, yields and employment are not presented here, a few comments can be made about the findings. Changes in irrigated area in each village between the pre- and post-project periods were recorded from official data, and differences across project categories were insignificant. In any case this information offers limited insight on project impact because of a lack of data on such key determinants as hydrological characteristics of the aquifer. Given the lack of useful data, respondents were asked their perceptions of the effectiveness of water-harvesting investments in each project category. Respondents in the NGO–GO category most frequently reported that water harvesting was effective, followed by the NGO projects, the JS projects, and finally the MoA projects.

Data on farmers' net returns to cultivation on rainfed land were also analysed. Instrumental variables regression analysis showed that farmers under NGO projects were likely to have higher net returns than those under other projects, followed by NGO–GO and MoA projects and then JS projects. Only the NGO project category was significantly different from the category of villages with no project. It was difficult to determine the reason behind this result, but it may be because NGO projects often helped farmers obtain higher market prices for their crops and put them in touch with extension officers and input sales people.

While investigation of such factors as increased irrigation and net crop returns suffered from difficulties in isolating the causal relationships, they are consistent with the findings presented in this chapter that show superior performance by the NGO and NGO–GO projects. As introduced earlier, the distinguishing feature of these projects is their greater emphasis on social organisation relative to purely technical interventions. The importance of such social organisation makes intuitive sense given the uneven distribution of benefits and costs in watershed management as described above.

On the other hand, the data on treatment of the upper catchment and accessibility of fodder production suggest a possible trade-off between productivity and conservation objectives, and equity or poverty alleviation objectives. In particular, the NGO–GO projects were most successful in rehabilitating and protecting the upper catchment, but also led to reduced access to fodder from the common lands in this area. This inverse relationship is not surprising, because improving the upper catchment typically involves reducing access to this area. The NGO projects avoided this problem by working in villages without common land.

Findings from qualitative investigations provide additional insight into this issue. In particular, women and livestock herders in many project villages complained that they had suffered from loss of access to common lands. Herders indicated that wage employment offered under all projects was insufficient to compensate for lost access to these lands.

Landless, low-caste people were a small minority in most villages and could not influence the decision to close the common lands, which was usually based on a majority-rule vote. Some NGO–GO or NGO projects required a consensus among villages to initiate a project, but landless people explained in interviews that they could not reasonably stand up to the will of a more powerful majority.

Herders in some villages complained that despite promises that access restrictions would be temporary while vegetation was allowed to regenerate, common lands remained off-limits even after successful regeneration. In fact, success in achieving productivity and environmental objectives raises the risk of such inequity. Elsewhere, herders protected their own livelihoods but undermined project objectives by ignoring grazing restrictions. These findings from qualitative discussions are consistent with the result in Table 10.3 that a high population of shepherds raised the extent of erosion but also raised access to grass fodder, compared to other villages.

A survey of 349 respondents in 13 of the study villages supported these findings. Respondents' landholding size was positively associated with the perception that projects benefited them, and negatively associated with the perception that projects harmed them. Among landless people the unanimous complaint was lost access to common lands.

Landless people in NGO project villages reported more favourable impressions of the project than did people from other project villages, but, as reported above, most NGO project villages had no common land. Limiting watershed development to villages without common land would exclude a large number of villages, but it may be an intelligent approach to minimise equity tradeoffs.

Some NGO and NGO–GO projects in areas with common land developed innovative solutions to the problem of uneven distribution of project costs and benefits between the upper and lower watershed. They tried to build the interests of different groups into the project design at the outset, for example by granting landless people fishing rights in runoff ponds. Landless and near-landless respondents in these villages unanimously reported having benefited from watershed projects; they had an incentive to protect the upper catchment. Many variations on this approach are possible and could help spread the benefits of watershed development and thus increase its chances of success.

Landless people may suffer in the short term from watershed development, but they may benefit in the long term if compensation mechanisms help them to survive the short term. For example, after 4 years of watershed management in one NGO–GO project village, labourers indicated that they could find 8 months of employment whereas previously they could only find 3 months of employment. Respondents in this study were asked whether they obtained more, less, or the same number of employment days than before the project period. Those in the NGO–GO and NGO project villages indicated with much greater frequency that employment opportunities had risen, whereas those under the JS, the MoA, and in non-project villages indicated more frequently that employment had declined.

Conclusions

To summarise, the key finding of the evaluation was that more-participatory projects were more successful in protecting upper catchments to promote water harvesting. On the other hand, too often protection of upper catchments came at the expense of landless people whose livelihoods relied heavily on them. Long-term sustainability of project outcomes is questionable under these inequitable conditions because the landless have every incentive to try to thwart access restrictions to common lands. A few projects have taken innovative steps to build landless people's interests directly into efforts to protect the commons, and others need to experiment with similar arrangements.

A broader question concerns whether watershed management projects' economic and environmental impacts justify the investments made. This study is not able to answer that question definitively due to the inability to identify causal links between watershed project interventions and increases in irrigation and returns to cultivation. This shortcoming results from the broad array of confounding factors that determine irrigated area and returns to cultivation, the lack of baseline data available for this study, which made a double difference analysis impossible, and the inadequacy of data related to groundwater. Lack of hydrological data also limits the ability to examine inter-village watershed externalities, although they are likely to be small in the study area.

The instrumental variables method used in this study provides a practical approach to evaluation in a situation of inadequate data where a quasi-experimental design is infeasible. However, as mentioned above, the estimates in such an approach are only as good as the instruments available for analysis.

The lack of information that hampered this research has other, more serious implications. In particular, it means that government planners lack sufficient data to draw firm conclusions about the returns to different kinds of watershed development investments. Given the vast size of the budget for watershed projects, better information about their performance would go a long way toward more cost-effective government planning. Currently too many funds are allocated on the basis of too little information, with too much potential for waste.

The data shortage takes two forms: firstly, a lack of baseline data against which to compare current conditions, and secondly, a lack of monitoring data for easy assessment of current conditions and comparison across projects. Most projects collect a small amount of baseline data, but not for the purpose of evaluation. It is difficult to obtain and, in any case, inadequate for the task. Furthermore, no effort is made to collect comparable baseline and monitoring data across different project categories. Of course this makes systematic comparison of projects difficult, and it makes a double difference analysis impossible for most indicators of project success.

Given the major place of watershed management in India's efforts to promote agricultural development in rainfed areas, it would make sense to establish an interministerial commission responsible for monitoring and evaluation. A common data collection protocol could be used for projects under all the different categories.

Endnotes

[1] This study did not include villages under the new guidelines of the Ministry of Rural Development, which called for more attention to social organisation. The projects were just getting underway at the time of the data collection for this study, so it was too soon to include them.

[2] Watersheds fall within village boundaries in all project categories except the Ministry of Agriculture, in which a watershed covers a few villages.

[3] Although this chapter reports only village-level data, information was also gathered at the individual and group levels. Individual data were gathered through individual interviews, group data were collected through focus groups and group interviews, and village-level data were gathered through key informant interviews with repeated cross-checking to ensure accuracy.

[4] The total amount spent per hectare under the watershed project to date is multiplied by the probability that a given project worked in that village. This value can be positive only for the project that worked in that village, since expenditure was zero under all the other projects.

References

Baker, J. (2000) *Evaluating the Impact of Development Projects on Poverty: A Handbook for Practitioners*. World Bank, Washington, DC, 217 pp.

Bound, J., Jaeger, D.A. and Baker, R.M. (1995) Problems with instrumental variables estimation when the correlation between the instruments and the endogenous variable is weak. *Journal of the American Statistical Association* 90(430), 443–450.

Campbell, D.T. and Russo, M.J. (1999) *Social Experimentation*. Sage Publications, Thousand Oaks, California, 407 pp.

Chopra, K., Kadekodi, G. and Murty, M.N. (1990) *Participatory Development, People and Common Property Resources*. Sage, New Delhi, India.

Farrington, J., Turton, C. and James, A.J. (eds) (1999) *Participatory Watershed Development: Challenges for the Twenty-First Century*. Oxford University Press, New Delhi, India, 382 pp.

Government of India (1992) WARASA: *National Watershed Development Project for Rainfed Areas (NWDPRA) Guidelines*, 3rd edn. Ministry of Agriculture (MoA), New Delhi, India.

Greene, W. (2000) *Econometric Analysis*, 4th edn. Prentice Hall, Englewood Cliffs, New Jersey, 1004 pp.

Indian Journal of Agricultural Economics (1991) Subject 1. Watershed Development. XLVI, 241–327.

Jyotsna Jalan and Martin Ravallion (2003) Estimating the benefit incidence of an anti-poverty program by propensity-score matching. *Journal of Business and Economics Statistics* 21(1), 19–30.

Kerr, J. and Sanghi, N.K. (1992) Indigenous soil and water conservation in India's semi-arid tropics. *Gatekeeper Series no. 34.* Sustainable Agriculture Programme, International Institute of Environment and Development (IEED), London, UK, 30 pp.

Kerr, J., in collaboration with Pangare, G. and Lokur, V. (2002) Watershed development projects in India. *Research Report 127.* International Food Policy Research Institute (IFPRI), Washington, DC, 90 pp.

Lal, R. (ed.) (2000) *Integrated Watershed Management in the Global Ecosystem*. CRC Press, Boca Raton, Florida, 395 pp.

Manski, C. (1995) *Identification Problems in the Social Sciences*. Harvard University Press, Cambridge, UK, 172 pp.

Singh, R.P., Vittal, K.P.R., Das, S.K. and Sanghi, N.K. (1989) Watershed technology stabilises yields in Andhra Pradesh. *Indian Farming* 39(9).

11 Assessing Economic and Environmental Impacts of NRM Technologies: An Empirical Application Using the Economic Surplus Approach

M.C.S. Bantilan[1], K.V. Anupama[1] and P.K. Joshi[2]

[1] *International Crops Research Institute for the Semi-Arid Tropics (ICRISAT), Patancheru, Andhra Pradesh, India*

[2] *International Food Policy Research Institute (IFPRI), New Delhi, India*

Introduction

This chapter addresses methodological and empirical complexities in assessing the impact of crop and resource management research through a concrete case study. It develops an applied economic surplus analysis of welfare gains, using farm survey data to measure farmer benefits from increased yields, reduced unit costs, and higher income. The environmental aspects of natural resource management (NRM) research impacts present special challenges in measurement across time and space. Farmers' perceptions of long-term environmental changes are highlighted as a means to augment or substitute for narrower quantitative indicators.

The case of groundnut production technology (GNPT) in central India illustrates the methodological and empirical issues in estimating research payoffs to NRM research investments. The GNPT was developed for the semi-arid tropics (SAT), a region usually characterised by water scarcity, low soil fertility and land degradation. Impact analysis of GNPT presents estimated costs and benefits using the principle of economic surplus and complements this with a detailed account of both quantitative and qualitative information provided by scientists and experts, including farmers.

Groundnut production technology (GNPT)

The research and development team that developed the GNPT package aimed to raise groundnut production by generating research information on various groundnut crop production components and integrating them into a 'package' of technology options. The technology package that was developed in 1986 integrates crop and resource management options detailed in Table 11.1. These components can be divided into five broad categories: land, nutrient, water, insect and pest management, and improved varieties.

Table 11.1. Technology components of the groundnut production technology (GNPT).

Component		Improved package (GNPT)	Local practice
C1	Land management Seedbed	Raised bed and furrow (RBF)	Flat
C2	Nutrient management		
	Farmyard manure	5–12 t/ha	10 t/ha
	Ammonium sulphate	100 kg/ha	Diammonium phosphate: 100 kg/ha
	Single superphosphate	300–400 kg/ha	Murate of potash: 100 kg/ha
	Zinc sulphate	10–20 kg/ha every 3 years	20 kg/ha every year
	Ferrous sulphate	2–3 kg/ha	–
	Gypsum	400 kg/ha	200 kg/ha
C3	Water management	Furrow or sprinkler to improve efficiency of water use	Flood
C4	Disease and pest management (effective control of insects, diseases and weeds, seed dressing/treatment)	Bavistin, dimethoate, monocrotophos	Need based
	Seed dressing	Thiram, Bavistin or Dithane M 45	Thiram
C5	Seed		
	Improved variety	ICRISAT varieties	Local varieties
	Seeding rate	125–150 kg/ha	120–125 kg/ha

During 1987–1991, International Crops Research Institute for the Semi-Arid Tropics (ICRISAT), through its Legumes On-Farm Testing and Nursery (LEGOFTEN) Unit, was an active partner with the Indian Ministry of Agriculture and other agencies in identifying and demonstrating appropriate technology options for increased groundnut production. The team reviewed all available and relevant research information and carefully identified production constraints in the major oilseed-producing regions of India. This package was thoroughly discussed with the national agricultural research service (NARS) and State Departments of Agriculture. This collaboration in a technology exchange programme provided ICRISAT with an opportunity to confirm the suitability and viability of the GNPT concept in farmers' fields. Although some components of the package (i.e. improved varieties, fertilisers, seed dressing) were already being used by farmers, ICRISAT's value addition took the form of information on appropriate timing and dosage rates of inputs.

The two new essential innovations introduced were land and water management. The land management component of the GNPT entails preparation of raised-beds and furrows (RBF) for groundnut production (Fig. 11.1). Compared to the practice of traditional farmers, who used 1–2 harrowings to sow groundnut on flat land, the RBF technologies were designed to reduce soil erosion, provide surface drainage, concentrate organic matter and fertiliser application, and reduce soil compaction around plants. Over a period of time, the concept of RBF was modified to suit the requirements of the farmers into a narrow-bed and furrow configuration, i.e. a bed of 75 cm, with ridge and furrow systems. The water management component was introduced to improve water use efficiency through furrow and sprinkler irrigation.

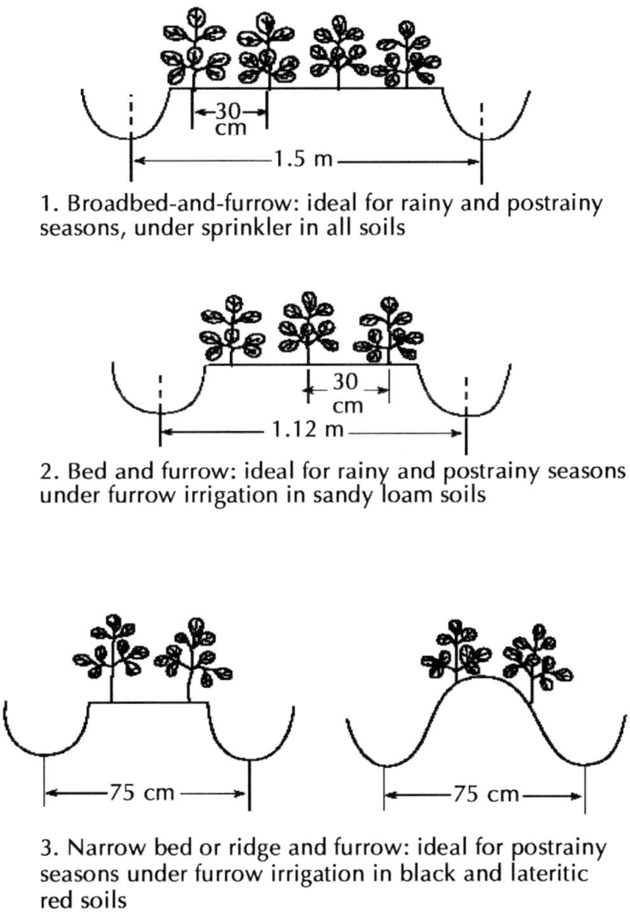

Fig. 11.1. The raised-bed and furrow (RBF) method of groundnut cultivation.

Groundnut Production and Markets in India: Background

India is the world's second largest producer of groundnut after China. Groundnut is one of the most important food and income-generating oilseed cash crops for smallholder farmers in semi-arid India. About 80% of the groundnut crop is rainfed, and is grown in southern, western, and parts of central India during the southwest monsoon. The remaining 20% is irrigated. Groundnut is mostly cultivated in red sandy soils (Alfisols) in many states, but it is also grown in shallow to medium-deep black soils in some parts of the country.

Groundnut yields in India vary widely depending on the production system (Freeman *et al.*, 1999). Rainfed groundnut yields roughly 0.9 t/ha, while the irrigated crop yields about 1.6 t/ha. Important improved groundnut cultivars include TMV 2, SB 11, CG 2, JL 24 and J 11, although these have never completely replaced the local cultivars. After the introduction of GNPT by LEGOFTEN, the area under groundnut production in India increased from 6.84 million ha in 1987/88 to 8.67 million ha in 1991/92 and groundnut production increased from 5.88 million t in 1987/88 to 7.07 million t in 1991/92. Rainy-season groundnut yields increased from 700–1000 kg/ha to 1.5 t/ha; postrainy season-yields rose from 2 to 4 t/ha, and summer yields rose from 1 to 3 t/ha after the introduction of GNPT.

Groundnut demand increases were driven by population growth, although the increase was moderated by rising prices. About 80% of Indian groundnuts are crushed for oil, and groundnut remains the vegetable oil of preference; but its share in the vegetable oil market is declining as consumers shift to such cheaper alternatives as rapeseed, sunflower, and imported palm oil. Large quantities of the groundnut meal produced in India are traded. Groundnut oil is thinly traded because in India substantial quantities of the oil produced are domestically consumed.

Methods for Research Evaluation

The unique empirical challenges of NRM impact assessment include both problems of measurement, and the attribution of research impacts. An impact analysis begins by measuring research benefits. Information on the actual cost of research and development (R&D) and technology transfer is combined with the stream of benefits based on the rate of technology uptake or levels of adoption. The approach quantifies those impacts that were amenable to quantification, while systematic documentation describes those that were difficult to quantify. For a five-component package like GNPT, the research evaluation includes measurement of the stepwise adoption of various technology options, estimates of on-farm benefits, and the relative significance of specific components among quantifiable variables. For the non-quantifiable impacts, researchers and farmers are important sources of detailed descriptions that may serve as a basis for evaluating as many effects as possible, or qualitatively understanding associated research impacts.

Research impacts documentation

The practical measurement of research impacts necessarily involves tracking and understanding the process based on detailed description by both researchers and research beneficiaries. In the absence of hard facts or documented data, detailed descriptions are an important way to understand the basis for estimates of costs and benefits associated with economic and environmental effects.

Because post-project long-term monitoring of GNPT was not undertaken, a systematic process of documentation was crucial for the evaluation process in order to carefully delineate various types of impacts: market and non-market, on-site and off-site, as well as intra- and inter-temporal effects. The implications of these aspects for impact assessment also require the analysis of counterfactuals for non-market effects. Additionally, the complexity of estimating impacts considering economic vs. environmental effects is recognised when some effects are already reflected in yield gains, but some environmental effects are non-quantifiable and do not relate to markets.

Data

Information was collected through farm interview surveys using a structured questionnaire, focus group meetings and participatory rapid rural appraisals, together with interviews with researchers on technical aspects of GNPT. Data on the following aspects were collected from farmers for the 1994/95 crop season:

1. Size of holding, total sown area, irrigated and non-irrigated areas
2. Land use and cropping pattern
3. Cost of groundnut production
4. Input and output data
5. Crop yields and prices
6. Farmer perceptions of sustainability issues and the constraints to adoption of GNPT.

Information on adoption trajectories for different technology options was collected, including:

1. Total groundnut area
2. First year of adoption of different GNPT components
3. Extent of adoption of different GNPT components in the first year
4. Extent of adoption during the period 1992–1994
5. Modification in technology components, if any.

District-level data for area and production were compiled from the Maharashtra State Department of Agriculture records, and disaggregated data below the district level were obtained from the Office of the Agricultural Development Officer (ADO) in each district. Rates of adoption obtained from the survey were also crosschecked with the ADO. Price data were re-collected from seed dealers and several traders dealing with the GNPT components. Estimates of elasticities used earlier estimates by Murty (1997),

Radhakrishna and Ravi (1990) and ACIAR (1992), and were validated using expert opinion.

The sample

Multi-stage stratified random sampling (using size of holding and intensity of groundnut cultivation as the basis for stratification) was used to select a representative group of groundnut farmers in order to assess the adoption and impact of different GNPT components. The technology was originally targeted at eight states in the Indian SAT: Andhra Pradesh, Gujarat, Karnataka, Madhya Pradesh, Maharashtra, Orissa, Tamil Nadu, and Uttar Pradesh. However, only in Maharashtra did government and non-government agencies follow up with the dissemination of technologies, and the State Ministry of Agriculture recommended the full GNPT package. Since the objective was to assess the adoption and evaluate the impact of the package, the evaluation of its impact therefore focused on Maharashtra.

The first and second stages of sampling involved stratification by the intensity of groundnut cultivation, while the last stage was stratified by size of holding. In the first stage of sampling, all districts growing groundnut were stratified into high and low intensity categories by the total area sown to groundnut. Two districts each from the top 50% and lower 50% intensity groups were selected at random. In the second stage of sampling, each selected district was stratified into three groups of *talukas* (sub-districts) by tercile of area sown to groundnut (high, medium, or low). Similarly, villages in each *taluka* were subdivided into three strata, also by tercile of groundnut sown area (details in Joshi and Bantilan, 1998). In the last stage of sampling, farm households were grouped into large (>4 ha), medium (1–4 ha) and small (<1 ha) categories according to size of farm holding. The final sampling units were identified through random selection of farmers in randomly selected villages in selected *talukas*. The final sample included 355 farm households.

Estimating the adoption pathway

Many crop and resource management technology packages that include several components are adopted component by component in step-wise patterns (Byerlee and Hesse de Polanco, 1986; Traxler and Byerlee, 1992). Establishing an accurate picture of adoption patterns among groundnut farmers can be complex. The five components of the GNPT package can be combined into ten pairs, ten triples, five quadruples, and one set of all five (Table 11.2). The adoption pattern can be established from the survey data by analysing farmers' responses when asked whether they practised different GNPT components. If the answer was yes, the farmer was asked to recall the first year of adoption for different components. Two additional questions were useful: 1. the extent of adoption of different GNPT components in the first year; and 2. the extent of adoption during the last 3 years ending in 1994. Several components of the technology package were already known and had been adopted even before the introduction of the package, and farmers were free to choose and adopt any of its subsets. Hence, adoption sequences were evaluated by tracking discrete subsets of options available to the farmer, for example, all subsets that included at least the land management option (shown as shaded components in Table 11.2). A systematic approach to tracking multiple technology adoption entailed measuring all subsets of technology components that included: 1. at least one option (say, land management); 2. two specific options (say, improved variety and land management); and 3. all options (full adoption).

Table 11.2. All possible combinations of the five components[a] of the groundnut production technology (GNPT) package.

One component adopted	Two components adopted	Three components adopted	Four components adopted	All components adopted
C1	C1C2	C1C2C3	C1C2C3C4	C1C2C3C4C5
C2	C1C3	C1C2C4	C1C2C3C5	
C3	C1C4	C2C3C4	C1C2C4C5	
C4	C3C4	C1C3C4	C2C3C4C5	
C5	C2C3	C1C3C5	C1C3C4C5	
	C2C4	C1C4C5		
	C2C5	C1C2C5		
	C1C5	C2C3C5		
	C3C5	C2C4C5		
	C4C5	C3C4C5		

[a]See Table 11.1 for a description of the components.

Farm survey data also served to estimate and project the adoption patterns of different GNPT components over time. By fitting a logistic function to data on the first year of adoption and data for the period 1989–95, the proportion of farmers affected by GNPT could be projected. The logistic function is defined as:

$$A_{it} = \frac{C_i}{(1 + e^{-(a+bt)})} \qquad (1)$$

where A_{it} is the percentage adoption of the ith component of the GNPT in the tth year; C_i is the adoption ceiling of the ith component; b is the rate of adoption; and a is the constant intercept term.

Research benefits and costs

Estimation of market benefits

Underlying the empirical application of the measurement of GNPT impacts is the principle of economic surplus, described in detail in Alston *et al.* (1995) and Swinton (Chapter 7, this volume). This principle is based on the idea that improved technologies enhance productivity or reduce the groundnut producers' unit cost of production, which translates into an outward shift in the producer's supply curve. Considering the conventional, comparative-static, partial equilibrium, closed economy model of supply and demand in the groundnut commodity market, and assuming simple linear demand and supply equations, a parallel supply shift (*k*) may be expected to occur due to a measurable reduction in unit cost of production when farmers adopt the GNPT technology package. As a point of reference, Fig. 11.2 shows the supply shift from S_0 (without GNPT) to S_1 due to measured unit cost reduction (*ae*) with the adoption of GNPT. For each cropping season, the change in the groundnut consumer surplus (ΔCS) and producer surplus (ΔPS) can be calculated using the formulae

$$\Delta CS = P_0 Q_0 Z (1 + ½ Z \eta) \quad (2)$$

$$\Delta PS = (J - Z) P_0 Q_0 (1 + ½ Z \eta) \quad (3)$$

where P_0 and Q_0 are the base groundnut price and quantity; $Z = -(P_1 - P_0)/P_0$; *k* is the unit cost reduction (equal to distance *ae* in Fig. 11.2); $J = k/P_0$; $(P_1 - P_0)$ is the change in market price; and η is the absolute value of the price elasticity of demand.

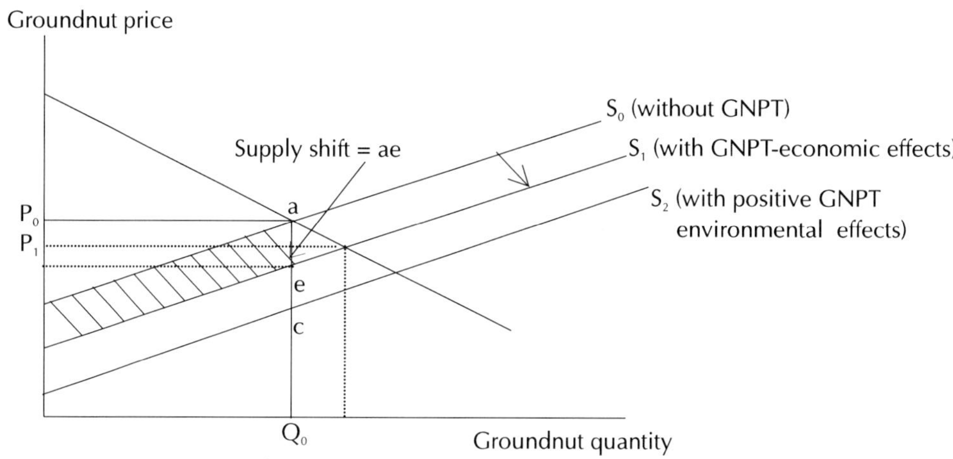

Fig. 11.2. Measurement of economic and environmental benefits due to adoption of groundnut production technology (GNPT) components.

Equations 2 and 3 can be used to calculate the empirical market benefits from adoption of the technology package. Annual gains are computed over the horizon the benefit is expected to accrue at actual adoption levels. The above estimation process only covers benefits accruing due to measurable market effects.

Computing the value of a supply shift

By custom, the magnitude of a supply shift (distance *ae* in Fig. 11.2) is measured by the change in unit cost of production and referred to as '*k*' (following Alston *et al.*, 1995). Establishing the actual supply shift (*k*) for adoption of GNPT involves understanding the unit cost reduction resulting from adoption of each of the possible GNPT options available to the farmer. This complex procedure can be overcome by categorising discrete subsets of options, among the whole range of 31 GNPT component mixes identified in Table 11.2.

Estimates of the *k*-shift in the supply function can be derived by using information available from on-farm trials. For analysis of the GNPT package, Pawar *et al.* (1993) provided results from trials managed by farmers and supervised by researchers. Different sets of technology options under on-farm trials presented alternative scenarios, namely:

- With and without improved package: This allowed comparison of the improved package of the GNPT, including improved varieties, RBF method, and other management practices, with the local package (full adoption)
- With and without RBF: This set compared only the effects of RBF with the flat land method of groundnut production, keeping the remaining components of the improved technology at their recommended level (at least RBF)
- With and without improved management practices: This option considered the use of improved varieties and compared the improved package of management practices with the traditional management package (i.e. partial adoption involving management practices only holding the effect of improved varieties).

The calculation of the supply shift *k* involves the use of the on-farm input and output data generated for each of the above scenarios. In particular, unit cost of production (Rs/t) was calculated based on total input cost and corresponding yield levels. Pairwise comparison of the unit cost incurred for the improved options versus the benchmark package generated a supply shift estimate for each scenario.

Inclusion of environmental impacts in the evaluation of NRM research benefits

In the process of examining the inclusion of environmental impacts in the evaluation of NRM research, it is useful to conceptualise specific scenarios detailing the nature of impacts by considering whether or not: 1. the effects of the technology intervention can be valued using conventional markets; 2. the effects are on-site or off-site or both; and 3. they have dynamic effects. Following this idea, Lubulwa and Davis (1997) identified four types of impact:

1. *On-site market impacts.* These impacts are specific to the site targeted by the technology intervention, do not have downstream effects, and can be evaluated using conventional markets. One example is exploitative farming systems that do not adequately replenish nutrients extracted during agricultural production. This activity has negative impacts as it reduces soil depth, degrades soil structure, decreases aeration, and increases salinity. The effects are on-site and may also have dynamic effects on crop productivity. These impacts are reflected in declining crop yields and can be valued using markets for the relevant crops.
2. *Off-site market impacts.* This represents off-site effects at locations different from where the technology impacts are targeted (e.g. downstream effects). Using the same example above, downstream effects that can be valued using markets include silting of rivers, reduced capacity for water storage, lowering water-table levels and the high costs of dredging irrigation canals.
3. *On-site non-market impacts.* This type of impact is specific to the site targeted but is not reflected in the marketplace. A good example is the slash and

burn practice used by farmers to expand cultivation area. A major impact of this practice is the loss of ecological biodiversity at the slash and burn site, but this impact cannot be valued using conventional markets. Contingent valuation or other similar techniques would be needed to value such an impact.

4. *Off-site non-market impacts.* This type reflects impacts that affect non-targeted locations as well as future generations. Water purification, carbon sequestration, and reduced flooding are all examples of downstream benefits resulting from upland watershed management.

Systematic process documentation of the research and impact pathways is necessary in order to understand the source of the impact and quantify the nature of the impact. More importantly, this process documentation enables identification of those variables that have market impacts and those that have non-market environmental impacts. The measurement of environmental effects in monetary terms within the context of the principle of economic surplus draws from changes in the social marginal cost of production (supply curve) and the demand for the marketed product. Figure 11.2 illustrates the measurement of a positive environmental effect as an additional supply shift resulting from the reduction in environmental damage or positive environmental effects caused by a specific option. In this case, cost-reducing research will shift the supply curve further from S_1 to S_2 thereby reducing the marginal cost by '*ec*'. The total cost reduction effect is represented by the sum of the supply shift due to cost reduction of the technology and a further shift caused by environmental effects. Thus, marginal environmental benefits are accounted for in the total unit cost reduction that is estimated as $ac = ae + ec$. This process adjusts the benefit calculations for implicit price changes. If, however, the effect of the resource management technology is negative, the supply curve S_2 shifts backwards reflecting the environmental damage and corresponding increase in cost. The following section details the analysis of market and non-market impacts of GNPT.

Research cost

Data on research costs can be based on project report documents and historical evidence, as well as on interviews and discussions with the scientists and extension staff who were directly involved in conducting research, on-farm trials, and technology transfer activities. The annual cost of developing and packaging the GNPT, plus the cost of its diffusion and dissemination were estimated by using the formula:

$$GNPTRC = C_{ic} + C_{nars} + C_{ext} \qquad (4)$$

where *GNPTRC* is the annual research and technology transfer cost of all components; C_{ic} is the annual research and overhead costs incurred at ICRISAT; C_{nars} is the annual research and other costs at the NARS; and C_{ext} is the annual cost of extension incurred by the technology transfer department of NARS.

Evaluation of Economic and Environmental Benefits

Farm-level benefits of the GNPT: quantitative estimates

Accounts of actual on-farm practices by representative farmers derived from the sample survey gave estimates of the benefits realised by farmers that include yield gains, cost saving and higher incomes (source: survey data of 1994/95 crop season):

1. The average groundnut yield among adopters was 2.2 t/ha, an increase of about 38% over the 1.6 t/ha among non-adopters
2. The unit variable cost of groundnut production under improved management was Rs3.86/kg in compared to Rs4.58/kg under local practices, a saving of about 16%; and
3. Net incomes among adopters averaged Rs21,470/ha in contrast to Rs15,580/ha among non-adopters, a gain of about 38% for the adopters.

Note that these estimates were obtained without accounting for the possibility of selection bias, an aspect that warrants consideration in future research.

On-farm trial data also provide estimates of the yield gain and unit cost reduction effects of GNPT. The value of the unit cost reduction is summarised for the three subsets chosen for this analysis based on on-farm trials detailed in Table 11.3:

a. k_1 = Rs1,198/t is achieved with the improved GNPT package (including improved varieties, RBF method, and other management practices), compared with the local package (full adoption)
b. k_2 = Rs564/t is achieved with the improved package of management practices compared with the traditional management package (with use of improved varieties in both cases), i.e. partial adoption involving management practices only, holding the effect of improved varieties.
c. k_3 = Rs270/t comparing the effects of RBF with the flat land method of groundnut production, keeping the remaining components of the improved technology at their recommended level (one component). This estimate is assumed to measure the unit cost reduction due to RBF.

Table 11.3. Cost of production and yield of groundnut under on-farm trials with different technology options, Maharashtra, India, 1987–91 (adapted from Pawar *et al.*, 1993).

Technology components		Yield	Cost	Unit cost
Management	Variety	(t /ha)	(Rs/ha)	(Rs/t)
Improved	Improved	3.49	6990	2002.86
Improved	Local	1.97	5990	3040.61
Local	Improved	2.56	6570	2566.40
Local	Local	1.74	5570	3201.15

By the nature of the measurable market effects listed above, the total value of the supply shift is only partially accounted for by taking these estimates of unit cost saving from adoption of the GNPT package instead of the existing practice.

Benefits as described by farmers in surveys and focused group interviews

Farmers described the additional benefits in a pilot survey (1999–2000), participatory rural appraisals and focus group interviews (Box 11.1).

Box 11.1. Welfare changes due to the adoption of groundnut production technology (GNPT) components, based on farm survey, participatory rural appraisals and focus group meetings (Bantilan *et al.*, 2003).

1. Raised-bed and furrow land configuration (RBF) improved soil moisture conservation (75% of survey respondents).
2. RBF was perceived to improve field drainage (75% of survey respondents).
3. RBF saved nutrients and water (28% of survey respondents).
4. Reinvestment in agricultural implements and inputs brought long-term stability to the farming system in the villages.
5. Stability of the farming system increased farmers' options in making decisions about cropping pattern (cash vs. subsistence crops) or investing in production vs. investing in schooling, housing, household assets.
6. The GNPT options were observed to have spillover effects beyond groundnut production. The RBF method was found applicable to such other crops as chillies, soybean, pigeonpea, chickpea, sunflower, mustard and some vegetables. Application of micronutrients to selected crops was also becoming popular where farmers had learned about the GNPT package.
7. Assets acquired for GNPT are being used for other crops, and have enabled cultivation in other seasons.
8. The community has become more socially inclusive, with greater interaction between members of different social categories. Respondents attributed this to a direct consequence of GNPT adoption, as it made landowner farmers more dependent on tribal and landless labour for longer periods throughout the year.
9. Credit rating of the village has risen.
10. Due to the newly found visibility conferred by GNPT adoption successes, the Maharashtra Government targeted the village for special development programmes (e.g. rural sanitation, wasteland development, integrated mother and child development).
11. Empowerment – a general improvement in self-esteem, confidence, ability to innovate were expressed in an increased diversity of crops cultivated, greater choice of investments, and greater access to credit, information, and government agents.
12. Higher pod yields with GNPT generated on-farm employment in shelling, especially for women. The overall labour requirement was about 12% higher with the GNPT than with the existing local practices.
13. For the marginalised groups (tribals and landless labourers), year-round employment ensured adequate food and nutrition for all members of the household.
14. Increased labour demand replaced out-migration of labour by in-migration.

Delineating market and non-market impacts

Table 11.4 summarises the overall impacts of GNPT adoption and delineates the market and non-market impacts in columns 2, 3 and 4. Yield-increasing or cost-reducing benefits cited in column 2 can be measured and directly included in the economic surplus calculations. Quantifiable measurements of these indicators give an initial basis for estimating the parallel *k*-shift in the supply function.

Table 11.4. Analysis of market and non-market impacts of groundnut production technology (GNPT).

Component	Market impacts	Non-market impacts	Environmental effects
C1 Land management			
RBF seedbed	Yield gains Saves 20% of input cost compared to conventional flat system	• Agricultural sustainability • Reduces soil erosion • Reduces water logging • Helps move salts to furrows, and from furrows to drains • Conserves soil moisture during deficit rain • Concentrates organic matter and fertiliser application • Reduces soil compaction, providing loose and well-aerated soil for growing crop • More soil depth for better development of root mass	+ (Greater yield stability, increased water availability off-site and in future, enable cultivation in other seasons) – (Off-site increase in soil salinity)
	Change in labour demand	• More labour required • Reduces drudgery for women in weeding operations (labourers sit in furrows and weed) • Efficient use of tractor and field machinery; interculturing with tractor/ bullock implements • Less power requirement for land preparation in successive years	
C2 Nutrient management			
Farmyard manure	Increase in groundnut yields	Improves soil physical properties and soil health	+ (Increase carbon content)

Continued

Table 11.4 Continued.

Component	Market impacts	Non-market impacts	Environmental effects
Ammonium sulphate	Increase in groundnut yields	Environmental effects	+ (Checks soil alkalinity) – (Causes water pollution)
Single super-phosphate	Increase in groundnut yields	Environmental effects	+, –
Zinc sulphate	Increase in groundnut yields	Environmental effects	+, –
Ferrous sulphate	Increase in groundnut yields	Environmental effects	+, –
Gypsum	Increase in groundnut yields	Environmental effects	+, –
C3 Water management			
Sprinkler irrigation	Reduced unit cost due to enhanced water use efficiency	Positive environmental effects due to reduced pest incidence Efficient water utilisation through GNPT offers potential long-term benefits, particularly in increasing water availability off-site and in the future	+
C4 Disease and pest management			
Fungicidal seed treatment	Good quality seeds reduce yield loss and increase employment potential		+, –
Herbicides and pesticides	Reduced yield losses	Negative health effects Adverse effects on water quality	– (Skin allergies)
C5 Seed			
Improved variety	Increase in yields	Conserves biodiversity, checks insect pest incidence	+
Seed rate	Increase in yields	Check insect pest infestation	+
Sowing–dibbling	Yield increase due to good and uniform plant population Increase in employment	Increase drudgery on women	–
Seed dressing	Increased yield	Check insect pest infestation	+

Some non-market impacts may also be indirectly reflected in the calculation of economic benefits to the extent that they affect improvement in yields or unit cost reduction. For example, improvement in the soil physical properties listed in column 3 may be reflected in enhancing groundnut yields. But, there are some indirect or long-term benefits that are difficult to measure as shown in columns 3 and 4 of Table 11.4. These include agricultural sustainability resulting from enhanced biodiversity and health effects. Ideally the value of these impacts can be obtained by seeking appropriate relationships between a chosen GNPT technological intervention and environmental effects. Finding a unique equation or a functional relationship that can be used to quantify, in physical terms, the effect on human health or air quality or other environmental impacts of each component could be difficult. For example, while soil health is believed to improve with the GNPT's land and nutrient management interventions, there are no data or models to measure the specific effects on soil health (J.V.D.K. Kumar Rao, personal communication, 2004). Nevertheless, descriptions of the likely environmental effects of GNPT interventions by Pawar *et al.* could help in impact assessment (1993; and C.S. Pawar, personal communication, 2004):

- The natural acidity of ammonium sulphate checks the alkalinity of the soil. This is a positive effect in alkaline soils, but excess applications of ammonium sulphate can also result in negative environmental effects
- Pollution levels are high with local practices of fertiliser application
- Water quality can be reduced when excess nitrogen is applied to crops
- Micronutrients like zinc sulphate and ferrous sulphate help maintain the yield potential of the soil. Zinc sulphate is used to rectify the zinc deficiencies of the crop. Ferrous sulphate is used to rectify iron deficiencies incurred by waterlogging
- Herbicides and pesticides, if used in large quantities, can cause severe damage to the environment; exposure can also trigger skin allergies in farmers. Prior to the introduction of GNPT, farmers applied excess quantities of pesticides due to lack of awareness. ICRISAT educated the farmers about appropriate dosages and safe handling procedures, thereby mitigating negative environmental effects and farmer health risks.

Listing the positive and negative effects, in Table 11.4, aids in the analysis of market and non-market impacts of the GNPT management options. It records the market impacts representing yield gains or reduced yield losses and changes in unit cost from adoption of GNPT components. The inventory of non-market effects is substantial. The RBF land management appears to have had significant positive environmental effects resulting to greater long-term yield stability, increased water availability off-site and in the future. Agricultural sustainability was enhanced through reduced soil erosion and reduced waterlogging during periods of heavy rain. The other components including nutrient management, disease and pest management and water management improved the soil physical properties and soil health. The environmental benefits included increased carbon content and checked soil alkalinity. Negative effects (environmental costs) from water pollution arose from the use of ammonium sulphate and other micronutrients and pesticide runoff. When pesticide use exceeded recommended levels, it also caused adverse health effects. Finally, although increased groundnut yields increased incomes, denser planting and groundnut shelling created added drudgery for women.

Table 11.4 illustrates how a qualitative understanding of the nature and direction of the impacts can provide a basis for determining the range of possible conditions that would simulate potential benefit levels. In this case it is important to understand the source of the impact, the nature of an impact, and the relationship between an impact and those variables that can affect current, potential, or future producers and consumers. Even though the effects on the environment are complex, the identification and understanding of GNPT effects narrows the field remaining for evaluation. Table 11.4 highlights how the conventional calculations that exclude environmental effects can skew measures of the full technology impact.

Approximations of Economic and Environmental Effects

This section applies the approach discussed above to estimate the total gains due to GNPT technology. Estimates of basic parameters are explained and procedures are illustrated.

Production, price and elasticities

a. The annual base level of groundnut production was 151,280 t in the four selected districts of Maharashtra (average during 1988–1990; source: ICRISAT District-Level Database)
b. The base groundnut price was Rs6533/t (average groundnut price in Maharashtra during 1988–1990, source: ICRISAT District-Level Database)
c. The price elasticity of demand was 0.5 and price elasticity of supply was 0.1 (Radhakrishna and Ravi, 1990).

Research lags

On the estimation of the research lag (i.e. the period of investment required before benefits were realised), the survey indicated that GNPT adoption first took place in 1989. A research lag of 12 years was measured from the time of initial research started in 1974 to the introduction of the technology in farmers' field in 1986 and a further lag of 3 years before first year of actual adoption.

Adoption estimates

Using the methodology introduced in the earlier section on adoption, the survey data covering the period 1989–1994 were used to develop the adoption pathway for GNPT (Fig. 11.3). The results above confirm the situations of partial adoption and step-wise adoption. They indicate that different technology components of GNPT are adopted in a step-wise process of adopting improved varieties, nutrient management, soil management, and other components of the package depending upon: 1. information about the technology, 2. the availability of necessary resources or inputs, 3. marginal returns to the technology, 4. risks, and 5. the suitability of technology traits.

The logistic function was used to estimate the adoption curve and predict the future path, e.g.:

$$A_t = \frac{40}{(1 + e - ^{(-2.6 + 0.69t)})} \quad \text{for adoption of at least RBF} \quad (5)$$

$$A_t = \frac{98}{(1 + e - ^{(-3.2 + 0.34t)})} \quad \text{for adoption of at least improved varieties} \quad (6)$$

Similar estimates can also be obtained for any selected component or subset of GNPT.

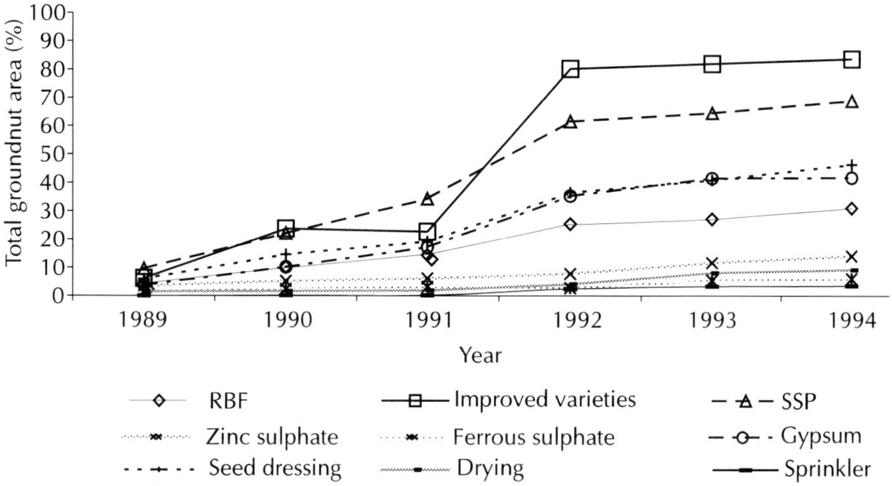

Fig. 11.3. Adoption patterns of groundnut production technology (GNPT) components in Maharashtra.

Figure 11.4 depicts the adoption path for the RBF component, estimated using the logistic function, showing a consistent increase in adoption of the RBF. Because this adoption path reflects those households adopting RBF (some of whom did not adopt other GNPT components), it overestimates adoption of the full package.

Farmers who adopted the concept of RBF but lacked appropriate implements did not strictly adhere to making beds 1.5-m wide. This illustrated an important dimension of crop and resource management technologies: farmers adapt technologies to meet special needs, changing the technologies in the process.

Among the other GNPT components, the adoption rate of improved groundnut varieties rose dramatically from 6% in 1989 to 84% in 1994. The adjusted rate of adoption of improved varieties was higher for those farmers practising the RBF method. The accelerated adoption of improved varieties may be attributed to the dissemination of information on GNPT. At the time of the survey in 1994 the sprinkler method of irrigation was yet to be

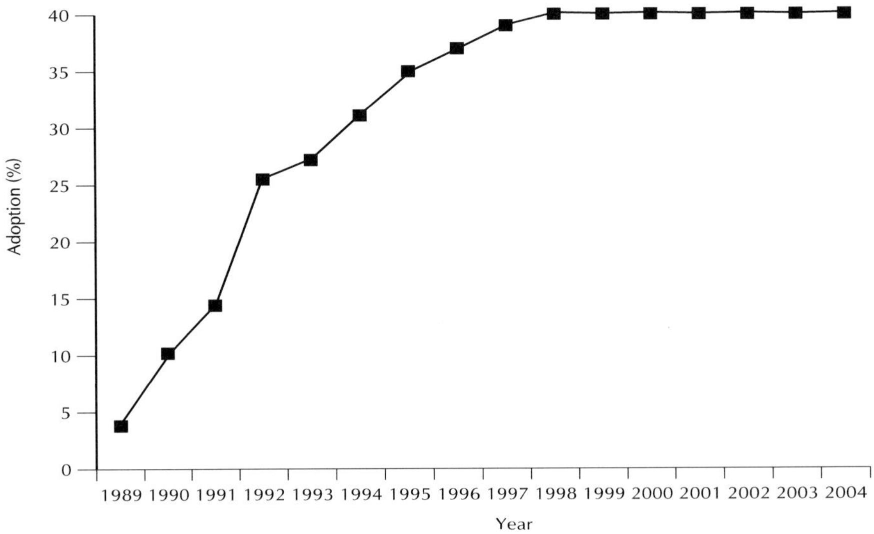

Fig. 11.4. Adoption of raised-bed and furrow (RBF) of groundnut production technology (GNPT) in selected districts of Maharashtra, India, 1989–1995 (projected to 2004).

adopted by the majority of groundnut cultivators. By the late 1990s, the use of sprinkler irrigation in Maharashtra had been substantially enhanced by government subsidies.

Research cost estimates

The estimated cost of research and technology transfer is detailed in Table 11.5. The annual cost of ICRISAT, C_{ic}, was estimated as:

$$C_{ic} = SAL_{ic} + OPR_{ic} + OVR_{ic} + OFD_{ic} \quad (7)$$

where SAL_{ic} is the annual salary of the research team; OPR_{ic} is the annual operational expenses required to undertake GNPT development, packaging, and diffusion; OVR_{ic} is the annual overhead cost at the Institute; and OFD_{ic} is the annual cost incurred to conduct on-farm trials and demonstrations in farmers' fields.

The salary of the research team at ICRISAT, SAL_{ic}, is considered to include the salaries of all those associated with the research project (SAL_i), each weighted by the proportion (w_i) of their time devoted to developing and packaging the GNPT, that is,

$$SAL_{ic} = \sum_{i=1}^{n} w_i * SAL_I \quad (8)$$

This annual salary cost was estimated at US$34,900. The operational cost (OPR_{ic} = US$12,215) of developing and packaging the GNPT was assumed to be 35% of the salary. This assumption is based on historical norms at ICRISAT. The overhead costs (OVR_{ic}) are usually considered to be half of the research expenses (Byerlee, 1996); this figure (US$47,115) was based on research resource allocations to different research projects at ICRISAT. Since the technology components were packaged and recommended for groundnut, pigeonpea, and chickpea, the research and packaging costs for GNPT was proportionately distributed. The share of groundnut in the total area of the three crops was used as a basis for allocating research costs to GNPT (US$45,600).

Table 11.5. Annual research and technology transfer cost (US$) of groundnut production technology (GNPT), 1974–2000.

Component	Year	Cost (US$)
Research		
Salary	1974–86	34,900
Operations	1974–86	12,215
Overheads	1974–86	47,115
NARS	1974–91	9,500
Technology transfer		
Packaging/on-farm trials	1987	24,000
On-farm trials	1988–90	20,000
On-farm trials	1991	10,000
State expenses	1992–2000	7,500

The NARS was involved in packaging the technology and conducting on-farm trials. To assess this cost, several researchers who worked for the NARS were consulted. It was determined that, on the basis of NARS participation in the development and packaging of the technology, the NARS incurred a cost of about US$4560 (approximately 10% of ICRISAT's total cost). Similarly the cost of on-farm research and technology transfer activities (OFD_{ic}) undertaken through the LEGOFTEN Technology Transfer Network, which started in 1987, was proportionately allocated. The expenses incurred in technology transfer (C_{ext}) through the Maharashtra Department of Agriculture during the post-LEGOFTEN period were calculated using the share of groundnut in total area in the State as no separate documentation exists on resource allocation for each commodity or technology.

The technology packaging and its transfer started from 1987 through the LEGOFTEN programme. The initial budget for this programme (1987 and 1988) was met through ICRISAT's core funds, and later (1989–1991) through financial assistance from the International Fund for Agricultural Development (IFAD). In the first year, when different components of technology were integrated, the cost of GNPT (US$24,000) was computed on the basis of the proportionate area under groundnut. In subsequent years, the total budget allocated to LEGOFTEN was distributed (US$20,000) to represent the GNPT package that was apportioned according to the number of on-farm trials conducted on groundnut. The budget of the State Department of Agriculture for GNPT extension activities during 1987–1991 was also met through the LEGOFTEN programme. The expenses incurred in technology transfer through the state departments of agriculture during the post-LEGOFTEN period were calculated using the share of groundnut in the total cropped area in the state, as no separate information on resource allocation to each commodity/technology is documented. On the basis of the salary, operations, and overheads, the annual technology transfer cost during the post-LEGOFTEN period was calculated to be US$7,500. This cost was considered from 1992 until 2000. Since the research and technology transfer costs incurred by ICRISAT, NARS, and the state departments of agriculture were rough estimates based on available ICRISAT Annual Reports and interviews with scientists involved in the project, a sensitivity analysis was also performed by increasing the cost of research and technology transfer by 10–20%. The results revealed that the internal rate of return (IRR) is rather insensitive to changes in costs of research and technology transfer.

Supply shift

The unit cost of production (Rs/ton) was calculated based on total input cost and corresponding yield levels. Pairwise comparison of the unit cost incurred by GNPT enhanced options vs. the traditional practice generated supply shift estimates for each scenario. For the three scenarios described in the previous section, three levels of on-farm unit cost reduction were taken: k_1 = Rs1,198/t, k_2 = Rs564/t, and k_3 = Rs270/t.

Table 11.6 presents the stream of research and technology transfer costs and market-based research benefits using the unit cost reduction estimates (k_1, k_2 and k_3) above, levels of adoption represented by Fig. 11.4, price, quantity and elasticity estimates. It also gives the estimated net present value, IRR, and benefit–cost ratio under three different scenarios. As noted earlier, the estimate using the adoption path for RBF gives an upper bound of the benefit levels. (A lower bound can be estimated using the adoption pathway of the GNPT component that has been adopted least, i.e. at a ceiling level of 15% based on the data.)

The analysis revealed that the IRR of GNPT was 25.3% if the total package of the GNPT is adopted. The total net present value of information from the research and technology transfer programme on GNPT was estimated to be US$3.45 million. The benefit–cost ratio was 9.37, which means that every US$1 invested in developing and disseminating GNPT produced an average benefit of US$9.37 throughout the period.

Given the environmental effects recorded from the analysis above (largely positive but also partially negative), two different scenarios of positive and negative net environmental effects were simulated. Because the major impacts were felt to be captured by the effects on marketable crop yields, the sensitivity analysis scenarios involved modest levels of change: a 10% increase in unit cost reduction from the base level of full GNPT package adoption, and a 5% decrease in unit cost reduction from the base level. The analysis revealed that positive environmental effects that might further increase the unit cost reduction could result in a benefit–cost ratio of 9.73 and an IRR of 26.17. The second scenario of a negative environmental effect by a marginal rate of 5% could reduce the benefit–cost ratio to 8.26 and result in reducing the IRR to 24.95. Negative environmental effects would have to increase the social value of unit production costs by 79% for the benefit–cost ratio to fall to the break-even level of 1.0. Such an increase in units costs is implausibly high, given the dominantly beneficial environmental effects reported by farmers and focus groups. None the less, these simulations show the sensitivity of research impacts when environmental effects are considered.

Table 11.6. Market-based cost and benefit streams for research and technology transfer of the groundnut production technology (GNPT) package.

	Cost (US$'000)		Benefits (US$'000)		
Year	ICRISAT	NARS	Full package	Partial package[a]	Land mgt (RBF)[b]
1974	45.6	4.56	0	0	0
1975	45.6	4.56	0	0	0
1976	45.6	4.56	0	0	0
1977	45.6	4.56	0	0	0
1978	45.6	4.56	0	0	0
1979	45.6	4.56	0	0	0
1980	45.6	4.56	0	0	0
1981	45.6	4.56	0	0	0
1982	45.6	4.56	0	0	0
1983	45.6	4.56	0	0	0
1984	45.6	4.56	0	0	0
1985	45.6	4.56	0	0	0
1986	24.0	4.56	0	0	0
1987	20.0	4.56	0	0	0
1988	20.0	4.56	0	0	0
1989	20.0	4.56	162.57	76.15	36.42
1990	10.0	4.56	460.62	215.75	103.19
1991	0.0	7.50	650.29	304.59	145.68
1992	0.0	7.50	1,151.56	539.39	257.97
1993	0.0	7.50	1,228.33	575.34	275.17
1994	0.0	7.50	1,404.45	657.84	314.63
1995	0.0	7.50	1,580.57	740.33	354.08
1996	0.0	7.50	1,670.89	782.64	374.31
1997	0.0	7.50	1,761.21	824.94	394.54
1998	0.0	7.50	1,806.37	846.09	404.66
1999	0.0	7.50	1,806.37	846.09	404.66
2000	0.0	7.50	1,806.37	846.09	404.66
2001	0.0	0.00	1,806.37	846.09	404.66
2002	0.0	0.00	1,806.37	846.09	404.66
2003	0.0	0.00	1,806.37	846.09	404.66
2004	0.0	0.00	1,806.37	846.09	404.66
2005	0.0	0.00	1,806.37	846.09	404.66
Internal rate of return (IRR) (%)			25.26	19.15	13.50
Net present value (US$ '000)			3,452.94	1,389.06	453.45
Benefit–cost ratio			9.37	4.39	2.10

[a]Partial = management practices only.
[b]Land mgt (RBF) = raised-bed and furrow.

Summary and Conclusions

This chapter principally illustrates an empirical estimate of economic surplus using the case of GNPT developed by ICRISAT and its partners in the Indian NARS. The case study illustrates the critical importance and use of qualitative information in understanding the additional environmental and long-term effects due to the adoption of NRM technologies.

To quantify the returns to investment on research and technology exchange, three aspects were examined:

1. Benefits (both economic and environmental) accruing from the research and technology exchange programme
2. Adoption rates and the spread of different components of GNPT
3. Research and technology exchange cost involving research partnerships among international and national research programmes as the extension sector.

Economic surplus and distribution of welfare gains were estimated by assuming a parallel shift in supply function due to investment in the research and technology development. Internal rates of return, net present values and benefit–cost ratios were computed under three options:

1. Full adoption of the GNPT package
2. Adoption of only management practices
3. Adoption of only land management (RBF) with other practices remaining the same.

Because environmental effects were not measured in monetary terms, two sensitivity analyses were carried out under scenarios related to net positive and negative environmental effects.

The survey results show that farmers initially adopted parts of the crop and resource management package, and adapted the technology options according to their needs, convenience, and resource endowments. Logistic growth functions were estimated to describe the rate of adoption of each GNPT component. The adoption analysis illustrates the nature and dynamics of adoption of NRM technologies.

The estimation of benefits accruing from GNPT involved computation of welfare gains based on yield gains and/or reduction in unit production costs. The inclusion of qualitative environmental effects encompassed impact dimensions not captured via the measurable reduction in unit cost or yield gains due to lack of quantifiable or long-term data. The difficulty of quantifying many environmental costs and benefits challenged the approach to incorporating these effects into cost–benefit analysis. The environmental effects were characterised by systematically tracking both individual and interaction effects of GNPT components. Thorough analysis is based on systematic documentation coupled with reasonable estimates of economic effects.

Environmental effects can have a large overall impact. The results show that if environmental effects reduced fully accounted unit costs by just 10% more than market effects, the net present value of the GNPT would increase by US$0.4 million and the IRR would increase by 1%. Clearly, environmental effects in the assessment of NRM options cannot be ignored. As stated by Winpenny (1991), the environment is not free, even though there may not be a conventional market for its services. In the context of decisions based on cost–benefit analysis, it is important to understand the source of the impact, the nature of an impact, and the relationship between an impact and those variables that can affect current, potential, and future consumers and producers. This means that valuing as many effects as possible and plausible, narrows the field remaining for pure judgement.

Acknowledgement

The authors would like to thank the contributions of Drs S.N. Nigam, C.S. Pawar and J.V.D.K. Kumar Rao.

References

ACIAR (Australian Centre for International Agricultural Research) (1992) *Research Priorities for International Research Database.* ACIAR, Canberra, Australia.

Alston, J.M., Norton, G.W. and Pardey, P.G. (1995) *Science Under Scarcity. Principles and Practice for Agricultural Research Evaluation and Priority Setting.* Cornell University Press, Ithaca, New York, 624 pp.

Bantilan, M.C.S., Parthasarathy, D. and Padmaja, R. (2003) Enhancing research-poverty alleviation linkages: Experiences in the semi-arid tropics. In: Mathur, S. and Pachico, D. (eds), with the collaboration of Jones, A.L. *Agricultural Research and Poverty Reduction: Some Issues and Evidence.* CIAT publication no. 335, Economics and Impact Series 2. Centro Internacional de Agricultura Tropical (CIAT), Cali, Colombia, pp. 173–188.

Byerlee, D. (1996) Returns to agricultural research investment. *Presented at a Training Program on Agricultural Research Evaluation and Impact Assessment, 16 June 1996, Indian Agricultural Research Institute, New Delhi, India.*

Byerlee, D. and Hesse de Polanco, E. (1986) Farmers' stepwise adoption of technological packages; evidence from the Mexican Altiplano. *American Journal of Agricultural Economics* 68(3), 519–527.

Freeman, H.A., Nigam, S.N., Kelley, T.G., Ntare, B.R., Subrahmanyam, P. and Boughton, D. (1999) *The World Groundnut Economy: Facts, Trends, and Outlook.* International Crops Research Institute for the Semi-Arid Tropics (ICRISAT), Patancheru, India, 52 pp.

Joshi, P.K. and Bantilan, M.C.S. (1998) Impact assessment of crop and resource management technology: a case of groundnut production technology. *Impact Series no. 2.* International Crops Research Institute for the Semi-Arid Tropics (ICRISAT), Patancheru, India, 60 pp.

Lubulwa, G. and Davis, J.S. (1997) Environmental and human health issues in the evaluation of agricultural research. In: Bantilan, M.C.S. and Joshi, P.K. (eds) *Proceedings of an International Workshop on Integrating Research Evaluation Efforts, 14–16 December 1994.* International Crops Research Institute for the Semi-Arid Tropics, Patancheru, India, pp. 66–73.

Murty, K.N. (1997) Trends in consumption and estimates of income and price elasticities of demand for major crops in the semi-arid tropics of India – a compendium. *Socioeconomics and Policy Division Progress Report 123*. International Crops Research Institute for the Semi-Arid Tropics (ICRISAT), Patancheru, India, 40 pp.

Pawar C.S., Amin, P.W., Jain, K.C., Kumar Rao, J.V.D.K. and Srivastava, M.P. (1993) On-farm research on groundnut, pigeonpea, chickpea, and transfer of technology to semi-arid tropics farmers in India, 1987–91. *Final Report to IFAD*. International Crops Research Institute for the Semi-Arid Tropics (ICRISAT), Patancheru, India, 66 pp.

Radhakrishna, R. and Ravi, C. (1990) *Food Demand Projections for India*. Centre for Economic and Social Studies, Hyderabad, India, 71 pp.

Traxler, G. and Byerlee, D. (1992) Economic returns to crop management research in a post-green revolution setting. *American Journal of Agricultural Economics* 74(3), 573–582.

Winpenny, J.T. (1991) *Values for the Environment. A Guide to Economic Appraisal*. Overseas Development Institute, London, UK, 277 pp.

12 Assessing the Economic and Environmental Impacts of Conservation Technologies: A Farm-level Bioeconomic Modelling Approach

B. Shiferaw[1] and S.T. Holden[2]

[1] *International Crops Research Institute for the Semi-Arid Tropics (ICRISAT), Nairobi, Kenya*
[2] *Agricultural University of Norway, Ås, Norway*

Introduction

Along with degradation of the productive resource base, widespread poverty and population growth are major concerns for sustainable intensification and agricultural development in many poor regions of the world. The relationship between population pressure, poverty and environmental degradation has been a subject of debate and controversies for many years, with an upswing in the debate over the last 30 years (Boserup, 1965; Cleaver and Schreiber, 1994; Tiffen *et al.*, 1994; Grepperud, 1996; Templeton and Scherr, 1999). Earlier studies on technology choice among smallholders in Ethiopia found that low or negative initial returns to conservation technologies could undermine investments in such practices (Shiferaw and Holden, 1998). Some evidence indicated that population pressure, poverty and land scarcity may even encourage removal of conservation structures (that occupy productive lands) introduced in the past through food-for-work programmes.

Although the empirical evidence is mixed and less conclusive (Templeton and Scherr, 1999), there is emerging consensus on the potential nexus between population density, poverty and land degradation in some less-favoured areas where poverty is associated with poor policies, and lack of access to markets and improved technologies (Reardon and Vosti, 1995; Heath and Binswanger, 1996). Under such conditions, poor land users often lack the wherewithal to invest in sustainability-enhancing options and could be caught up in a self-reinforcing nexus that may lead to worsening poverty and resource degradation. However, significant research and development effort is directed towards evolving options for improved natural resource management (NRM) to enhance the productivity and sustainability of production systems. Nevertheless, the basic question remains about the potential of technological and policy options to lift the poor out of poverty, and to what extent these options could

(eds B. Shiferaw, H.A. Freeman and S.M. Swinton)

actually contribute to sustaining the resource base and livelihoods under conditions of high population density and high risk of land degradation.

Bioeconomic models are suited to evaluating the potential impact of new technologies and policy options on rural livelihoods and the environment (resource conditions) at different temporal and spatial scales (Holden, Chapter 8, this volume). The integration of biophysical and socio-economic conditions into the local economy is an enhancement of earlier econometric approaches, since it allows more-precise simulation of household investment decisions and simultaneous assessment of the welfare and environmental impacts in a dynamic setting – a more suitable approach to assessing NRM impacts. The objective of this chapter is to illustrate how a multiperiod bioeconomic household-level model, in which changes in resource quality have feedback effects on future land productivity, can be used to explore the economic and environmental impacts of NRM technologies and policies. This model is used to test the influence of land scarcity and asset poverty (e.g. oxen and labour) on incentives to undertake sustainability investments. The integration of agroecological and socio-economic information has provided useful insights regarding the potential of alternative policy instruments and the impacts of new technologies. The model incorporates important features of the biophysical system and its dynamics along with market characteristics in the rural economy. The choice of crop and livestock production activities and NRM technology investments are jointly determined. The model is developed in Generalised Algebraic Modelling System (GAMS) using data from Andit Tid, in the central highlands of Ethiopia, an area inhabited by poor smallholder farmers and characterised by high population density, rugged topography, steep slopes, and severe problems of soil degradation.

The results show how land scarcity could drive conservation investments, while poverty in vital assets such as oxen and labour could deter investments in land and water management. The welfare and environmental impacts are very modest but are highest when the conservation technology does not reduce short-term crop yields. Otherwise, the level of adoption of these technologies and their effects on poverty and soil degradation are significantly reduced even when family labour is not limiting. This contributes to worsening the conditions of the poor and continued degradation of the resource base. For credit-constrained households the increased fertiliser use associated with improved credit availability may substitute for conservation effort. The following part of the chapter offers an overview of the case study area and important biophysical and socio-economic aspects included in the model, then the basic structure of the bioeconomic model is presented. This is followed by presentation and discussion of the simulation results. The final part highlights the major findings and policy implications.

The Biophysical and Socio-economic System

The study area (Andit Tid) is located in North Shewa, in the central highlands of Ethiopia, approximately 60 km north of Debre Berhan, along the main road from Addis Ababa. This implies that market access is fairly good. The area is characterised as a low-potential, cereal–livestock zone and suffers from severe soil degradation. Given the high altitudes, the land falls in two altitude zones: *Dega* zone (<3200 m asl) and *Wurch* zone (>3200 m asl). There are two distinct rainfall and growing seasons, the *Meher* (June–December, 1056.8 mm rainfall), and the *Belg* season (January–May, 315.4 mm rainfall).

Barley is the main crop, followed by wheat, horse bean, and field peas. Lentils and linseeds are also commonly grown. The cropping pattern depends on the local agroclimatic zone (see Table 12.1). Crop production mainly depends on organic fertilisers, while the use of mineral fertilisers is limited by lack of credit and the low profitability of applying it to some crops. Most of the production takes place in the low altitude zone but barley is grown also in the higher altitude zone in the *Belg* season. The major crops during the main growing season (the *Meher*) are barley, wheat, faba beans, field peas and lentils, in the low-altitude zone. In the *Belg* season, barley is grown in the high altitude zone, and lentils and field peas in the low altitude zone. Droughts are not common during the *Meher* season but can occur in the *Belg* season. Hailstorms and frost may damage crops during the *Meher* season.

Table 12.1. Crops grown in the different seasons and local agroclimatic zones.

	Cropping zone	
Season	Low altitude	High altitude
Main season (*Meher*)	Barley, wheat, faba beans, field peas, lentils, linseed	Fallow
Short-rainy season (*Belg*)	Field peas, lentils	Barley

The two dominant soil types are Andosols and Regosols. Andosols are dominant in the high-altitude zone while Regosols are common in the lower-lying areas. The Regosols are the most important and intensively cultivated soils. Andosols are mainly used to grow barley and are relatively rich in organic matter. Steep slopes and intensive cultivation increase the risk of soil degradation. An estimated 75% of the land area is steeply sloped (>25%). Soil erosion rates are very high and an estimated 21% of the agricultural land has shallow soils (<30 cm) and 48% medium-deep (30–60 cm) soil (Yohannes, 1989).

Cattle and sheep are the predominant types of livestock but goats, equines and chickens are also common. There are strong crop–livestock interactions in the system. Crop residues are typically used as animal fodder. Oxen provide traction power to cultivate land and thresh crops. Animal manure is used to enhance soil fertility and for fuel. Fodder shortage is a constraint to livestock production. High population density and land scarcity increase competition between crop and livestock production. Sale of small stock (sheep, goats and chickens) complements both household consumption and crop–production activities.

Some conservation technologies were introduced through food-for-work programmes in the early 1980s. With the termination of programme benefits in the early 1990s, farmers have been selectively removing soil conservation structures from their plots (Shiferaw and Holden, 1998). The removal seems to be accelerated when structures occupy productive land and increase land scarcity, or when they do not contribute to increasing short-term yields. How poverty affects this process and the potential economic and environmental impacts from such NRM investments are not well understood. Farm households possess usufruct rights to land. Following the land reforms of 1975 and frequent land redistributions thereafter, landlessness is uncommon, and land is fairly distributed according to family size (see Table 12.2). This means that livestock wealth is often a better indicator of household wealth and wealth differentiation. The oxen rental market is underdeveloped (Holden and Shiferaw, 2004) and ownership of traction power is an important asset that determines crop income. When the necessary traction power is lacking and rental markets are imperfect, land ownership by itself may not necessarily translate into better living conditions for the household. Typically, households lacking traction power either rent out land to households with two or more oxen, or depend on relatives with oxen to cultivate their lands. Hence, local communities often use oxen ownership as a wealth indicator. Therefore oxen ownership along with farmland and family labour endowments were used as proxy indicators for household poverty. Future work will need to extend this through use of other more-relevant poverty indicators.

Production remains largely subsistence based. The small towns in the vicinity, inhabited mainly by local traders, serve as markets in the area. Owing to the difficult terrain, there are significant transportation costs to some of the accessible output markets in the area. Farm-gate prices are adjusted for such transaction costs based on local estimates. Some markets for labour, land and livestock exist within the village or in the nearby towns. The labour market is largely inactive, but the local wage rate varies seasonally depending on local demand. Labour may be hired in cash, in kind (fixed output share) or in exchange for traction power. Formal institutional credit is largely unavailable. Hence, the basic model does not include credit, but this assumption was relaxed to assess the effect of credit policy on welfare and sustainability investments. Off-farm income options are mainly limited to local agricultural wages and self-employment in petty trade within the vicinity.

Along with biophysical and experimental data collected by the Soil Conservation Research Project (SCRP), socio-economic data mainly collected in 1994 and complemented in 1998 were used to formulate and develop the model. The availability of on-site biophysical and socio-economic data made it possible to assess technology and policy impacts using a multi-period bioeconomic model. In 1994, about 26% of the households had no oxen, 15% had one ox, and 56% had two oxen. Less than 5% of households were landless, mainly young families awaiting land allocation by the State. Table 12.2 shows the basic characteristics and resource endowments of the different household groups. For better simulation of the biophysical system and variations in land quality, land was classified into eight different soil depth and slope classes (Table 12.3 and Fig. 12.1).

Table 12.2. Basic farm household characteristics in Andit Tid, 1994.

	Household type[a]			
Variables	No ox	One ox	Two+ oxen	Average
Family size	2.80	5.80	7.20	6.10
Consumer units	2.60	5.17	6.58	5.55
Labour units	1.53	2.78	3.98	3.23
Own farm size (*Timad*)[b]	5.55	7.68	11.00	9.05
Operated crop area	3.30	5.08	8.84	6.73
Own cultivated area	3.00	4.79	7.80	6.07
Rented-in land	0.30	0.28	1.04	0.66
Rented-out land	1.55	0.18	0.10	0.31
Tropical livestock units (TLU)	1.45	3.52	7.10	5.10
Oxen	0	1.00	2.30	1.53

[a] The sample size was 10 households with no oxen, 30 households with one ox, 40 households with two or more oxen.

[b] Land areas are measured in *Timad*, approximately 0.25 ha.

Table 12.3. Land area (in *Timad*) by farm household category, soil type, soil depth and slope classes.

				Household category		
Soil type	Codes	Soil depth class (cm)	Slope class (%)	No ox	One ox	Two+ oxen
Andosols (A)		All	All	2.03	2.82	4.02
	A0–30 cm (1)	0–30	0–20	0.91	1.26	1.80
	A30–60 cm	30–60	0–20	0.57	0.78	1.12
	A>60 cm	>60	0–20	0.32	0.44	0.63
	A0–30 cm (2)	0–30	>20	0.24	0.33	0.48
Regosols (R)		All	All	3.52	4.88	6.98
	R0–30 cm (1)	0–30	0–20	1.62	2.25	3.21
	R30–60 cm	30–60	0–20	0.86	1.19	1.69
	R>60 cm	>60	0–20	0.31	0.44	0.62
	R0–30 cm (2)	0–30	>20	0.73	1.01	1.44

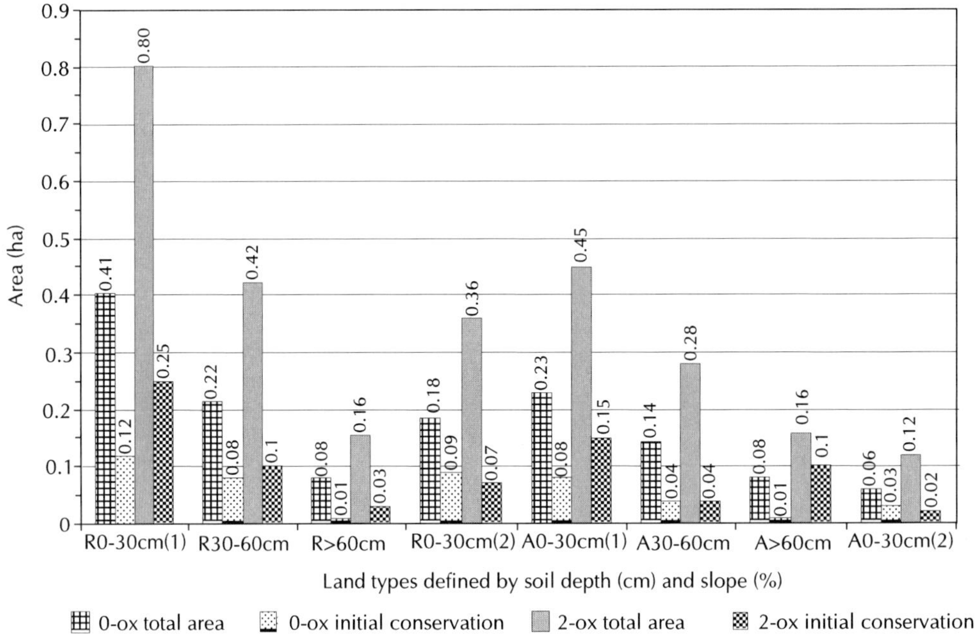

Fig. 12.1. Cultivable land and initial level of conservation by household group and land type (as defined in Table 12.3).

The Bioeconomic Model

Understanding farm households' incentives and constraints to intensification of land use, technology choice and investment behaviour, and analyses of the resulting pathways of development requires integration of biophysical and economic modelling approaches at the household level (Ruben *et al.*, 1998). The bioeconomic model developed here uses a non-separable farm household model (de Janvry *et al.*, 1991) as a basis. Production, consumption and investment decisions are jointly determined in each period. This results from imperfections in input and output markets that introduce divergence between selling and buying prices (price bands). In such situations, decisions are constrained by market imperfections, and household attributes and asset endowments will affect production and investment choices. The on-site costs of soil erosion and nutrient depletion are endogenous in the model and their future land productivity impacts influence the choice of land management practices. Off-site effects are not accounted for, but the model allows quantification of soil erosion and runoff that may also affect water bodies and other farmers in the vicinity.

The farm household maximises the discounted utility (*DU*) subject to resource supply, market access and subsistence consumption constraints:

$$DU = \sum_{t=1}^{T} U_t (1 + r_t)^{-(t-1)} \quad (1)$$

The utility function is specified as:

$$U_t = \frac{1-\mu}{\left(\frac{FY_t}{SY_t}\right)^{\mu-1}} + \mu - 1 \qquad (2)$$

where FY_t is the full income of the household (as defined in Equation 4). SY_t is the subsistence (poverty line) level of full income estimated based on the annual poverty line income (Dercon and Krishnan, 1996) of Birr 528 (US$1 = Birr 6 in 1993/94) per consumer unit (CU) and minimum consumption of leisure time in the area. The utility function (U_t) has an elasticity of marginal utility of income (also called flexibility of money) equal to $-\mu$. The curvature of the utility function has a relative risk aversion coefficient equal to μ. The marginal utility of income estimated for different countries ranges from –3 at low levels of per capita income to –1.1 at higher levels (Bieri and de Janvry, 1972). A value of $-\mu = 3$ was used. As defined, the utility function attains a negative value when income is less than subsistence, a zero value when income is just equal to subsistence, and a positive value when income is higher than subsistence consumption. This provides a good indicator of the welfare impacts of conservation investments

Based on Holden *et al.* (1998) the rate of discount r is endogenous in the model and is determined by the level of income and asset endowments:

$$r_t = \frac{FV_{t+1}}{PV_t} - 1 \qquad (3.1)$$

$$PV_t = z + \beta \left\{ \frac{(FY_t - \Sigma_s w_{st} L^c_{st})}{YCU_t} \right\} \qquad (3.2)$$

where PV_t is the present value equivalent of future income (FV_t) the household is willing to accept instead of waiting for one more year. The PV_t is assumed to be dependent on the level of income per CU in each period (YCU_t); the value of β is determined from an econometric model estimated for farmers in Ethiopia (Holden *et al.*, 1998). The upper and lower bounds of PV_t are estimated based on survey data and the highest and lowest discount rates found for households. Based on average incomes, the value of z is calibrated at levels consistent with the highest and lowest bounds for different household groups. In this way, an increase in household income increases the present value equivalent of future income, and reduces the rate of discount, indicating the household's ability to trade-off current consumption to improve future livelihoods. If the income level falls, the opposite would occur. The effects on technology choice and investments are estimated by solving the model for upper and lower bounds on the discount rates.

Household full income is given by:

$$FY_t = \sum_{g=1}^{G} \sum_{c=1}^{C} A_{cgt} \left\{ p_{ct} y_{cgt} \left(x_{cgt} \right) - \sum_{i=1}^{I} e_{icgt} x_{icgt} \right\} + \sum_{V=1}^{V} L_{vt} \left\{ p_{vt} y_{vt} (x_{vt}) - \sum_{i=1}^{I} e_{ivt} x_{ivt} \right\} + \sum_{s}^{S} w_{st} (L_{st}^{c} + L_{st}^{of}) \quad (4)$$

A_{cgt} is the area of crop c produced on land type g in year t. L_{vt} is production of units of livestock v in each period. x_t is a vector of inputs used in production of a unit of crop c in land type g and livestock v in year t. p is the per unit price of crops or livestock and e is the per unit input cost. y_{cgt} is the yield function for the production of crop c and y_{vt} is the yield function for livestock v. In year t, family home time (leisure) in each season (s) is L^c_s while w_s is the seasonal reservation wage (after transactions costs). The seasonal off-farm labour supply is L^{of}_s.

Linkages between the economic and biophysical system

The key equations that link the biophysical system with the economic behavioural model are embedded through the production functions that include the effect of changes in soil quality. Change in the soil nutrient stock is the cumulative outcome of positive and negative processes. Use of organic and mineral fertilisers adds soil nutrients, while soil erosion depletes both rooting depth and soil nutrients. The cumulative change in the available nutrient stock affects crop yields in the following years. Depending on the cost of abating soil degradation through conservation and/or fertiliser use, this creates the economic incentive to adopt new sustainability-enhancing practices. The change in the soil nitrogen (N) stock is given by:

$$N_{t+1} + N_t - \delta \left[N_t - \eta \left(SE_t \right) \right] - \eta \left(SE_t \right) \quad (5)$$

where SE_t is the period t rate of soil erosion, δ is the share of soil N mineralised in each period and η is the N composition of the soil. Based on the advice of agronomists, an enrichment ratio of 2 for eroded soil and an annual mineralisation rate of 1% for soil N were used. The change in plant-available soil-N due to soil erosion and nutrient depletion from period to period (dN) is computed as:

$$dN = \delta (N_t - N_{t+1}) \quad (6)$$

where δ is as defined above. The cumulative reduction in plant-available N is included in the production function (Equation 7.2) to influence crop yields in each period. Since incorporating the effect of phosphorus (P) depletion on land productivity requires additional data on P-fixation, conversion of stabile P to labile P, and the total P-stock in the soils, the model currently includes only the effects of depletion of rooting depth and soil-N on crop yields.

Crop yield (y_{cgt}) for crop c on land type g in period t is estimated in two steps. Firstly, the intercept term (y^{int}) representing the depth–yield relationship without fertiliser use was estimated econometrically as a function of soil depth (SD_t) and soil type (ST) based on the SCRP time-series collected at the site (Shiferaw and Holden, 2001). Secondly, responses to N and P were estimated from Food and Agriculture Organization of the United Nations (FAO) fertiliser-response studies (Ho, 1992) and the soil productivity calculator (Aune and Lal, 1995) as a function of fertiliser nutrients and the cumulative change in the available soil-N (dN_t). Hence, the intercept term and the yield function are given as:

$$y_{cgt}^{int} = f(SD_t, ST) \tag{7.1}$$

$$y_{cgt} = f(y_{cgt}^{int}, dN_t, N_t, P_t) \tag{7.2}$$

where N_t and P_t are nitrogen and phosphorus available to plants.

The rate of soil erosion (SE_t), and hence the change in soil depth for each land type, in each period depends on the soil type (ST), slope (SL), rainfall (RF), land management or conservation technology used (K), and the type of crop grown (c):

$$SE_t = f(ST, SL, RF_t, K_t, c_t) \tag{8}$$

The parameters of Equation 8 were obtained from the SCRP experiments at the site or were estimated based on plot-level survey data. In return, soil erosion affects soil depth in each period such that:

$$SD_t = SD_{t-1} - \varphi SE_t \tag{9}$$

where φ is the conversion parameter. Hence, the soil depth trajectory depends on the initial soil depth and the cumulative level of soil erosion. Most of the model *parameters* were exogenously determined. These parameters include input and output prices, wage rates, seasonal working days (excluding religious holidays), population growth rate, activity-wise resource requirements, nutrient content of local foods, and household asset endowments. Given the objective function and a set of resource availability and market constraints, the model determines optimal values of *variables* that represent crop–livestock production, consumption and conservation investments.

Other model variables and constraints

Major activities in the model include production of six crops on eight land types with ten levels of fertiliser use [diammonium phosphate (DAP) and urea]; two land management options; two cropping seasons; consumption, storage and selling of crops; allocation of family labour (over ten seasons) for production, conservation, off-farm employment (constrained) and leisure;

seasonal labour hiring; production, selling and consumption of livestock; buying of agricultural products for consumption; buying of livestock feed (crop residues); and constrained local markets for renting in/out land and oxen. The model constraints include limits that the use and sale of available resources (e.g. land, seeds, labour, fertiliser, oxen power, food, animal feed and liquidity) could not exceed total household endowments:

$$\Phi A_t - X_t^b \leq X_t^w \tag{10.1}$$

$$X_t^s \leq X_t^w - \Phi^w A_t \tag{10.2}$$

where A_t is a vector of the level of activity, Φ is a vector of total and Φ^w owned resource requirement per unit of activity A, X^w is a vector of owned resources, X^b is a vector of bought (hired) resources, and X^s is a vector of sold or out-rented resources. Available resource supplies can be increased through participation in markets (10.1). According to local norms, the model assumes that labour may be hired in cash, in kind (fixed output) or in exchange for traction power. Land can be in-rented in cash or in kind (fixed output), the price depending on its quality. The model also allows in-renting or out-renting of oxen in exchange for labour or cash. When the family resource stock is nil (e.g. fertiliser), all the demand will be met from markets. When markets exist, resources not used in production can also be sold, but the amount used and sold cannot exceed available supplies (10.2). The overall cash and credit constraint is specified as:

$$P_t^b X_t^b + (1+\gamma)X_{t-1}^{cr} - P^s X_t^s \leq X_t^{lq} + X_t^{cr} \tag{11}$$

where P^b is the buying and P^s selling price, X^{lq} liquidity at hand and X^{cr} is the level of credit (with interest rate γ) received during each period. When liquidity is non-existent, all purchases will be financed from available credit and sale of resources (inputs or products). When credit is not available, cash expenditures cannot exceed cash income from sales. The interest and the principal from the credit used in the previous period [$(1+\gamma)X_{t-1}^{cr}$] should be paid back during the next period. Consumption requirements were specified as:

$$\lambda \left[X_t^w + X_t^b\right] \geq \Omega \tag{12}$$

where λ is a vector of nutrient composition of owned (X_t^w) and purchased (X_t^b) foods and Ω is the biologically determined nutritional requirement for carbohydrates, fats and proteins. Households can use markets to meet resource demand (10.1) and consumption requirements (12) but buying activities for inputs and products include a price band of 5–10% over farm-gate selling prices. All purchases are also subject to a cash constraint given in Equation 11. The model also allows for the import of commonly consumed crops not grown in the area. Taste and food diversity constraints reflecting observed consumption choices were also imposed. Consumption of grains could also include savings from previous production. The consumption requirements depend on family size and CUs. The production balance in each year for consumed products is given as:

$$Q_{cons} + Q_{seed} + Q_{sold} + Q_{stored} = Q_{Tot} \quad (13)$$

This indicates that the total production is consumed (Q_{cons}), used as seed (Q_{seed}), sold (Q_{sold}) and/or stored (Q_{stored}) for subsequent periods. Likewise, family labour is allocated to different activities seasonally as follows:

$$L_{st}^{c} = L_{st} - (L_{st}^{f} + L_{st}^{of}) \quad (14)$$

This shows that family labour in year t and season s (L_{st}) is used on-farm (L_{st}^{f}), off-farm (L_{st}^{of}), and the residual consumed as leisure (L_{st}^{c}). Off-farm employment is constrained to average levels estimated from the survey for different household groups. Other constraints include restrictions on crop rotations such that cereals follow land sown to legumes in the previous period. Accounting equations include land, crop and technology-specific soil erosion; cumulative changes in soil depth; and cumulative changes in conservation investments. Changes in the stock of animals was specified for each type as:

$$LV_t = (1 - \theta - m)LV_{t-1} + LVR_{t-1} + LV_t^b - LV_t^s \quad (15)$$

where LV_t is adult livestock in period t, θ is the culling rate, m is the mortality rate, LV_{t-1} is the closing stock in the previous period, LVR_{t-1} is young stock of certain ages in the previous period reared into adult animals in period t and LV^b and LV^s are animals bought and sold during the period. Production and rearing of young stock is given as:

$$(1 - m)kLV_t^f = LVR_t + LVR_t^c - LVR_t^s \quad (16)$$

where LV_t^f is female animals of reproductive age, and k is the litter size per reproductive female. The total number of newborns, adjusted by the mortality rates (m), is reared (LVR), consumed (LVR^c) or sold (LVR^s) within the year. The detailed structure of the model is presented in Holden and Shiferaw (2004).

Scenarios for analysis of technology and policy impacts

The bioeconomic model was used to simulate the adoption and potential impact of two types of land and water management options introduced into the area by the SCRP and the Ministry of Agriculture. These technologies were developed based on graded soil–stone bunds to enhance water infiltration, and drainage of excess water, and to reduce soil erosion. Farmers indicated that the structures occupy productive land and reduce yields in the initial period, especially on steeper slopes. In order to assess how this will affect adoption of these technologies, we specified two stylised versions of the technology. Type I is when the initial effect of area loss from adoption of the conservation methods is negligible and short-term yields are unaffected, and Type II is when loss of productive land and other undesirable effects may reduce initial yields with conservation by 5–10% depending on the slope. The Type I situation may arise if conservation improves soil fertility or raises relative returns to fertiliser use and offsets the negative effect of area loss. The Type II situation may arise when positive effects are negligible or when negative outcomes are dominant. Both are very likely and valid scenarios. Even if Type II conservation has a short-term yield penalty, it could still be attractive in the long term as crop yields exceed those without conservation. The length of time needed for this to occur will depend on the interaction between existing soil depth, the level of soil erosion and the type of crop grown. However, with a positive discount rate, delayed benefits may not create incentives for small-scale farmers to adopt these technologies. The model captures these relationships and impacts on welfare outcomes and the condition of the resource base.

Furthermore, depending on slope, adoption of these technologies is estimated to require 100–120 working days/ha while annual maintenance requires 15–20 days/ha. The model also allows removal of some of the existing conservation structures installed through food-for-work programmes and mandatory polices of the past. Figure 12.1 shows the area of land under different categories and the existing level of conservation in the initial year. Removal is assumed to require 25% of the labour need for construction. The decision to remove will depend on the availability and opportunity cost of family labour, the ability to pay for hired labour, the scarcity of land, and the expected returns from removal or maintenance of the structures. The expected return will in turn depend on the crop grown, the soil type and the slope of the land.

The two variants of the technology (Type I and II) are nested in the model for two household groups: without oxen (poor households), and with a pair of oxen (less-poor households). Since farm and non-farm employment opportunities are limited, it is hypothesised that the relative availability of land and oxen assets will be crucial for household welfare while the relative abundance of family workforce relative to land will contribute to increased conservation investments. In order to capture this complex relationship, each of the two household groups are further disaggregated into two sub-groups depending on the relative endowment of land and labour resources within the household at the initial period. Hence, four modelling scenarios are developed: without oxen and land-scarce, without oxen and land-abundant, with two oxen and

Table 12.4. Household sizes in the selected scenarios at the initial period.

	Households with two oxen		Households without oxen	
	Land-scarce[a]	Land-abundant[b]	Land-scarce[b]	Land-abundant[a]
Family size	7.2	4.2	7.2	2.8
Worker units	4.0	1.5	4.0	1.5
Consumers units (CU)	6.6	3.0	6.6	2.6
Land (Regosols)[c]	6.98	6.98	3.52	3.52
Land (Andosols)	4.02	4.02	2.03	2.03
Total farm size	11.00	11.00	5.55	5.55
Total farm size per capita	1.53	2.62	0.77	1.98

[a] These are average values for the group from the study area.
[b] Labour endowments are adjusted to explore the effect of changes in land–labour ratios.
[c] The land areas are in *Timad* (approximately 0.25 ha).

land-scare, and with two oxen and land-abundant. Table 12.4 shows the major attributes and cumulative asset endowments of these four household groups. The model uses the detailed land classification shown in Table 12.3. The multi-period model, written in GAMS, is solved for $t = 5$ years. The 5-year model has about 25,700 variables and solves within 1–2 hours using present-day Pentium-4 computers.

Simulation Results

As stated earlier, the bioeconomic model allows a simultaneous evaluation of the level of technology adoption and the associated effects on productivity, human welfare and sustainability. The optimised model provides extensive results on the crop–livestock economy, marketed surplus, conservation investments, consumption levels and changes in soil depth and soil erosion. The main focus here is on adoption of NRM technologies and productivity and environmental impacts. The differential conservation adoption patterns and the resulting livelihood and resource conservation outcomes for the different household groups are discussed. The level of conservation investments is reported for the different land types at varying endogenous rates of discount.

Adoption of NRM technologies

Households with a pair of oxen

Boserup (1965) hypothesised that intensification of land use and investments to enhance land productivity will be limited when land is more abundant than labour. This suggests that labour-scarce families with large farms will have lower incentives to increase the intensity of labour use and other inputs per unit of land to enhance its productivity. This may particularly be the case if land markets are imperfect and surplus land cannot be sold or leased out to others. These simulations also indicate that when land is more abundant than labour, the land users lack sufficient incentives to make significant erosion control investments (see Tables 12.5 and 12.6). The level of investment in conservation and soil fertility management is much larger for land-scarce households than for land-abundant households. When conservation does not incur a short-term yield penalty (Type I), the land-scarce households make significant conservation investments in all land types except the steep slopes that are mainly used for grazing. While labour-scarce households adopt conservation practices on a maximum of one-third of the different land types, the land-scarce (labour-rich) households are able to adopt conservation on up to 97% of the area of some land types (Table 12.5).

Compared to the land-scarce household, the short-term welfare impact of soil degradation in terms of future productivity decline is relatively less for the land-abundant household. Even if soil erosion increases on untreated lands, households with relatively abundant land will have enough land to

Table 12.5. Livelihood and environmental impacts for households with two oxen: when conservation technology does not take land out of production.

Household type	Household welfare	Average net income (Birr/consumer unit (CU))[a]	Adoption of conservation practices (% total area) terminal period: Regosols 0–30 cm (1)	Regosols 30–60 cm	Regosols >60 cm	Regosols 0–30 cm (2)	Andosols 0–30 cm (1)	Andosols 30–60 cm	Andosols >60 cm	Andosols 0–30 cm (2)
Land-abundant[b]	4.521	1305.8	32.4	26	25.8	22.2	2.6	17.9	0	0
Land-scarce[b]	2.076	708.6	78	97.6	93.5	97.2	0	96.4	57.1	0
Land-abundant[c]	2.68	1296.4	32.4	26.0	25.8	22.2	2.6	17.9	0	0
Land-scarce[c]	1.208	702.8	78.1	97.6	93.5	95.9	0	96.4	57.1	0
Land-abundant[d]	2.68	1296.4	100	100	100	100	1.3	100	0	0
Land-scarce[d]	1.208	702.8	70	57	100	100	0	100	80	0

[a] In 1993/94, US$1 = Birr 6. Current rates are about US$1 = Birr 8.6.
[b] Low discount rate (ranges: 0.25 to 0.26 for land-scarce, and 0.21 to 0.22 for labour-scarce households).
[c] High discount rate (ranges: 0.57 to 0.58 for land-scarce, and 0.50 to 0.51 for labour-scarce households).
[d] Percentage of the initial area of treated land maintained at the end of the terminal period (high discount rate).

Table 12.6. Livelihood and environmental impacts for households with two oxen: when conservation takes 5–10% land out of production.

Household type	Household welfare	Average net income (Birr/consumer unit (CU))[a]	Adoption of conservation practices (% total area) terminal period: Regosols 0–30 cm (1)	Regosols 30–60 cm	Regosols >60 cm	Regosols 0–30 cm (2)	Andosols 0–30 cm (1)	Andosols 30–60 cm	Andosols >60 cm	Andosols 0–30 cm (2)
Land-abundant[b]	4.511	1304.0	0	0	0	0	0	0	0	0
Land-scarce[b]	2.03	700.8	84.6	0	0	13.2	3.6	0	0	0
Land-abundant[c]	2.674	1293.8	0	0	0	0	0	0	0	0
Land-scarce[c]	1.1857	695.0	64.5	0	0	13.1	0	0	0	0
Land-abundant[d]	2.674	1293.8	0	0	0	0	0	0	0	0
Land-scarce[d]	1.1857	695.0	0	0	0	0	0	0	0	0

[a,b,c,d] Refer to footnotes to Table 12.5.

maintain their current welfare levels. The limited effect of degradation on their welfare reduces the incentive to mitigate the externality, especially when the rental value of land does not increase with conservation investments. A labour-scarce household with relatively abundant land will cultivate some of the land and rent out the rest. The incentive to treat out-rented land with conservation investments depends on the expected economic benefits. It was found that village land rentals markets do not reflect the value of conservation investments but do reflect land quality aspects that affect its productivity. This means that land of the same quality (whether or not treated with conservation measures) has the same rental value and that there is no short-term economic incentive for the land 'owner' to invest in conservation. Therefore the model does not choose conservation on out-rented plots. This result would have changed if the rental value of land decreases due to soil degradation as in share-tenancy arrangements. Future work may need to assess such effects. Shortage of labour relative to land also means that the labour-scarce household may have to hire-in labour in order to install labour-intensive conservation investments. The cumulative effect of scarcity of labour and land abundance is lower soil conservation effort for the labour-scarce household.

For Type I conservation technologies, it was also found that the labour-scarce households maintain much of the initial conservation (except those on deep soils where erosion effects are low or on marginal lands used for grazing) previously installed on their lands through programme benefits, while the land-scarce households dismantle most of the initial conservation (Table 12.5).

The investment gap and resulting impacts on the welfare of households and the resource base are even more pronounced for Type II conservation technologies that could take some land out of production and reduce initial crop yields (Table 12.6). In this case, both types of households quickly dismantle the existing conservation structures, especially in plots where the perceived risk of erosion is low. However, land-scarce households eventually install them on shallow soils where their effect on productivity is high and hence conservation benefits are large (Table 12.6). The re-investment on some plots seems to occur in later years as welfare levels improve from livestock production and storage of surplus grains. This may not be the case if risk were to be included in the model (Holden and Shiferaw, 2004). Compared to Type I technology, in the 5-year period considered here the overall conservation investment is highly reduced. The households may not, however, have removed the initial conservation investments if a longer planning horizon and a lower discount rate were used (although this may not be a realistic assumption). Moreover, since the discount rates are high and a longer time period is required for conservation benefits to have appreciable effects on productivity, the upper and lower bound discount rates in both cases did not show significant differences in household conservation investments.

Table 12.7. Livelihood and environmental impacts for households without oxen: when conservation technology does not take land out of production.

Household type	Household welfare	Average net income (Birr/consumer unit (CU))[a]	Adoption of conservation practices (% total area) terminal period							
			Regosols				Andosols			
			0–30 cm (1)	30–60 cm	>60 cm	0–30 cm (2)	0–30 cm (1)	30–60 cm	>60 cm	0–30 cm (2)
Land-abundant[b]	3.371	1049.4	22.2	37.7	13.7	0	35.2	28.7	13.7	0
Land-scarce[b]	–2.937	434.0	67.6	95.3	87.5	0	39.1	62.9	87.5	0
Land-abundant[c]	1.901	1027.2	22.2	37.7	13.7	0	35.2	28.7	13.7	0
Land-scarce[c]	–1.921	425.0	22.2	95.3	87.5	0	39.1	65	87.5	0
Land-abundant[d]	1.901	1027.2	74	100	100	0	100	100	100	0
Land-scarce[d]	–1.921	425.0	0	0	0	0	0	0	0	0

[a,b,c,d] Refer to footnotes to Table 12.5.

Table 12.8. Livelihood and environmental impacts for households without oxen: when conservation technology takes 5–10% land out of production.

Household type	Household welfare	Average net income (Birr/consumer unit (CU))[a]	Adoption of conservation practices (% total area) terminal period							
			Regosols				Andosols			
			0–30 cm (1)	30–60 cm	>60 cm	0–30 cm (2)	0–30 cm (1)	30–60 cm	>60 cm	0–30 cm (2)
Land-abundant[b]	3.364	1048.4	0	0	0	0	0	0	0	0
Land-scarce[b]	–3.06	430.0	22.2	0	0	0	39.1	0	0	0
Land-abundant[c]	1.901	1025.6	0	0	0	0	0	0	0	0
Land-scarce[c]	–1.99	421.8	22.2	0	0	0	39.1	0	0	0
Land-abundant[d]	1.901	1025.6	0	0	0	0	0	0	0	0
Land-scarce[d]	–1.99	421.8	0	0	0	0	0	0	0	0

[a,b,c,d] Refer to footnotes to Table 12.5.

Households without oxen

The corresponding results for the two household groups without oxen are presented in Tables 12.7 and 12.8. Under Type I technology, it was also found that the relative abundance of labour and scarcity of land improves the likelihood of sustainability investments. However, compared to households with a pair of oxen, the level of adoption of conservation is reduced, so the productivity and sustainability impacts of improved NRM options are relatively diminished. When the household is poor both in oxen and land, large family sizes put high pressure on the household's ability to meet subsistence needs. While the lack of oxen for ploughing compels the household to rent out land, imperfections in food markets and the presence of price bands work in the opposite direction. Under pressure from conflicting market influences, the household in-rents some traction power to grow a portion of its subsistence needs and rents outs some of its land. It spends about 15% of the available working time on hiring-in oxen for traction. However, meeting the consumption requirements of a large household becomes difficult unless the surplus labour finds some employment off-farm; the household allocates the allowable maximum 25% of the available labour time in activities that include petty trade and employment within and outside the village to earn supplemental income. If the labour market is missing, the model becomes infeasible, indicating that the household is simply unable to meet its subsistence needs unless external assistance (e.g. food aid) is provided. If sufficient off-farm employment is available, labour-rich households without oxen are more likely to reduce on-farm labour and work more off-farm, which may further depress investment in conservation. When off-farm employment is limited (as in this case) the household invests labour to install Type I conservation technologies (see Table 12.7). These investments occur on prime agricultural land where conservation benefits are high while steep slopes [R0-30cm(2) and A0-30cm(2)], mainly used as pasture for livestock, are left without conservation.

When the household is poor in both oxen and labour, the relative abundance of land and shortage of labour discourages conservation investments. The household will hire-in some traction power and labour seasonally to produce part of its subsistence, but will rent out the remaining land annually without conservation. Since fewer workers also mean smaller CUs, the household with relatively abundant land is able to meet its subsistence needs although it invests relatively less in conservation practices. Imperfections in land, oxen, labour and credit markets jointly constrain labour- and oxen-poor households from investing in conservation while compelling them to rent out part of their land assets to labour- and oxen-rich households within the village. If the revenue from land rentals declines because of soil degradation (i.e. rental markets reflect the value of soil conservation), and if labour, oxen and credit markets function well, the labour-scarce household is likely to use much of its land for itself or rent out it after undertaking conservation investments. Currently there is no credit for conservation, and fertiliser credit is minimal and unreliable (see below on the effect of credit). Both selling and long-term leasing of land are illegal in Ethiopia. Along with productivity-enhancing technical change, lifting such restrictions could enhance the value of land and the efficiency of land rental markets. Empirical evidence in Africa and elsewhere shows that under favourable policies (e.g. secure land rights) and market conditions, and when sustainability investments provide high relative returns, smallholders are unlikely to ignore the sustainability impacts of current land-use decisions (user costs) (Tiffen *et al.*, 1994; Heath and Binswanger, 1996; Templeton and Scherr, 1999; Holden *et al.*, 2001). These are important policy constraints that need to be tackled to encourage land investments in Ethiopia.

As expected, labour-scarce households maintain more of the initial conservation measures than land-scarce households. The situation is very different for Type II conservation technologies (Table 12.8). In this case, both households remove the conservation structures on their plots. Only land-scarce and labour-endowed households allocate some labour for conservation. Hence, the level of conservation adoption is minimal and the attained impact on the quality of the resource base is very limited mainly because exploitative traditional agricultural practices with high levels of soil erosion (up to 40 t/ha) continue (Shiferaw and Holden, 2001).

Economic and sustainability impacts

The above results have clearly shown the roles of land and labour scarcity in household conservation investment decisions. It was hypothesised that the endowment of traction power and farmland will largely determine the welfare impacts of new technologies. Households that are poor in land and oxen can therefore be expected to attain the lowest level of welfare. The discounted utility (welfare) and the average net income per CU for the different scenarios are presented in Tables 12.5–12.8. The results show that adoption of NRM practices is very minimal for Type II technologies. This means that the farmer will largely use existing practices and the welfare and environmental impacts from such interventions will be minimal. Comparison of the welfare and income differences under Type I and II technologies can therefore reveal the economic impacts associated with adoption of improved NRM practices. For example, the land-abundant household attains a welfare level of 4.521 under Type I, which declines to 4.511 under Type II where no adoption has occurred, representing a discounted welfare gain in 5 years of 0.22%. Similarly the average net annual income per CU has shown a slight increase of about Birr 2 (0.15%), which amounts to about Birr 10 in 5 years. These are direct benefits associated with the reduction in soil degradation from adoption of the conservation technology. It is to be noted that the best NRM technology simulated (Type I) does not enhance yield; it only reduces soil erosion while yields remain unchanged in the initial years. The economic gain would have been more pronounced had the NRM technology also contributed to growth in crop yields. Moreover, in all the scenarios simulated, the better-off households with two oxen attain the poverty line level of welfare ($U_t>0$ and $DU>0$) under both technology alternatives. Oxen-

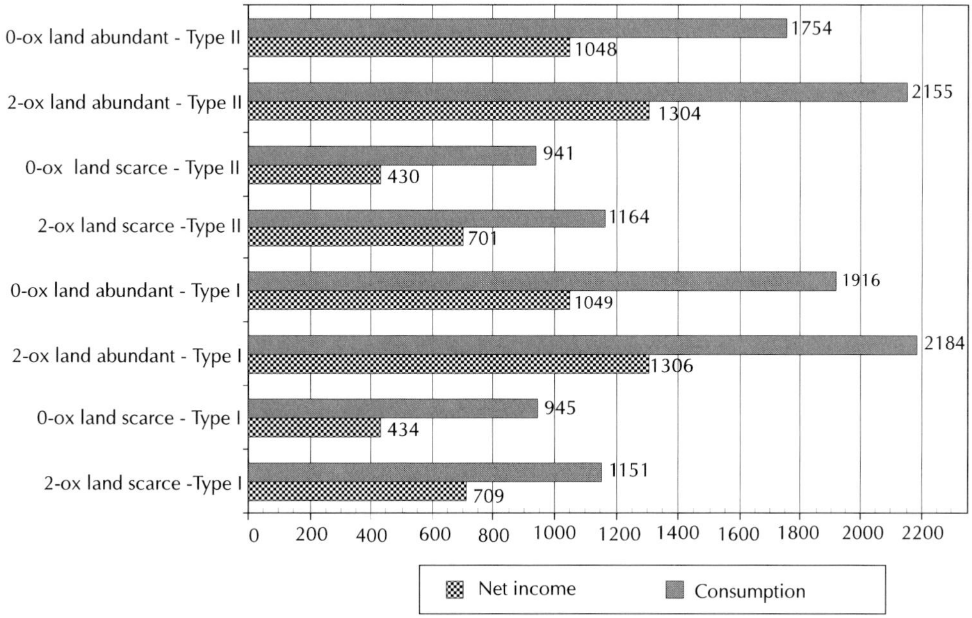

Fig. 12.2. Average annual net income (Birr) and consumption (kcal) per consumer unit by household group (US$1 = Birr 6 in 1993/94).

owning and land-rich households with fewer consumers attain the highest level of welfare. Households without oxen attain the poverty-line welfare level only when land is not scarce and the number of consumers is limited. A combination of land and oxen poverty along with insufficient off-farm employment opportunities makes the household unable to attain the poverty-line full income. Hence, these households are unable to escape poverty ($U_t<0$) even when Type I conservation is used. This probably explains why many poor households in the area also have small-sized families.

The income and consumption outcomes (at low discount rates) are depicted in Fig. 12.2. Using the annual poverty line income of Birr 528/CU and subsistence calorie requirement of 840 kcal/CU, the results show that all household groups attain the subsistence level of consumption but not the poverty-line net income. Land-scarce households without oxen fall far short of this level of income even though they meet their subsistence level of calorie consumption.

It will be useful to assess the level of economic gain from adoption of improved conservation practices. The gain in household net income attained per unit of land area conserved can be estimated from comparison of the net income with and without adoption of Type I technologies. For example, the average household annual net income for land-scarce and two-oxen households with adoption of Type I technologies is about Birr 51 higher than that without adoption. This amounts to about Birr 36/ha/year of conserved land. If irreversibility in soil degradation is assumed, the perpetual on-site net gain from adoption of conservation practices amounts to Birr 72 to 180/ha

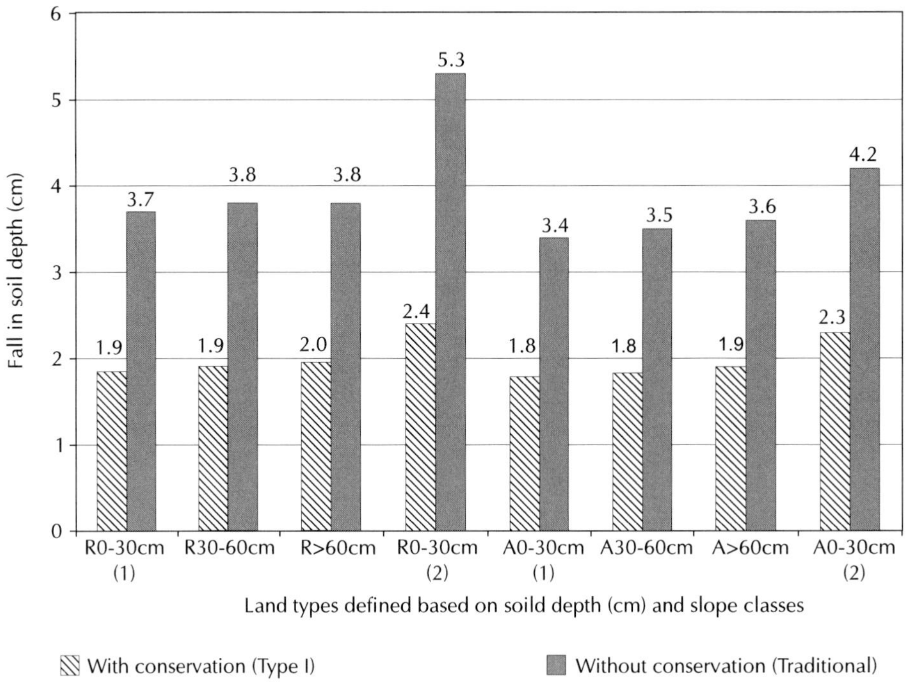

Fig. 12.3. Reduction in soil depth in 10 years: land-scarce 2-ox household (land types as defined in Table 12.3).

using the farmer's high (50%) and low (20%) discount rates. Adoption of high-yielding varieties and other options is likely to increase the net farmer benefits from conservation.

In order to show the long-term environmental or sustainability impacts of adopting improved management practices, the model was solved for a planning horizon of 10 years under Type I and traditional practices. The results are shown in Fig. 12.3. The fall in soil depth under Type I conservation technology is about half of that under traditional management. Depending on the soil and land type, soil depth declines by about 1.8–2.5 cm with conservation, but this increases to 3.2 to 5.4 cm under traditional management. As was shown in Equation 7.2, crop yields depend on many variables including the use of organic and inorganic fertilisers. Figure 12.4 shows the effect of soil degradation on crop productivity under differing levels of fertiliser use. If farmers do not use chemical fertilisers, barley yields decline by about 175 kg/ha without conservation (No Cons), while this loss falls to less than 50 kg/ha with conservation (Cons). This indicates that, depending on the relative returns, farmers have the option of using fertilisers to replace lost nutrients or of investing in conservation practices to mitigate the effect of soil degradation. Policies for fertiliser or conservation subsidies have been used to achieve productivity and/or sustainability objectives. Since fertiliser price subsidies are no longer popular policy options, it could be useful to investigate how the credit constraint might affect farmers' conservation choices. This is explored further in the following section.

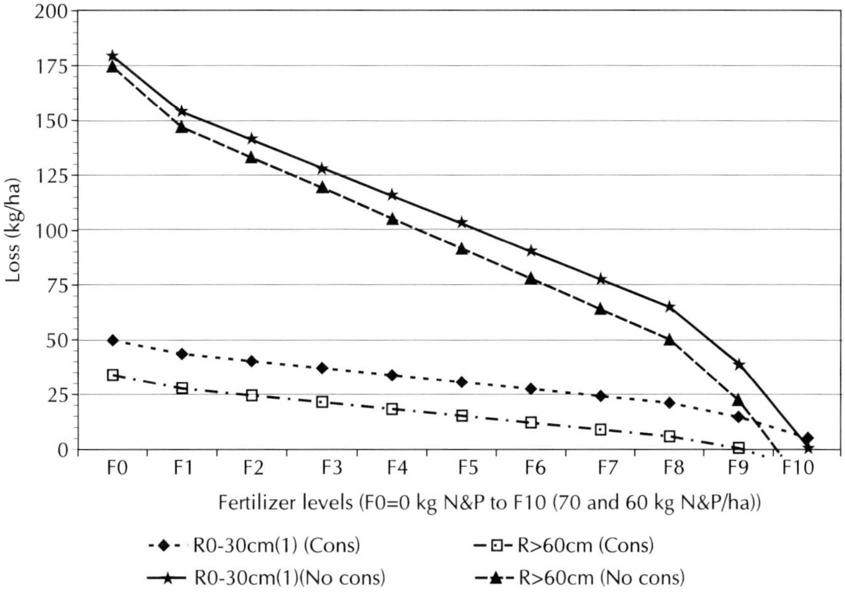

Fig. 12.4. Fertiliser use and decline in barley (*Meher* season) yields in 10 years under alternative land management practices (Type I vs. traditional).

Effect of credit for fertiliser

As discussed earlier, credit was not included in the base model presented thus far. What happens if the Ethiopian government increases allocation of credit for conservation and production purposes in the future? Availability of credit and fertiliser use are critical ingredients for stimulating adoption of improved technologies. Several earlier studies have shown that subsidised credit may increase fertiliser use (especially when profitable varieties are available) and may discourage investment in soil conservation (Shiferaw *et al.*, 2001; Holden and Shiferaw, 2004). When cheap credit is available, high levels of fertiliser use can easily replace lost nutrients and reduce the need for soil conservation. The same effects can be expected from fertiliser price subsidies. As Fig. 12.4 shows, under high levels of fertiliser use, the relative productivity benefits of conservation disappear and crop yields will be similar to those without conservation. We find that for credit-constrained households, increased availability of input credit could discourage investment in conservation. This is demonstrated using results for the poor and land-scarce household group (Table 12.9). As the availability of input credit improves, the level of conservation investment declines progressively, even when Type I conservation technology is available. With Type II conservation technologies, access to credit seems to entirely wipe out all the incentives for conservation. In this case fertiliser use becomes more economical than soil conservation to counter soil degradation. The decrease in sustainability investments occurs while short-term welfare improves because of increased fertiliser use and improved land productivity. It is not clear, however, for how long fertilisers

Table 12.9. The effect of access to credit on smallholder conservation investments (land-scarce households without oxen).

Techno-logy type	Credit level[a]	Fertiliser use (kg)[b]	Regosols				Andosols			
			0–30 cm (1)	30–60 cm	>60 cm	0–30 cm (2)	0–30 cm (1)	30–60 cm	>60 cm	0–30 cm (2)
I	No credit	D=150, U=62	67.6	95.3	87.5	0	39.1	62.9	87.5	0
	Limited	D=235, U=134	65.2	84.8	25	0	39.1	35	62.5	0
	High	D=250, U=147	32.1	41.9	25	0	39.1	35	12.5	0
II	No credit	D=150, U=62	22.2	0	0	0	39.1	0	0	0

[a] The 'limited' credit level is specified as Birr 400 from formal (12% interest) and Birr 250 (60% interest) from informal sources (US$1 = Birr 6 in 1993/94). The 'high' level of credit is twice that of the 'limited' level.
[b] D = diammonium phosphate (DAP), U = urea.

can be used to mitigate the effect of soil degradation. Agronomists argue that a minimum soil depth is essential for crop production and that once soil erosion reduces the rooting depth below a given threshold level, the marginal productivity of fertiliser use may decline. This indicates that as soil degradation increases, more fertiliser may be required to compensate for losses and to sustain crop productivity. This trade-off could be tackled through interlinkage of credit supply with conservation requirements (Holden and Shiferaw, 2004), a policy that could foster win–win economic and environmental outcomes.

Conclusions and Policy Implications

In resource-poor regions with high population pressure, sustainable use of land and other resources has become an important policy and development problem. Improved NRM interventions are important to reverse soil degradation and sustain agricultural productivity. Several recent studies have posited a nexus between poverty and the ability to undertake sustainability investments, especially when markets are imperfect. Bioeconomic models that interlink biophysical information with behavioural economic models at different spatial scales in a dynamic perspective are most suited to the analysis of NRM impacts and to determine how poverty in certain assets affects investment decisions. Using data from the Ethiopian highlands, it has been shown how a non-separable bioeconomic household model can be used to track these relationships and impacts, and how the effect of technology and policy changes affecting NRM can be evaluated simultaneously in terms of economic efficiency (the incentive to adopt the technology), welfare (poverty effects) and sustainability (resource conditions). The model is formulated for four stylised household groups and captures production, biophysical diversity and market conditions in the area. The results show that when land is relatively abundant, households are unlikely to carry out labour-intensive conservation investments. An increase in family labour coupled with scarcity of land, however, increases the incentive to invest in conservation, especially when opportunities for off-farm employment are limited and profitable conservation technologies are available. In this case, higher adoption of resource management practices leads to positive impacts on household welfare and sustainability of resource use.

It is also found that poverty in labour and traction power forces households to rent out land to other relatively better-off households. Under the existing system of usufruct rights to land in Ethiopia, sustainability investments that do not affect short-term crop yields do not affect the rental value of land. In this case, the oxen- or labour-poor households rent out land without conservation because the returns from renting are the same. This points to the need for new policies and interventions that would improve the efficiency and effectiveness of land rental markets and create incentives for land users to consider the future productivity impacts of current land-use decisions (user costs).

The economic incentive to invest in conservation drastically decreases when the new technologies increase scarcity of land and decrease crop yields in the short term (Type II). This scenario seems to explain the extensive removal in the study area of conservation measures introduced in the past. Unfortunately, better options that provide short-term benefits to the poor are rarely available and the only reasonable way to encourage investments in such practices is to provide some targeted subsidies (e.g. cost-sharing). However, when farmers are able to perceive the consequences of soil degradation and use-rights are secure, they are able to adopt Type I conservation technologies without additional incentives. Only labour-scarce households and those without the necessary traction power are unable to make significant investments due either to the relative abundance of land or to the high opportunity costs of labour.

The direct economic gains from the adoption of Type I technologies are quite modest. The average annual income gain is estimated at about Birr 36/ha, which translates to an increase in annual income per consumer of Birr 10 in 5 years. This is partly because the nature of the technology simulated in this case does not improve yields. Higher benefits can be expected if conservation also enhances land productivity. But the low return to available conservation technologies is a major factor that makes conservation investments less attractive than competing alternatives (e.g. off-farm employment or livestock production). This suggests the need to develop NRM technologies that provide attractive economic gains along with sustainability benefits. Land-scarce households without oxen even failed to attain the poverty-line income. The level of conservation adoption and its impact is lowest for land-abundant households. Adoption of conservation measures did not arrest soil degradation, but did provide substantial benefits in terms of maintaining soil depth and improved crop productivity. The decline in soil depth with conservation is half of that under traditional practices, but the yield reduction is less than one-third of that without conservation. Fertiliser use could also reduce yield losses. There is some evidence that increased fertiliser credit may substitute for conservation effort. This may require cross-compliance types of policies that link fertiliser credit with conservation requirements.

Nevertheless, evaluation of economic and environmental impacts will not be complete until the added social benefits are compared with the research and development (R&D) costs incurred in generating and delivering these technologies on a larger scale. When these costs are low and the associated economic and sustainability benefits are high, improved social efficiency from such NRM investments can be expected.

References

Aune, J. and Lal, R. (1995) The tropical soil productivity calculator: a model for assessing effects of soil management on productivity. In: Lal, R. and Stewart, B.A. (eds) *Soil Management Experimental Basis for Sustainability and Environmental Quality.* CRC Press, Lewis Publishers, Boca Raton, Florida, pp. 499–520.

Bieri, J. and de Janvry, A. (1972) *Empirical Analysis of Demand under Consumer Budgeting*. Gianini Foundation Monograph 30. University of California, Berkeley, California, 60 pp.

Boserup, E. (1965) *The Conditions of Agricultural Growth. The Economics of Agrarian Change under Population Pressure*. Earthscan Publications, London, UK, 124 pp.

Cleaver, K.M. and Schreiber, G.A. (1994) *Reversing the Spiral. The Population, Agriculture and Environment Nexus in Sub-Saharan Africa*. The World Bank, Washington, DC, 227 pp.

de Janvry, A., Fafchamps, M. and Sadoulet, E. (1991) Peasant household behaviour with missing markets: some paradoxes explained. *Economic Journal* 101, 1400–1417.

Dercon, S. and Krishnan, P. (1996) A consumption-based measure of poverty for rural Ethiopia in 1989 and 1994. In: Kebede, B. and Taddesse, M. (eds) *The Ethiopian Economy: Poverty and Poverty Alleviation. Proceedings of the Fifth Annual Conference on the Ethiopian Economy*. Addis Ababa University, Addis Ababa, Ethiopa, pp. 77–101.

Grepperud, S. (1996) Population pressure and land degradation: the case of Ethiopia. *Journal of Environmental Economics and Management* 30, 18–33.

Heath, J. and Binswanger, H.P. (1996) Natural resource degradation effects of poverty and population growth are largely policy induced: the case of Colombia. *Environment and Development Economics* 1, 64–84.

Ho, C.T. (1992) *Results of NPK fertilizer trials conducted on major cereal crops by ADD/ NFIU (1988–1991)*. ADD/NFIU Joint Working Paper 43, Ministry of Agriculture, Addis Ababa, Ethiopia, 85 pp.

Holden, S.T. and Shiferaw, B. (2004) Land degradation, drought and food security in a less-favoured area in the Ethiopian highlands: a bioeconomic model with market imperfections. *Agricultural Economics* 30(1), 31–49.

Holden, S.T., Shiferaw, B. and Wik, M. (1998) Poverty, credit constraints, and time preferences: of relevance for environmental policy? *Environment and Development Economics* 3, 105–130.

Holden, S.T., Shiferaw, B. and Pender, J. (2001) Market imperfections and land productivity in the Ethiopian highlands. *Journal of Agricultural Economics* 52(3), 53–70.

Reardon, T. and Vosti, S.A. (1995) Links between rural poverty and the environment in developing countries: asset categories and investment poverty. *World Development* 23(9), 1495–1506.

Ruben, R., Molla, H. and Kuyvenhoven, A. (1998) Integrating agricultural research and policy analysis: analytical framework and policy applications for bioeconomic modeling. *Agricultural Systems* 58, 331–349.

Shiferaw, B. and Holden, S. (1998) Resource degradation and adoption of land conservation technologies in the Ethiopian highlands: A case study in Andit Tid, north Shewa. *Agricultural Economics* 18(3), 233–248.

Shiferaw, B. and Holden, S.T. (2001) Farm-level benefits to investments for mitigating land degradation: empirical evidence from Ethiopia. *Environment and Development Economics* 6, 335–358.

Shiferaw, B., Holden, S.T. and Aune, J. (2001) Population pressure and land degradation in the Ethiopian highlands: a bioeconomic model with endogenous soil degradation. In: Heerink, N., van Keulen, H. and Kuiper, M. (eds) *Economic Policy Reforms and Sustainable Land Use in LDCs: Recent Advances in Quantitative Analysis*. Springer-Verlag, Berlin, Germany, pp. 73–92.

Templeton, S.R. and Scherr, S.R. (1999) Effects of demographic and related microeconomic change on land quality in hills and mountains of developing countries. *World Development* 27, 903–918.

Tiffen, M., Mortimore, M. and Gichuki, F. (1994) *More People—Less Erosion: Environmental Recovery in Kenya.* John Wiley and Sons, Chichester, UK, 311 pp.

Yohannes, G. (1989) *Land-use, agricultural production and soil conservation methods in the Andit Tid Area, Shewa Region.* Research Report 17. Soil Conservation Research Project, Ministry of Agriculture Soil and Water Conservation Development, Addis Ababa, Ethiopia, 151 pp.

13 Assessing the Impacts of Natural Resource Management Policy Interventions with a Village General Equilibrium Model

S.T. Holden[1] and H. Lofgren[2]

[1] *Agricultural University of Norway, Ås, Norway*
[2] *International Food Policy Research Institute (IFPRI), Washington, DC, USA*

Introduction

Rural economies in developing countries are characterised by significant transactions costs and imperfect information (Hoff *et al.*, 1993; Sadoulet and de Janvry, 1995). The resulting gap between (higher) buying prices and (lower) selling prices causes farm households to be only partially integrated into markets. Missing markets or non-participation in markets may cause production decisions of farm households to become non-separable from consumption decisions (Strauss, 1986; de Janvry *et al.*, 1991). Transaction costs cause isolation of markets and inter-spatial and inter-temporal price variation. Many prices therefore become local and interdependent (general equilibrium prices) as well as dependent on external and local conditions. Such economies may be characterised as constrained Pareto-inefficient and the implication is that there may almost always be room for Pareto-improving interventions (Greenwald and Stiglitz, 1986, Stiglitz, 1986).

Land degradation may be the most serious environmental problem requiring prompt attention in sub-Saharan Africa (SSA), and is particularly severe in the densely populated areas in eastern Africa (Stoorvogel and Smaling, 1990; World Bank, 1996). Poverty is particularly severe in the Ethiopian highlands where Holden and Shiferaw (2002) found that poverty in combination with market imperfections affected the ability and willingness of poor households to invest in conserving their own land. The challenge to researchers and policy makers is to identify technologies and policies that can reduce the problems of environmental degradation while at the same time improving the welfare of the poor.

 Natural Resource Management in Agriculture: Methods for Assessing Economic and Environmental Impacts (eds B. Shiferaw, H.A. Freeman and S.M. Swinton)

This chapter explores some of these alternative natural resource management (NRM) interventions and their impacts using a village computable general equilibrium (CGE) model to capture welfare and relevant environmental effects. The specific objectives are to:

- Explain why and when an economy-wide modelling approach is preferred for impact assessment
- Demonstrate how a village bioeconomic CGE model can be developed for the purpose of evaluating the impacts of NRM technology and policy interventions
- Present and discuss the key findings from the case study.

The case study focuses on assessing the welfare and environmental impacts of such NRM policy options as input and output taxes and subsidies and the welfare impacts of changes in levels of land productivity. The model, developed for a rural economy in the Ethiopian highlands, is used to track the associated impacts on efficiency and equity, and on the level of soil degradation. The economy is characterised by transaction costs in both the internal village markets and in the markets linking the village economy to the external world. Compared to other village economies in Ethiopia the village economy that is modelled here has high agricultural potential and good market access. The model contains a representation of the crop–livestock system in the area through nested (i.e. inputs are combined at different levels as shown in Table 13.2 later) constant elasticity of substitution (CES) production functions with more realistic elasticities of substitution[1], multiple inputs and multiple outputs. Imperfections in the local markets[2] are captured through market-specific transaction costs[3] and price bands for some commodities and factors, and missing markets for others. The level of land degradation, represented by a decline in soil productivity, is a function of soil type, crop type, and crop management (fertiliser use). The CGE model is calibrated to a 1993 village social accounting matrix (SAM) for the study area.

Rapid land degradation (soil erosion and nutrient depletion) has been documented in the East African Highlands (FAO, 1986; Stoorvogel and Smaling, 1990; Grepperud, 1996). This degradation is due to erosive cropping and low levels of fertiliser use. Low fertiliser use is attributed to poor market access (high marketing costs), lack of capital/purchasing power, poor access to credit, and production and price uncertainties. The policy reforms of the Ethiopian Government have involved devaluation and removal of subsidies on fertilisers and other inputs. It has been questioned whether the fertiliser subsidy removal causes more rapid land degradation and whether a subsidy on fertilisers could be defended on environmental grounds (Holden and Shanmugaratnam, 1995).

We therefore use the village CGE model to assess the impacts of: 1. removal of output price tax[4] leading to output price increases of 10%, 2. reduction of fertiliser subsidy from 20% to 10%, 3. increase in fertiliser subsidy from 20% (1994 level) to 30%, and 4. soil degradation in the following year, on social welfare (household income) across different household groups, production, export, and import. The first three scenarios are 'with and without policy change' comparisons, while the fourth one is a two-period scenario where we see the consequences of 1 year of land degradation on household welfare in the following year.

In the next part of the paper, we provide the rationale for use of village CGE models for NRM policy impact assessment. Policies and NRM in Ethiopia are then discussed, followed by a description of the case study area. The structure of the village CGE model is summarised, followed by a presentation and analysis of the simulation results and the authors' conclusions.

Village CGE Models for NRM Policy Impact Assessment

CGE models are economy-wide models that can be applied at different levels of aggregation, from local (village, watershed, community), to a district, region, country, group of countries, or the whole world.

A SAM is needed for the area for which the model is to be built. The SAM typically includes production activities, commodities, factors of production, institutions (including household groups, firms and government), savings–investment and possibly capital accounts, and 'rest of the world'. The SAM is a database that provides input (structure of the economy and starting values) for the CGE model.

The SAM is used to give a complete map of resource, commodity and service flows in an economy. It requires that all sources and sinks for the transactions together with their associated prices are identified. Market imperfections that lead to non-separability and price differences between sellers and buyers of factors of production create a need to have household group-specific activity accounts (for activities, commodities and factors of production) in the SAM. Price bands due to transaction costs also cause a need to operate with different prices for net sellers and net buyers. The grouping of households should be based on wealth/resource characteristics to obtain relatively uniform groups. Such groups also tend to have similar market participation behaviour. Furthermore, a CGE model consists of a system of equations that captures the behaviour of agents in the model, price formation, market characteristics, material balances, and trade relations.

CGE models have the capacity to capture general equilibrium (GE) effects that cannot be captured in household models. General equilibrium effects may occur due to changes in policies, shocks (e.g. droughts), or improvement of technologies.[5] They may also come gradually due to population growth or land degradation. The beauty of these models is that they are flexible in terms of identifying which outputs and inputs should have an endogenous or exogenous price. The scale at which the model is constructed and the nature of markets (degree of isolation or integration) determine which prices become endogenous and which remain exogenous. Holden (Chapter 8, this volume) presents a typology of village economies and village economy models. The scale of analysis has to be fitted to the issue that should be analysed. Village (watershed or community) models are relevant when there are significant local GE effects (Holden, Chapter 8, this volume). Imperfections in markets may also cause producer–consumer households to make non-separable production and consumption decisions. The simultaneous existence of endogenous shadow prices and local GE effects causes a need to nest non-separable household models into micro-CGE models.

The links between macro-policy changes and rural micro-economies and the environment may be complex. The main links between the economy and the environment in agriculture-based poor rural economies go through the agricultural production activities of farm households. Natural resources are depleted in the production process, reducing the production potential of the farming system unless a sufficient amount of productivity-raising investments are made. The production and investment decisions of farm households are endogenous responses to exogenous changes in policies and other external factors, conditioned by household and farm characteristics. This implies that land degradation becomes dependent on household characteristics, i.e. the poverty of households may affect how they manage their land.

Holden *et al.* (1998b; and also Holden, Chapter 8, this volume) developed a typology of village economies and village economy models based on the size of transaction costs in relation to trade and the degree of differentiation in asset ownership within villages. This typology indicates that it is relevant to use village CGE models only when significant transaction costs lead to endogenous price determination in village markets. This means that the market for some factors or commodities clears within the village such that prices are determined through interplay of supply and demand within the local economy. Some differentiation is required in order for farm households to have incentives to trade with each other, given that there are also transaction costs related to local trade as well. Holden *et al.* (1998b) found that a remote Zambian village, in which local trade was insignificant, could be modelled as a number of non-separable farm household models. This shows that in the absence of local trade, economy-wide models are not required. Taylor and Adelman (1996) modelled village economies as consisting of a number of separable farm household models, allowing GE-effects but not allowing poverty and other household characteristics to affect production and investment decisions. Lofgren and Robinson (1999) developed a CGE model that allowed household groups to endogenously choose between participation and non-participation in markets.

CGE models at the micro-level (village, watershed, community) have the strength to capture important aspects of the local structure of production, decision-making, market characteristics (including local GE effects), and natural resource linkages. GE effects at a higher level, like price effects due to technical change in a wider area, cannot be captured, however, unless such prices are endogenised. Village CGE models will typically treat such prices as exogenous because local production and imports and exports are considered to be too small to affect prices in the broader economy. However, many output prices for tradable commodities may also be exogenous in national CGE models (small open economies). This is therefore not a limitation of the CGE modelling approach as such, but rather a reflection of the structure of the economy and the price-formation process.

How much do we then gain or lose by using village-level CGE models to evaluate technology and policy impacts associated with NRM? CGEs capture important economy-wide effects that household models cannot capture. Household models may on the other hand capture more detailed technology specifications and seasonal variations and constraints than CGE models. One may therefore say that the CGE models should contain simpler versions of the household models. The two types of models can be seen as complementary rather than as substitutes in relation to NRM technology and policy impact assessment.

Policy Changes and NRM in Ethiopia

Market imperfections are common in rural markets in Ethiopia. This may partly be due to historical factors since economic policy only recently (in 1991) was changed from a socialist, top-down planning system to a more market-friendly regime. It may take time before more efficient markets develop. The poor infrastructure also causes transaction costs to be high. Ethiopian land reform in 1975 resulted in an egalitarian distribution of land among farm households. While all land is state-owned, user rights have been allocated to individual households through the land reform in 1975 and several land redistributions in the years that followed. Land sales are illegal but there are active land rental markets. The distribution of other rural assets, most importantly livestock, is less egalitarian and this creates incentives for trade within villages, including the renting of land and oxen.

The most serious environmental problem in Ethiopia is land degradation, primarily caused by soil erosion and nutrient depletion. This leads to on-site and off-site external effects in the sense that the level of soil degradation may be higher than what is socially optimal. The on-site external effects can be external because of high discount rates due to market imperfections, poverty, and insecure or unspecified private property rights (Holden *et al.*, 1998a). One consequence of imperfections in markets and tenure regimes is land degradation, manifested in declining productivity, as users lack incentives to make sufficient investments in the land that they operate. Such degradation may be irreversible. The net present value of this permanent productivity loss that may be considered an inter-temporal externality (Holden and Shiferaw, 2002) can be considerable. In this study we estimate the size of this productivity loss and assess how it is affected by various policy reforms.

With the regime change in 1991 and its replacement by a more market-friendly government, Ethiopia embarked on structural adjustment policy reforms. These reforms included devaluation of the exchange rate, removal of fertiliser subsidies, removal of price controls for agricultural commodities (pan-territorial pricing), and privatisation of public enterprises.

The development strategy, called 'Agricultural Development-Led Industrialisation,' is focused on the development of labour-intensive industries that rely heavily on domestic raw materials and inputs from smallholder agriculture. The new strategy aims to stimulate market development, competition, and efficiency. Consequently, the Government has dissolved producer cooperatives, reduced the role of state farms, abolished compulsory food grain quotas, and removed price controls on agricultural commodities and domestic trade restrictions.

Like many African countries, Ethiopia followed a pan-territorial fertiliser pricing policy and provided subsidies to smallholder farmers. These subsidies are often blamed for creating wrong incentives to farmers although the universality of this claim has been questioned (Holden and Shanmugaratnam, 1995). The fertiliser subsidy in Ethiopia was 15% in 1993, 20% in 1994, 30% in 1995, and 20% in 1996. Following the devaluation in 1992, fertiliser prices increased sharply, causing a decline in fertiliser consumption in that year. Fertiliser subsidies were therefore introduced, but later reduced and then eliminated starting from 1997. Fertiliser use has remained low in Ethiopia after the reforms. In terms of nutrients the average rate of fertiliser application is 7 kg/ha in Ethiopia against 9 kg/ha for SSA, and 65 kg/ha worldwide. A new fertiliser distribution policy was introduced in 1997. It called for elimination of fertiliser subsidies and pan-territorial pricing system for fertiliser. The involvement of the private sector in importation and distribution of fertilisers was also stimulated.

Although dependence on rainfed agriculture and frequent droughts continue to pose serious concerns, most macro indicators suggest that economic performance has been strengthened since the introduction of the reforms. On average, growth seems to have accelerated, both in agriculture and other parts of the economy, while overall inflation has remained moderate. Both exports and imports have grown much more rapidly than gross domestic product (GDP), thus drastically increasing the openness of the economy.

The environmental and poverty impacts of the changes in output tax and input subsidy policies have not yet been carefully analysed. Policy changes took place based on general assumptions that were not empirically verified. This study attempts to provide a careful assessment of the efficiency, distributional and environmental impacts of some of the policies directly affecting NRM in the country.

Case Study Area

The study area consists of the Hidi, Hora Kilole and Borer Guda peasant associations in the Ada-Liben district in Showa region in the central highlands of Ethiopia. This area is favourably located approximately 20 km from Debre Zeit, which is near the main highway and only 50 km east of the capital, Addis Ababa. In addition to good market access, the area enjoys a high agricultural potential. Ada-Liben district is a surplus producer of teff (*Eragrostis tef*), the main crop (both in terms of consumption and market sales) and the preferred cereal among Ethiopian consumers. The production system is an integrated crop–livestock system where oxen provide traction power for land cultivation, and straw from grain production is the main source of animal fodder. Very little communal land exists as most of the land has been distributed to individual households.

Land rental markets are active, particularly given that land cannot be sold or purchased. Usually, fixed-rent contracts are used. Livestock, most importantly oxen, are the most important privately owned asset in the study area. The distribution of this resource is less egalitarian and is a good indicator of household wealth. Oxen ownership is also an important indicator of farming capacity due to the crucial role of oxen. Rental markets for oxen are less important because of moral hazard problems in relation to oxen management and because proper timing of ploughing is crucial on the dominant heavy black soils in the area. In this setting, characterised by a non-egalitarian distribution of oxen and impediments to oxen rental, households without oxen rent out much of their land. Typically, households with two or more oxen rent in land as they have excess ploughing capacity. Households with one ox tend to exchange oxen with other one-ox households as a pair of oxen is needed for ploughing. Average household and farm characteristics for different oxen ownership groups of households are presented in Table 13.1. In 1993/94, 25% of the households had no oxen, 17% had one ox, 34% had two oxen and 24% had more than two.

Table 13.1. Basic farm household characteristics in the survey area of Ethiopia, 1993.

	Household category by number of oxen				
Variable	0	1	2	>2	All
Share of population (%)	25	17	34	24	100
Female headed households (%)	27	7	7	3	11
Farm size (kert)[a]	4.48	7.4	8.18	11.25	7.83
Total income (Birr)[b]	2,992	4,893	5,792	12,279	6,489
Male work force (adult equivalents)	0.71	1.34	1.57	2.84	1.62
Female work force (adult equivalents)	0.91	1.01	1.12	1.68	1.18
Consumer units (adult equivalents)	2.47	3.76	4.12	6.47	4.2
Tropical livestock units	0.31	2.46	4.46	9.12	4.09

[a]1 kert = 0.3 ha.
[b]In 1993/94, US$1 = Birr 6; current rates are about US$1 = Birr 8.6.

The local economy is highly agriculturally oriented, as the diversification into non-farm activities is limited. However the area is a net importer of unskilled labour (seasonal demand in crop production) but exports some skilled labour.

Holden *et al.* (1998a) found that the subjective discount rates of farm households in the study villages were high and that they were influenced by wealth as poorer households had higher subjective discount rates. This indicates that households are credit constrained and that poorest households suffer most. Credit in kind (in the form of fertiliser) was provided in the area but this credit also appeared to be rationed in 1993/94 when the study was undertaken.

Holden and Shiferaw (2002) estimated the farm households' willingness to pay (WTP) to sustain land productivity in the area as almost all farm households stated that land productivity was declining over time. They also estimated farmers' perceived average rates of land productivity decline. Shiferaw and Holden (1999) used the Universal Soil Loss Equation adapted to Ethiopian conditions to estimate soil erosion in the area, and production functions adapted from experimental studies at other locations in Ethiopia to estimate the impact of erosion on crop yields. The resulting estimates of average rates of land productivity decline were about twice as high as those estimated based on farmers' judgements.

Model Description

General model structure

As the starting point for our village CGE model, we use a standard CGE model developed by Lofgren *et al.* (2002). This model uses General Algebraic Modeling System (GAMS) software that aims to make CGE analysis more cost-effective and more accessible to a wider group of analysts. The model has been applied to a large number of countries. Its flexibility is based on two main features: 1. It separates the model from the database, making it easier to apply the model to new settings, and 2. It permits the user to choose among alternative assumptions for how factor markets and macro constraints operate (Lofgren *et al.*, 2002). Box 13.1 shows the main steps in formulating a village CGE model to evaluate technology and policy impacts.

Box 13.1. How can a village CGE for NRM impact assessment be built?

Building a CGE model for NRM impact assessment can be quite demanding in terms of data and skills. The major steps in the construction of a village CGE (using the standard model) can be summarised as follows:

- The required data need to be collected from the relevant village/watershed. The data to be collected should include the source (origin) and sink (destination) and price data for all transactions, and the relevant biophysical data (depending on the environmental aspects that need to be incorporated into the model)
- In order to capture the local market conditions, it is important to determine the market characteristics of the economy. This requires answers to such basic questions as which factor or output markets have endogenous prices and which of these markets are missing (have shadow prices)
- Then the households need to be classified into uniform categories based on their important resource characteristics
- The SAM can then be constructed by balancing income and expenditure data for households and household groups, balancing supply and demand for endogenous markets, including transaction costs in relation to trade where relevant, and finally balance the overall SAM
- One can then start to prepare the input file with the SAM for the CGE model. This requires defining sets for the CGE model including the structure of production and how changes in NRM will affect output
- It will also require incorporating the SAM in the input file, and the different elasticities for the model in relation to production, consumption and trade need to be defined
- At this stage the standard village CGE model can be adjusted and modified in terms of setting the proper market characteristics for all factors of production and commodities at the village level
- Once this is done the base simulation can be run and errors corrected until the model satisfies the basic requirements
- The simulation file can now be prepared to run the relevant policy experiments. At this final stage, it will be possible to run simulations with the new policy issues, compare results with the base values, and identify the efficiency, equity and sustainability impacts.

The modelling structure has also been adapted for village-level analysis. The general structure of the modelling system is kept separate from the specific structure and inputs needed for a specific economy to be modelled. Relatively few changes are therefore needed in its general structure. The specific structure is entered through an input file or database containing most of the relevant elements and quantitative inputs for modelling the specific economy. One important distinction for a village model from a country model is that a village does not have a separate currency with an exchange rate and this has implications for how the interactions of the village with the surrounding economy are balanced. Another distinction is that 'the rest of the world', export, import, and 'foreign' have different meanings in village and country models. For the village model these terms refer to outside the village, be it inside or outside the country.

The database of the model consists of the SAM, elasticity data for production, consumption and trade, and possibly physical factor quantities. The model is developed as a set of simultaneous equations, many of which are non-linear. The model must be exactly identified, i.e. the number of equations must be equal to the number of endogenous variables in the model. The equations capture the behaviour of agents in the model (producers, consumers, households, government), e.g. by incorporating first-order optimal conditions for profit and utility maximisation. Furthermore, the model contains material balance equations, market characteristics and price equilibria, government policy (taxes, subsidies, and transfers), savings–investment balances, and 'import–export' balances.

A simplified picture of the model and its building blocks is provided in Fig. 13.1 (Lofgren *et al.*, 2002). The arrows represent payment flows, while the material flows go in the opposite direction. All transactions in an economy (and in SAMs and CGEs) involve a flow of goods or services (material flows) and payment flows in the opposite direction. Refer to Lofgren *et al.* (2002) for a detailed mathematical description of the standard CGE model.

Specific model structure

The general model was modified to accommodate the characteristics of the economy in the study area. These modifications included fitting a village SAM to the requirements of the CGE model. Market imperfections (in markets for: land, labour, oxen ploughing services, manure, crop residues, and crop outputs) in the study area caused production and consumption-decisions of households to be non-separable. The prices for land and oxen services became endogenous to villages (closed village market only) and household groups depending on these factors as net sellers or net buyers of these factors of production. Shadow prices for non-traded factors (crop residues and manure) became endogenous to each of the household groups. Transaction costs in the local factor markets caused household group-specific factor prices for factors traded within the village to depend on whether the household group was a net seller or net buyer of the factor in the village

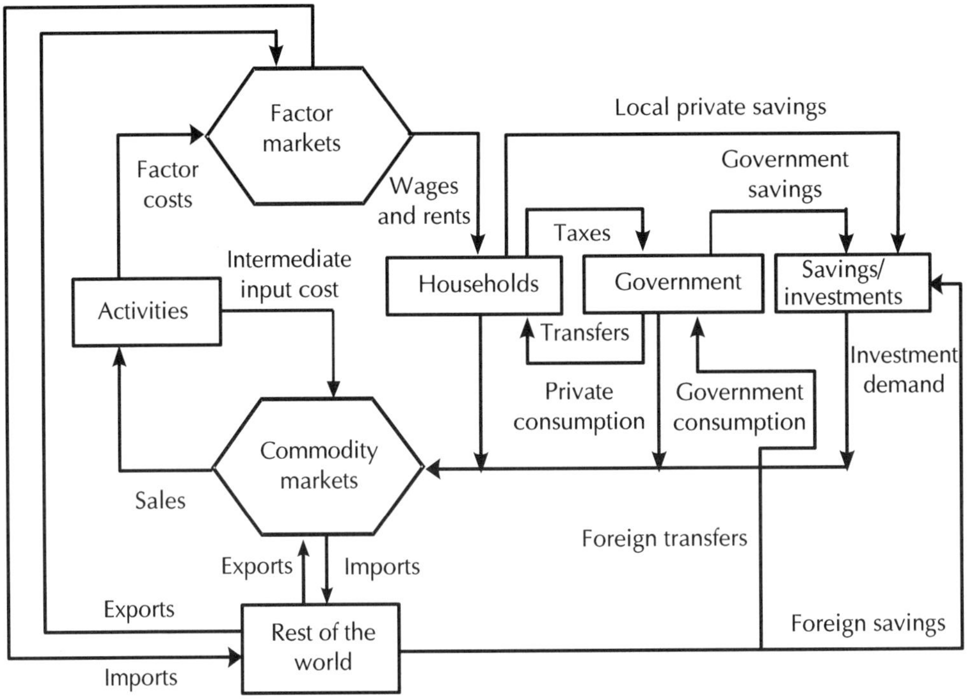

Fig. 13.1. Structure of payment flows in the standard CGE model (Lofgren *et al.*, 2002).

market. Household group-specific and general commodity accounts were used to capture transaction costs in relation to local and external trade.

As discussed earlier, oxen ownership is used as a basis for household group classification. One additional reason for this is that oxen ownership tends to drive the participation in land rental markets. In the CGE model and the underlying village SAM, the households were therefore divided into three groups on the basis of oxen ownership. Households with two oxen or more were pooled into one group, while households with zero and one ox were kept as separate groups.

The modelling structure is quite flexible and this has been further developed in order to handle a relatively complex farming system (crop–livestock system) and market imperfections induced by transactions costs. Households typically produce a variety of crops and livestock types. Agricultural production activities may have multiple inputs and multiple outputs. Outputs from one activity may be inputs in another activity. Typically crop residues are used as livestock fodder, while oxen are used for land preparation, and animal manure is used as fuel. Production technology is captured by nested, two-level CES-production functions, allowing substitution elasticities to be different at different levels of the nest (Table 13.2). At the bottom level, substitution elasticities between oxen, land, capital and labour may be quite low ($\sigma = 0.2$) relative to the elasticities between land and fertilisers at the higher level of the nest ($\sigma = 0.9$). This reflects the

Table 13.2. Basic structure of the village social accounting matrix (SAM).

	Activities	Commodities	Factors	Households	Government	Savings–Investment	Transaction costs	RoW	Total
Activities		Outputs							Activity income (gross output)
Commodities	Intermediate inputs			Private consumption	Government consumption	Investment	Transaction costs	Exports	Demand
Factors	Value-added, transaction costs								Factor income
Households			Factor income to households		Transfers to households			Transfers to households from RoW	Household income
Government	Producer taxes, value-added tax	Sales taxes	Factor income to Government, factor taxes	Transfers to Government, direct household taxes				Transfers to Government from RoW	Government income
Savings–investment				Household savings	Government savings			Foreign savings	Savings
Transaction costs		Transaction costs							Transaction costs
Rest of the World (RoW)		Imports			Government transfers to RoW				Foreign exchange outflow
Total	Activity expenditures	Supply	Factor expenditures	Household expenditures	Government expenditures	Investment	Transaction costs	Foreign exchange inflow	

relatively fixed relationship between land, oxen, capital and labour in relation to land preparation and the much higher flexibility that exists in relation to fertiliser application.

As a result of transaction costs, selling prices are typically lower than buying prices. In the base year, each household group is a net seller, self-sufficient, or a net buyer of various factors of production (inputs) and commodities. This also determines the choice of price when value added is imputed to the different factors in the SAM. Markets for crop residues and manure are missing. Manure is used for fuel (an input into the chores activity, representing miscellaneous tasks carried out inside the households). SAM cell values for manure are based on the nutrient contribution of manure and the cost of nutrients if they were bought as fertiliser (for nitrogen and phosphorus). The values of crop residues and fodder from grazing land were determined residually after subtracting the value of labour and animal stock from the value of livestock production.

A village SAM that was structured to match the requirements of the CGE model was constructed. The SAM is based on a survey of the economy in 1993, carried out in 1994. The basic structure of the village SAM is presented in Fig. 13.2. Transaction costs in relation to village exports and imports are captured by separate accounts while transaction costs in relation to internal village trade are primarily represented by the labour time needed to carry out the transactions. This implies a considerable expansion in the number of rows and columns in the SAM, making it too big for reproduction here (a copy of the SAM used in this case study can be obtained from the authors upon request).

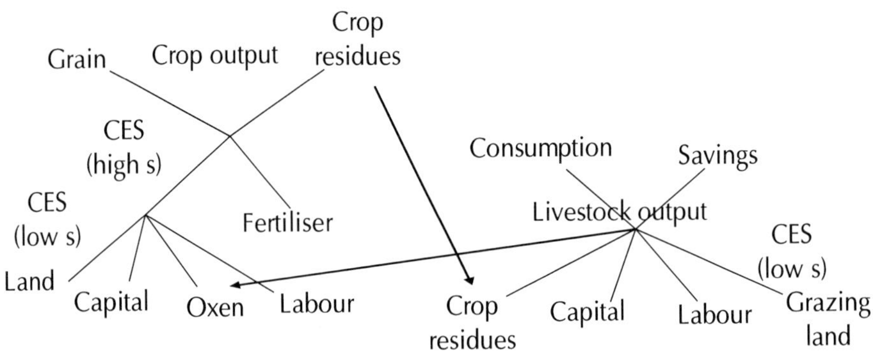

Fig. 13.2. Technology tree in crop and livestock production.

Land productivity declines have been estimated for the area for different types of soils and crops (Shiferaw and Holden, 1999). Information about farmers' perceptions on the rates of land degradation is available from Holden and Shiferaw (2002). The rate of productivity decline is reduced by the use of fertilisers that replace lost nutrients. The model is calibrated such that the estimated rates of productivity decline (annual mean, 1.1%) are taken as an indicator of the rate of productivity decline when no fertilisers are used, while the average rate of fertiliser use in the area is assumed to reduce the rate of productivity decline to the rate that farmers perceived (annual mean, 0.55%).

Some of the equations that capture the link between the land productivity and land degradation are described below. The land productivity decline per unit of land is a function of land type (A), crop choice (C), and household type (H) such that:

$$\psi_{ach} = \psi(F_{ach}, \psi_{MAX}(A, C)) = \psi_{MAX}(A, C)/(100\,(1 + \beta F_{ach} / Q_{ach})) \quad (1)$$

where F_{ach} is the household, land and crop specific fertiliser use, ψ_{MAX} is the maximum land productivity decline that takes place when no fertiliser (F) is added to the land, Q_{ach} is the current output of crop type c, on land type a by household type h, and β is a calibration parameter. The rate of productivity decline without fertiliser use is specified at two levels – high and low. The actual values used $\psi_{MAX}(A, C)$ for different crop types and land types in the model are given in Table 13.3. The intensity of fertiliser use (F_{ach}) is household group (H) specific and depends on the price of fertiliser (P_F), crop choice (C), crop price (P_C) and land type (A);

$$F_{ach} = F(H, P_F, P_C, C, A) \quad (2)$$

while the fertiliser price (P_F) depends on the import price (including transportation costs) (P_{FI}), and the level of subsidy (S_F);

$$P_F = P_{FI} - S_F \quad (3)$$

Table 13.3. Maximum land productivity decline rates used in the model.[a]

	Level of land degradation, annual yield decline (%)	
Land and crop type	High	Low
Upland, teff	4.1	1.2
Lowland, teff	0.38	0.25
Upland, other cereals	3.5	0.9
Lowland, other cereals	0.3	0.1
Upland, pulses	3.5	0.5
Lowland, pulses	0	0

[a] Based on Shiferaw and Holden (1999, 2000) and Holden and Shiferaw (2002).

Optimal fertiliser use is determined through the first-order conditions for the production functions; Equation 2 captures this. The first order conditions equate marginal value products to the prices of the respective inputs. An output price change will similarly affect the first-order conditions and affect both input use and output supply. Other things remaining the same, reduction in the fertiliser subsidy will cause the fertiliser price to increase, the level of fertiliser use to go down, and the level of productivity loss from land degradation to increase. This land degradation externality is aggregated across areas (L_{ach}) of different land types, crop types, and household groups, assuming that the process is irreversible,[6] with a social discount rate δ and that land use and output prices (P_{QC}) are constant over time:

$$LDEXT = \sum_H \sum_A \sum_C \psi_{MAX} L_{ach} P_Q / \delta \tag{4}$$

Household consumption in the model is captured by a Stone-Geary Linear Expenditure System that can handle broad commodity groups. Leisure is one of the commodities that is included in the model, consistent with theoretical household models but unlike typical macro CGE models. Agricultural production for home consumption is also included in the expenditure system. All household groups are net sellers of agricultural commodities, indicating the importance of the area as a surplus producer of food grains. The agricultural production activities are given in Table 13.4.

Table 13.4. The impact of alternative policies on land use and crop and livestock production activities for the three household groups[a] defined in Table 13.3.

Production activities by household group	Base results	Output price increases by 10%	Fertiliser subsidy decreases to 10%	Fertiliser subsidy increases to 30%	Land degradation 1 year (high)
Upland teff					
H0	36.5	5.5	–1.4	1.5	–2.1
H1	99.5	1.5	–2.1	2.4	–2.2
H2	1098.7	1.2	–1.6	1.8	–2.7
Lowland teff					
H0	30.2	5.6	–1.3	1.5	0.0
H1	52.0	–0.8	–2.4	2.6	–0.0
H2	823.9	0.8	–1.5	1.6	0.3
Upland other cereals					
H0	8.5	3.2	0.1	–0.1	–3.6
H1	19.8	1.5	0.1	–0.2	–2.3
H2	193.9	0.7	–0.5	0.6	–2.2
Lowland other cereals					
H0	5.6	7.4	–2.5	2.8	3.5
H1	7.4	6.4	–3.3	3.6	2.3
H2	53.2	7.5	–3.8	4.4	1.8
Upland pulses					
H0	2.9	1.5	2.1	–2.2	–1.9
H1	9.5	–2.4	3.1	–3.1	–3.6
H2	108.6	2.2	–0.3	0.4	–2.9
Lowland pulses					
H0	6.2	2.2	0.5	–0.4	–1.2
H1	12.5	–0.4	3.1	–3.0	–0.6
H2	147.9	0.0	0.5	–0.5	–0.7
Livestock					
H0	8.2	4.4	–0.9	1.0	–0.8
H1	53.0	1.2	–0.7	0.7	–0.6
H2	622.3	0.9	–0.4	0.5	–0.4

[a] Base results in '000 Ethiopian Birr; non-base simulation results as % change from base results.

The model is used to simulate the effects of an increase or a reduction in fertiliser subsidies, and an output price increase (tax reduction).

Along with other policy issues that affect NRM, fertiliser is used as an important NRM technology because it can reduce the rate of nutrient depletion. The primary NRM policy evaluated is a subsidy for fertiliser. In principle, it may be possible to think of funding of such a subsidy to come from a tax on crop output, but the output tax also has an impact on the land degradation externality. The model assesses the impact of these policies on household welfare, input use, output supply, and on land degradation. Cost–benefit analysis (CBA) is used to evaluate the overall efficiency of the policy change by incorporating the environmental outcomes into the cost–benefit calculations. The CBA includes the direct household income/expenditure effects of policy changes, the change in government tax income/subsidy cost, and the change in the environmental externality induced by the policy change. The CBA for the output and fertiliser price policies is done for high and low levels of degradation for three different social discount rates (3, 5 and 10%) (Tables 13.5 and 13.6).

Table 13.5. The social efficiency of a 10% output price increase (tax reduction).[a]

	Level of land degradation and social discount rates					
	High			Low		
Effects	3%	5%	10%	3%	5%	10%
Income effects	412.9	412.9	412.9	412.9	412.9	412.9
Productivity loss effects	–89.1	–53.4	–26.7	–24.1	–14.5	–7.3
Tax loss effect[b]	–357.2	–357.2	–357.2	–357.2	–357.2	–357.2
Net benefit	–33.8	1.9	28.6	31.2	40.8	48.0
Net benefit to cost ratio	–0.09	0.005	0.08	0.087	0.11	0.13

[a] In '000 Ethiopian Birr (except for the net benefit to cost ratios). The effects are calculated in relation to the baseline values before the price change.

[b] The tax loss effect is based on the assumption that output prices are controlled and taxed by the government.

NRM Policy Simulation Results

For each simulation experiment, in the following sections we present the impacts of NRM policies (fertiliser subsidies and output taxes) on household incomes, crop and livestock production, village exports, fertiliser use, land degradation, and the costs and benefits of the policy interventions themselves.

Because of the influence of market imperfections, the impacts of specific policies vary by land type, crop type and household group. This is a consequence of the non-separability of production and consumption decisions, making land use and land degradation a function of household characteristics. This has been one of the major limitations of many of the standard CGE models, which assume that production decisions are unaffected by poverty or equity.

Table 13.6. The social efficiency (CBA) of fertiliser subsidy changes.[a]

Difference from base (Fertiliser subsidy = 20%)	Policies at different social rates of discount: 3% Fertiliser subsidy 10%	3% Fertiliser subsidy 30%	5% Fertiliser subsidy 10%	5% Fertiliser subsidy 30%	10% Fertiliser subsidy 10%	10% Fertiliser subsidy 30%
Household income effect	–59.3	67.5	–59.3	67.5	–59.3	67.5
Land degradation externality effect (high level)	–42.3	48.8	–25.4	29.3	–12.7	14.6
Land degradation externality effect (low level)	–13.2	15.4	–8.0	9.2	–3.9	4.6
Government subsidy cost saving	47.1	–60.1	47.1	–60.1	47.1	–60.1
Net benefit (high level)	–54.5	56.2	–37.6	36.6	–24.9	22.0
Net benefit (low level)	–25.4	22.7	–20.2	16.5	–16.1	11.9
Net benefit to Government cost ratio (high level)	–1.16	0.93	–0.80	0.61	–0.53	0.37
Net benefit to Government cost ratio (low level)	–0.54	0.38	–0.43	0.28	–0.34	0.20

[a] In '000 Ethiopian Birr (except for the benefit to cost ratios). The government cost effect is the direct cost to the government of the fertiliser subsidy.

The model developed here demonstrates how this assumption can be relaxed to make CGE models capture poverty–NRM linkages more effectively.

The impact of increase in output prices

The results from a 10% increase in the prices for all outputs (reduction of output taxation) are examined first. The income effects for the different household groups are shown in Table 13.7. The 10% price increase leads

Table 13.7. The impact of alternative policies on real household incomes by household group.[a]

Household group	Base results	Output price increases by 10%	Fertiliser subsidy decreases to 10%	Fertiliser subsidy increases to 30%	Land degradation 1 year (high)	Land degradation 1 year (low)
Without oxen (H0)	263.4	14.8	–1.2	1.4	–1.3	–0.3
With one ox (H1)	281.1	13.4	–1.6	1.8	–1.5	–0.4
With two or more oxen (H2)	2972.1	11.3	–1.7	2.0	–1.5	–0.4

[a] Base results in '000 Ethiopian Birr; non-base simulation results as % change from base results.

to an increase in household real incomes of 11.3–14.8% for the different household groups. The poorest group responds more (14.8%) to the output price increase than any other group. This is mainly because the price increase makes production more profitable thereby allowing them to cultivate more of their land themselves (rent out less). This is supplemented by increased fertiliser use and livestock production. The impacts on production activities are shown in Table 13.4. An increase in output prices has a positive impact on production of most commodities although there is some variation across commodities and household groups. The price policy had a stronger effect on production of lowland cereals (wheat and barley). The lion's share of the production comes from households who own at least a pair of oxen.

Perhaps a surprising result is the impact on marketed surplus from the village (Table 13.8). The output price increase reduced the marketed surplus of grains because the income effect increases the demand for self-consumption more than it stimulates local production. A relatively high income elasticity for food consumption among the poor households explains the increase in food consumption. Market imperfections and low elasticities of input substitutions may also explain this low supply response. Similar results were found by Bardhan (1970) and de Janvry and Kumar (1981) in parts of India.

Table 13.8. The impact of alternative policies on village exports.[a]

Factor/trade/ activity	Base results	Output price increases by 10%	Fertiliser subsidy decreases to 10%	Fertiliser subsidy increases to 30%	Land degradation 1 year (high)
Teff	794.9	–1.2	–1.9	2.0	–1.5
Other cereals	9.7	–3.1	1.2	–1.5	–0.6
Pulses	34.4	–8.6	9.9	–10.0	–3.6
Livestock	211.4	0.7	–0.3	0.2	–0.3
Business	113.7	–13.3	1.9	–2.1	1.7
Skilled labour	13.8	–59.9	20.9	–19.2	19.6

[a] Base results in '000 Ethiopian Birr; non-base simulation results as % change from base results.

The effect of output prices on fertiliser demand is shown in Table 13.9. The price increase led to increased fertiliser use for cereal crops. At the aggregate level a 10% price change led to an almost equivalent (9.7%) increase in fertiliser demand.

Table 13.9. The impact of alternative policies on village import of labour and commodities.[a]

Commodity	Base results	Output price increases by 10%	Fertiliser subsidy decreases to 10%	Fertiliser subsidy increases to 30%	Land degradation 1 year (high)
Unskilled labour	91.4	2.7	–0.2	0.2	–0.2
Fertiliser	341.1	9.7	–10.7	13.6	–0.4
Agricultural commodities	246.9	7.1	–1.2	1.3	–1.1
Other commodities	1,031.3	3.7	–0.9	1.0	–0.8

[a] Base results in '000 Ethiopian Birr; non-base simulation results as % change from base results.

The impact of fertiliser subsidies

Two experiments with fertiliser subsidies are included – a reduction from the 1994 level of 20 to 10% and an increase to 30%. Although these two policy scenarios are mirror images of each other, the resulting impacts are not necessarily symmetrical, indicating that the policy effect is non-linear.

Table 13.7 shows that the reduction in fertiliser subsidy reduced household incomes by 1.2–1.7% while the increase led to an improvement of household incomes by 1.4–2.0%. The strongest relative change was for the wealthiest household group (2%). This is opposite to the distributional impact of output price changes. The wealthiest benefited relatively more from fertiliser price subsidies, while output price increase seemed to be more pro-poor.

Table 13.4 shows that a reduction in the fertiliser subsidy caused a reduction in cereal production in most cases, while it had mainly positive effect on the growing of pulses (legumes). This is mainly because the price rise causes a shift to crops that are less fertiliser-intensive or encourage planting of legumes usually grown without fertilisers. The subsidy removal also had a negative effect on livestock production because fodder production (crop residues) became more costly. The reduction in fertiliser subsidy caused a decrease in the export of teff and an increase in the export of other cereals and of pulses (Table 13.8). There was a small reduction in the export of livestock products and an increase in out-migration and outputs from small businesses.

The reduction in fertiliser subsidy from 20 to 10% caused a fall by 9.5–12.9% in the demand for fertiliser in the different cereal production activities (Table 13.10). The aggregate demand decreased by 10.7%. The impact of an increase in fertiliser subsidy from 20 to 30% caused a slightly stronger response in the form of an increased demand for fertiliser of 12.0 to 16.8%. This is also illustrated in Table 13.9 where it can be seen that the first experiment led to a reduction in the aggregate fertiliser import of 10.7% and the second experiment led to an increase of 13.6%. The reduction in the subsidy also leads to a slight reduction in the local demand for unskilled labour and in the demand for consumer goods.

The impact of price policies on the land degradation externality

The impact of alternative policies on the land degradation externality, assuming that land degradation is irreversible and causing permanent productivity losses, is shown in Table 13.11. The net present value of these permanent productivity losses (referred to here as land degradation externalities) were calculated for high and low levels of degradation and at varying social rates of discount (3, 5 and 10%). This is basically to explore how sensitive productivity losses (referred to here as land degradation externalities) were to the impacts of changes in assumptions. The reported values are the village-wide land degradation externalities as computed by Equation 4.

Table 13.10. The impact of alternative policies on fertiliser use by production activity and household group.[a]

Production activities by household group	Base results	Output price increases by 10%	Fertiliser subsidy decreases to 10%	Fertiliser subsidy increases to 30%	Land degradation 1 year (high)
Upland teff					
H0	5.5	15.1	−10.1	12.8	−0.9
H1	14.9	10.4	−11.1	13.9	−1.0
H2	161.4	9.5	−10.7	13.6	−1.4
Lowland teff					
H0	4.6	15.2	−10.0	12.7	0.8
H1	7.7	7.9	−11.2	14.2	0.6
H2	121.8	9.2	−10.6	13.4	1.0
Upland other cereals					
H0	0.7	13.2	−9.5	12.0	−2.4
H1	1.5	11.3	−9.7	12.3	−1.2
H2	14.4	9.9	−10.5	13.3	−1.1
Lowland other cereals					
H0	0.8	16.7	−11.4	14.6	3.5
H1	1.0	15.7	−12.1	15.6	2.6
H2	6.7	16.2	−12.9	16.8	2.2
Aggregate change		9.7	−10.7	13.6	−0.4

[a] Base results in '000 Ethiopian Birr; non-base simulation results as % change from base results.

Table 13.11. Sensitivity analysis of the impact of alternative NRM policies on the village-wide land degradation externality.[a]

	Level of land degradation and social discount rates					
	High			Low		
Policies	3%	5%	10%	3%	5%	10%
Base results	1226.9	736.2	368.1	353.1	211.8	105.9
Output price increase 10%	1316.0	789.6	394.8	377.2	226.3	113.2
Fertiliser subsidy 10%	1269.3	761.6	380.8	366.3	219.8	109.8
Fertiliser subsidy 30%	1178.1	706.9	353.4	337.7	202.6	101.3

[a] In '000 Ethiopian Birr.

Compared to the base scenario, the output price increase leads to more rapid land degradation because of more intensive cultivation. This is associated with the increase in production on erodible upland soils (Table 13.4). However, a reduction of the fertiliser subsidy also causes the land degradation externality to increase, while an increase in the fertiliser subsidy has the opposite effect. This may indicate that a fertiliser subsidy can be defended on environmental grounds (see below).

The short-term impact of soil degradation

These simulations were undertaken to explore the following-year impacts resulting from soil degradation in the current year. The model was run as a two-period model where land degradation takes place in the first year and the impacts occur in the following year. In order to see the upper and lower ranges in the resulting effects, the high and low level of land degradation were used in the analysis. The real land degradation should be somewhere between these two levels. The effect of land degradation during such a short period of time is fairly linear. It was found that the magnitude of the low level of degradation relative to the impact of the high level of degradation is constant across the different policy scenarios. Therefore, the values for high and low levels of degradation are reported only for the income effects (Table 13.7) and on the land degradation externality (Tables 13.5 and 13.6). For the other scenarios only the impacts of the high level of degradation are shown.

Table 13.7 shows that a high level of degradation leads to a loss in household net income of 1.3–1.5% in the following year while for a low level of degradation the losses are around 0.3–0.4%. In other words, this shows that a NRM intervention that arrests the level of degradation has the potential to increase household incomes by the respective levels. Table 13.4 shows how land degradation affects production activities. The land degradation impacts on crop production are high and negative for the uplands where the rates of erosion are high. However, it seems that land degradation leads to a shift of resources towards growing of less-erosive cereals (teff is most erosive) in the lowlands. There is also a negative impact on livestock production due to the negative effects on availability of fodder. As a consequence, Table 13.8 shows that the export of cereals and livestock from the village decreases while the migration of skilled labour and small businesses increases.

Table 13.10 shows that smallholders are likely to shift fertiliser use from upland soils to lowland soils where degradation is lower and the returns from fertiliser use are higher. Table 13.9 shows that there is a slight reduction in total fertiliser use due to land degradation. This implies that land degradation is not compensated for by increased fertiliser use. Land degradation also leads to a fall in demand for consumer goods due to the fall in household income.

The social efficiency of NRM policies

The blanket removal of input subsidies as part of structural adjustment programmes can raise some important questions. Fertiliser subsidies may sometimes be defended on environmental grounds (Pigouvian subsidies) as fertiliser use may be necessary to sustain land productivity (Holden and Shanmugaratnam, 1995). In this case, fertiliser subsidies could be used to stimulate land conservation if they were linked to conservation requirements (inter-linkage policies). Shiferaw and Holden (2000) used a partial-equilibrium farm-household analysis to explore these linkages. Although such effects are not explored here, whether or not certain kinds of fertiliser and output price policies can be defended on environmental grounds is examined.

The CBA for the output price increase (reduction in output tax) is considered first. Table 13.11 shows that the income effect of the output tax reduction is larger than the tax loss effect. The land degradation effect is negative as the increase in output prices increases the land degradation externality. However, Table 13.5 shows that only in the case of high level of land degradation and low social rates of discount does this policy become inefficient. This implies that an output tax policy can only be defended as an instrument on its own to internalise the land degradation externality when the social rate of discount is low and land degradation is high. The net returns to a tax reduction are fairly low, with a maximum of 13% when the level of land degradation is low and the social rate of discount is high (10%).

A CBA of the fertiliser subsidy policy experiments is given in Table 13.6. The CBA includes the household income effects, on-site land degradation externality effects due to loss in land productivity (high and low levels), and subsidy costs to the government. The table shows the net benefits with high and low levels of degradation and the respective net benefit to government subsidy cost ratios at three different social rates of discount.

The results show that the household income effects are stronger than the government subsidy cost effects and that the two effects go in opposite directions, i.e. a fertiliser subsidy generates income benefits to the poor, while the subsidy reduction generates cost savings to the government. The fertiliser subsidy provides positive benefits in terms of reductions in the externality while the opposite is true when the subsidy is decreased. Therefore when the future productivity loss effects are accounted, the net social benefits of an increase in the fertiliser subsidy are positive. The results also show that the subsidy reduction is associated with negative social net benefits. The net social returns to fertiliser subsidies are large and vary from 37 to 93% when land degradation is high and from 20 to 38% when land degradation is low. When the sustainability effects are accounted, this clearly indicates that policies for fertiliser subsidies can be justified.

Conclusions

This chapter has demonstrated how a village bioeconomic CGE model can be used to evaluate the impacts of alternative NRM policy interventions. In particular, the use of Pigouvian subsidies and taxes for internalising the land degradation externality has been assessed. The model has allowed assessment of the relative impacts of alternative policies on different socio-economic groups while taking into account the interaction of the socio-economic groups through local markets. Policy changes have both direct and indirect impacts that are captured by the model. The model was developed based on a village SAM constructed for the case study area. The policy scenarios are assumed to provide better predictions as they are based on an underlying SAM and market and technological relationships that give a better representation of supply responses and imperfections in local markets, caused by commodity and market-specific transaction costs, than models that show partial equilibrium and than CGE models that ignore market imperfections and the interrelatedness of production, consumption and NRM.

There are positive impacts on household welfare (poverty reduction) of reducing output taxes, but the output supply response was weak and the response in marketed surplus was even negative. This may be explained by the relatively strong income and profit effects due to the price increase, imperfections in factor markets and low elasticity of substitution between factors. Even though the response in production was weak, the increase in output prices contributed to increasing the (negative) land degradation externality.

The simulations showed that a reduction in fertiliser subsidies would reduce household incomes and would also increase the future productivity loss from land degradation, while it would also reduce government spending on subsidies. Accounting for these effects, the results showed that returns to removal of fertiliser subsidies are negative. The larger the land degradation effect, the more negative the returns from subsidy reduction will be. This seems to strongly justify the policy to subsidise fertiliser use because it represents a win-win option that contributes to poverty reduction while also reducing land degradation. The social returns to fertiliser subsidies were high even under assumed low levels of land degradation or high social rate of discount.

The impact of alternative policy interventions on land degradation is represented by the impact on short-term land productivity and through a permanent loss of productivity. Land productivity in the following year was affected by crop choice and fertiliser use in the first year. The on-site land degradation externality was calculated as the net present value of the loss in output due to a land productivity decline, assuming that this decline is irreversible. An increase in the output price or a reduction in fertiliser subsidies increased the land degradation externality while an increase in the input subsidy had the opposite effect. The impacts of land degradation on the income of the poor in the following year range from 0.3% (low level) to 1.5% (high level). NRM interventions that reduce land degradation can be expected to provide comparable income benefits to the poor.

An increase in output prices and an increase in fertiliser subsidies may be seen as alternative policies to reduce the bias against the rural sector. The analysis indicated that the social returns to increasing the fertiliser subsidy may be higher than reducing the output tax by the same amount. Actually, from an environmental perspective, a tax-subsidy regime that combines output taxation and fertiliser subsidisation may be socially optimal. This conclusion holds even at a low level of land degradation and at a high discount rate of 10%.

Endnotes

[1] Inputs like land, traction power (oxen), labour and seeds are of complementary nature in production of crops and they can, to a limited degree, substitute for each other. This is captured in this case using relatively low elasticities of substitution (between 0 and 1) in the CES production functions.
[2] Market imperfection refers to the situation where markets do not work perfectly competitively making prices exogenous to producers and consumers. Such imperfections include missing markets, imperfect competition (e.g. monopolistic traders), transaction costs causing price bands, rationing, and interlinked markets (e.g. share tenancy, credit in kind, barter trade).
[3] Transaction costs are additional costs that buyers and sellers face beyond the actual market price for a product. These may include time and other costs incurred in relation to travel, searching and negotiation. These costs create a 'price band' between the effective selling and buying prices for the same product.
[4] Output taxation has been a common policy in Africa and has been blamed for stagnation of African agriculture (see Krueger *et al.*, 1991). Output prices were controlled in Ethiopia up to the early 1990s. Although this study did not examine the institutional arrangements that may be needed to implement them, value-added taxes on farm output might be justified for generating government revenue.
[5] Technological change may be introduced as e.g. Hicks-neutral technological change, labour-saving technological change or land-saving technological change. Hicks-neutral technological change is a yield-enhancing technical change that does not alter the mix of inputs (see e.g. Angelsen *et al.*, 2001).
[6] Although some soil degradation may be reversible to some degree, soil formation is a slow process and erosion rates have been found to be more than ten times higher than the soil formation rates in Ethiopia.

References

Angelsen, A., van Soest, D., Kaimowitz, D. and Bulte, E. (2001) Technical change and deforestation: a theoretical overview. In: Angelsen, A. and Kaimowitz, D. (eds) *Agricultural Technologies and Tropical Deforestation.* CAB International, Wallingford, UK, pp. 19–22.

Bardhan, K. (1970) Price and output response of marketed surplus of foodgrains: a cross-sectional study of some North Indian villages. *American Journal of Agricultural Economics* 52, 51–61.

de Janvry, A. and Kumar, P. (1981) The transmission of cost inflation in agriculture with subsistence production: a case study in Northern India. *Indian Journal of Agricultural Economics* 36, 1–14.

de Janvry, A., Fafchamps, M. and Sadoulet, E. (1991) Peasant household behaviour with missing markets: some paradoxes explained. *Economic Journal* 101, 1400–1417.

FAO (1986) *The Highlands Reclamation Study – Ethiopia. Final Report.* Volumes 1 and 2. Food and Agriculture Organization of the United Nations, Rome, Italy.

Greenwald, B. and Stiglitz, J.E. (1986) Externalities in economies with imperfect information and incomplete markets. *Quarterly Journal of Economics* 101, 229–264.

Grepperud, S. (1996) Population pressure and land degradation: the case of Ethiopia. *Journal of Environmental Economics and Management* 30, 18–33.

Hoff, K., Braverman, A. and Stiglitz, J.E. (eds) (1993) *The Economics of Rural Organization. Theory, Practice, and Policy.* Oxford University Press, Oxford, UK, 590 pp.

Holden, S.T. and Shanmugaratnam, N. (1995) Structural adjustment, production subsidies and sustainable land use. *Forum for Development Studies* 2, 247–266.

Holden, S.T. and Shiferaw, B. (2002). Poverty and land degradation: peasants' willingness to pay to sustain land productivity. In: Barrett, C.B., Place, F.M. and Aboud A.A. (eds) *The Adoption of Natural Resource Management Practices: Improving Sustainable Agricultural Production in Sub-Saharan Africa.* CAB International, Wallingford, UK, pp. 91–102.

Holden, S.T., Shiferaw, B. and Wik, M. (1998a) Poverty, market imperfections, and time preferences: of relevance for environmental policy? *Environment and Development Economics* 3, 105–130.

Holden, S.T., Taylor, J.E. and Hampton, S. (1998b) Structural adjustment and market imperfections: a stylized village economy-wide model with nonseparable farm households. *Environment and Development Economics* 4, 69–87.

Krueger, A.O., Schiff, M. and Valdés, A. (1991) *The Political Economy of Agricultural Pricing Policy.* Volume 3: *Africa and the Mediterranean.* The Johns Hopkins University Press, Baltimore, Maryland, and London, UK, 340 pp.

Lofgren, H. and Robinson, S. (1999). Nonseparable farm household decisions in a computable general equilibrium model. *American Journal of Agricultural Economics* 81, 663–670.

Lofgren, H., Harris, R.L. and Robinson, S. (2002) A Standard Computable General Equilibrium (CGE) Model in GAMS. *Microcomputers in Policy Research,* Vol. 5. International Food Policy Research Institute, Washington, DC.

Sadoulet, E. and de Janvry, A. (1995) *Quantitative Development Policy Analysis.* The Johns Hopkins University Press, Baltimore, Maryland, 397 pp.

Shiferaw, B. and Holden, S.T. (1999) Soil erosion and smallholders' conservation decisions in the highlands of Ethiopia. *World Development* 27(4), 739–752.

Shiferaw, B. and Holden, S.T. (2000) Policy instruments for sustainable land management: the case of highland smallholders in Ethiopia. *Agricultural Economics* 22, 217–232.

Stiglitz, J.E. (1986) The New Development Economics. *World Development* 14(2), 257–265.

Stoorvogel, J.J. and Smaling, E.M.A. (1990) *Assessment of Nutrient Depletion in Sub-Saharan Africa: 1983–2000.* Volume I: *Main Report.* The Winand Staring Centre, Wageningen, The Netherlands.

Strauss, J. (1986) The theory of comparative statics of agricultural household models: a general approach. In: Singh, I., Squire, L. and Strauss, J. (eds) *Agricultural Household Models. Extensions, Applications, and Policy.* The Johns Hopkins University Press, Baltimore, Maryland, pp. 71–91.

Taylor, J.E. and Adelman, I. (1996) *Village Economies: the Design, Estimation, and Use of Villagewide Economic Models.* Cambridge University Press, Cambridge, UK, and New York, 264 pp.

World Bank (1996) *Towards Environmentally Sustainable Development in Sub-Saharan Africa: A World Bank Agenda.* The World Bank, Washington, DC.

Part V.

Towards Improved Approaches for NRM Impact Assessment

14 The Concept of Integrated Natural Resource Management (INRM) and its Implications for Developing Evaluation Methods

B. Douthwaite[1], J.M. Ekboir[2], S.J. Twomlow[3] and J.D.H. Keatinge[4]

[1] *Centro Internacional de Agricultura Tropical (CIAT), Cali, Colombia*
[2] *Consultant, Centro Internacional de Mejoramiento de Maïz y Trigo, Mexico*
[3] *International Crops Research Institute for the Semi-Arid Tropics (ICRISAT), Bulawayo, Zimbabwe*
[4] *ICRISAT, Patancheru, Andhra Pradesh, India*

Introduction

Agriculture in developing countries faces a huge challenge. In the next 50 years the number of people living in the world's poorer countries will increase from 5 billion to nearly 8 billion (Population Reference Bureau, 2001). Moreover, per capita food consumption needs to increase to adequately feed the 1.1 billion underfed people in the world (Gardner and Halweil, 2000). This means that in 2050 farmers will need to produce at least 50% more food from a natural resource base that is already damaged by human activity to the point where further degradation could have devastating implications for human development and the welfare of all species (World Bank, 2000).

The Green Revolution is widely credited with having averted a similar crisis when large-scale famines were predicted to threaten Asia in the 1970s and 1980s. The research component of the Green Revolution was largely based on the genetic improvement of a few commodity crops to enhance their productivity and improve their resistance to pests and diseases. The gains were largely confined to areas of high agricultural potential, and they often benefited the more prosperous farmers. In many cases, this research yielded large production gains at the expense of soil degradation, water, biodiversity, and non-cultivated land (Sayer and Campbell, 2001).

 Natural Resource Management in Agriculture: Methods for Assessing Economic and Environmental Impacts (eds B. Shiferaw, H.A. Freeman and S.M. Swinton)

A second Green Revolution is now needed. However, the situation today is dramatically different from when the first Green Revolution began and different research and development approaches are required. Old, top-down ways of working, in which international agricultural research centres (IARCs) see themselves as the main sources of agricultural innovations that are transferred to national agricultural research and extension systems (NARES) and downward to farmers, are no longer valid (Biggs, 1990; Clark, 1995). There is now a much more sophisticated understanding of how rural development occurs, which recognises that innovation has multiple sources and results from the action of a broad network of actors, of which IARCs and NARES are just a part (Hall *et al.*, 2003a). Research is now seen as part of a collective effort to create new technical and social options that rely more on local knowledge and less on a 'one size fits all' application of simple technologies and chemical inputs. Hence, working in partnerships has become much more important, as has grassroots participation of farmers and their organisations (Hall *et al.*, 2002). A second important area of change is that farmers are increasingly exposed to global markets, and while the information and communication revolution offers exciting opportunities for them to benefit, it also threatens to create a 'digital divide' between rural and urban areas (Malecki, 2003). Over-all, IARCs and NARES need to become much more nimble and responsive in the face of an ever-faster rate of change (Watts *et al.*, 2003).

Integrated natural resource management (INRM)[1] is an attempt to build a new agricultural research and development paradigm to meet the challenges and opportunities outlined above. Campbell *et al.* (2001) define INRM as 'a conscious process of incorporating the multiple aspects of natural resource use (be they bio-physical, socio-political or economic) into a system of sustainable management to meet the production goals of farmers and other direct users (food security, profitability, risk aversion) as well as the goals of the wider community (poverty alleviation, welfare of future generations, environmental conservation)'. Campbell *et al.* (2001) go on to say that evaluation has a crucial role in helping to build and support INRM. The objective of this chapter is to investigate the types of evaluation that are needed to build and support INRM.

Integrated Natural Resource Management (INRM)

INRM has grown out of farming systems research (FSR), which had its heyday in the mid-1980s and then all but disappeared from the list of research programmes by the early 1990s (Ravnborg, 1992). This was because FSR attempted, just as INRM is attempting today, to carry out research with complicated technologies in complex settings. Research on complex agricultural systems is difficult because of the multiple scales of interaction and response within and between physical and social subsystems, uncertainty, long time lags, and multiple stakeholders who often have contrasting objectives and activities (Campbell *et al.*, 2001).

Early FSR failed because by engaging with this complexity it was criticised for generating excessive amounts of data, being very costly to conduct, and yielding few results of immediate practical value. The other major cause of the failure of FSR was a lack of understanding of the role of farmers and other stakeholders in technology development (Röling, 1988; McCown, 2001). In many instances, researchers conducted their experiments in farmers' fields but failed to interact sufficiently with the farmers themselves; in other words, they continued their traditional research methods only this time outside the experimental station. The participation of private firms, consumers and farmer associations in the planning and execution of research was almost nil.

Early FSR learnt from its mistakes, evolved, and INRM is a result of this process. The term INRM was first coined in 1996 by the Consultative Group on International Agricultural Research (CGIAR) system, a coalition of 15 international research centres (CGIAR/TAC, 1998). INRM moved to centre stage in the CGIAR as a result of the 3rd CGIAR Systemwide External Review (CGIAR/TAC, 1998) recognising that a paradigm shift had occurred in 'best practice' NRM, in which 'hard' reductionist science was being tempered by 'softer' more holistic approaches. Specifically, the review identified a move from classical agronomy to ecological sciences, from the static analysis of isolated issues to systems' dynamics, from top-down to participatory approaches, and from factor-oriented management to integrated management. The CGIAR subsequently set up a task force to coordinate work on INRM [CIFOR, 1999 (The Bilderberg Consensus)].

One of the major outputs of the INRM initiative has been a special edition of the electronic journal *Conservation Ecology*, describing INRM concepts and practice. In a synthesis paper, Sayer and Campbell (2001) flesh out the definition given above, which emerges as a road map of how institutions might modify their way of doing business rather than by a set of tried and trusted approaches already in use. The guiding perspective of 'best practice' INRM is that standardised, generally applicable technologies or truths are unlikely because small-scale producers generally have multiple objectives, and achieving change involves the interplay of multiple stakeholders. Rather, research efforts should be directed at improving the capacity of agroecological systems to adapt to changes and to continuously supply a flow of products and services on which poor people depend, i.e. to improve systems' 'adaptive capacities'. In practice this means helping farmers and other managers of natural resources to acquire the skills and technologies to better control their resources, i.e. improving their 'adaptive management' abilities (Holling *et al.*, 1998; Hagmann and Chuma, 2002). INRM's way of working is to develop practical, local solutions in partnership with farmers together with an array of local and international partners. In deriving the solutions the best science is blended with local and specialised technological knowledge. The underlying principles learned in the local process can then be an ingredient used to develop solutions for similar conditions in different locations and environments. Sayer and Campbell (2001) describe five key elements of INRM.

Learning together for change

INRM must be based upon a continuous dialogue, negotiation and deliberation amongst stakeholders. Like jazz – NRM needs constant improvisation, so that each band member knows the weaknesses and strengths of the other players and that they all learn how to play together. Researchers cannot therefore remain exclusively outside: they need to engage themselves in action research to develop appropriate solutions together with resource users. In this process researchers and resource users: 1. define subsystems; 2. reflect and negotiate on future scenarios; 3. take action; and 4. evaluate and adapt attitudes, processes, technologies and practices. This learning cycle is the basis of resource management that can evolve.

Multiple scales of analysis

INRM attempts to integrate research efforts across spatial and temporal scales. This is because ecological and social processes are taking place over different time scales ranging from minutes to decades (Fresco and Kroonenberg, 1992). Slow-changing variables operate as restrictions to the dynamics of more rapidly-cycling processes. At the same time, fast changing variables affect the dynamics of the slow changing processes. As the system evolves, the dynamics of the different variables may experience sudden changes that reorganise the system. Usually these changes arise when the system reaches specific thresholds. In these reorganisation points, it is impossible to predict how the system will self-organise (Nicolis and Prigogine, 1989). Understanding a system, rather than just describing it, usually requires studying that system together with the other systems with which it interacts. Systems modelling is a practical approach to deal with variables that change more slowly than the length of a project. Modelling can also help farmers and other natural resource managers explore different scenarios, identify preferred ones, and then negotiate how to achieve them (van Noordwijk *et al.*, 2001).

Plausible promises

INRM needs to maintain a practical problem-solving approach that delivers tangible outputs. There needs to be some motivation for farmers to want to work together with researchers to develop technologies and processes. This motivation comes from ideas and technologies that make a 'plausible promise' to farmers of being of benefit to them. Working together builds trust and leads to further learning, from which other possibilities flow. Monitoring and evaluation and impact assessment can help identify and improve what is working effectively.

Scaling out and up

INRM runs the risk of being criticised for only producing local solutions. However, if natural resource systems are characterised adequately, for example, according to exogenous drivers as in the IITA Benchmark Area Approach,[2] then INRM can yield results that have application across broad ecoregional domains. While most INRM technologies cannot be scaled-out, INRM technologies together with the learning processes that allow rural people to identify and adapt new opportunities to their environments can be scaled-out. INRM recognises a difference between scaling-out where an innovation spreads from farmer to farmer, community to community, within the same stakeholder groups, and scaling-up which is an institutional expansion from grassroots organisations to policy makers, donors, development institutions, and other stakeholders key to building an enabling environment for change (Douthwaite *et al.*, 2003a). The two are linked: scaling-out occurs faster if INRM projects plan and invest in engaging with stakeholders who can help promote project outputs and create an enabling environment for them. Iterative learning cycles that take place in participatory technology development processes can also help create an enabling environment through interaction, negotiation and co-learning amongst different stakeholders.

Evaluation

Evaluation is key to adaptive management because it provides the real-time feedback necessary for constant improvisation in implementing INRM projects, and for learning and improving the performance of those involved. Evaluation also provides data for further negotiation between stakeholders, and for resource-allocation decisions. Stakeholders should agree on plausible strategies on how research will contribute to developmental change and then undertake regular monitoring of the implementation of these strategies to feed into the learning cycle. Success criteria and indicators, agreed early on in a project, are the basis for impact assessment and negotiation amongst stakeholders for resource-allocation decisions.

The discussion so far shows that INRM is based on a paradigm that is better able to cope with complexity than the top-down conceptual framework which underpinned much of the IARCs and NARES earlier successes with plant breeding.[3] New paradigms require new ways of looking at the world and new conceptual models for understanding it. These conceptual frameworks are important because they influence the ways that research and development interventions are conceptualised, planned and implemented. The authors contend that INRM would be well served by adopting an Innovation Systems (ISs) perspective, and that this perspective will help clarify the needs and roles for evaluation in INRM. The ISs framework has a long track record, has been widely adopted outside of agriculture, and is based on evolutionary economics (Nelson and Winter, 1983), institutional economics (Freeman, 1987), and stochastic processes and theories of complexity (Rycroft and Kash,

1999; Ekboir, 2003). The ISs framework has also been employed successfully in the analysis of post-harvest systems in South Asia (e.g. Hall *et al.*, 2003b) and is providing the conceptual framework for the emergent Institutional Learning and Change (ILAC) Initiative in the CGIAR (Watts *et al.*, 2003). The ILAC Initiative is being supported by the International Fund for Agricultural Development (IFAD), the Rockefeller Foundation and the German Deutsche Gesellschaft für Technische Zusammenarbeit GmbH (GTZ) and Bundesministerium für Wirtschaftliche Zusammenarbeit und Entwicklung (BMZ). It was born out of a frustration that conventional evaluation methods used in the CGIAR were not supporting the learning and change needed for the CGIAR centres to adapt to an ever-faster changing world. In explaining Rockefeller's support for the ILAC Initiative, Peter Matlon of the Rockefeller Foundation said: 'There is an urgent need for impact assessment and evaluation to play more self-critical learning roles. Impact assessment studies need to begin to address more systematically and rigourously the – "why?" questions – that is, not only what works, but also what doesn't, under what circumstances and, most importantly, what are the drivers that determine success or failure' (Mackay and Horton, 2003).

The types of development practice, including evaluation practice, being proposed by the ILAC Initiative (Watts *et al.*, 2003) are fully consistent with those required by INRM, as shown in Table 14.1.

At its simplest, an innovation system has three elements (Watts *et al.*, 2003): 1. the groups of organisations and individuals involved in the generation, diffusion, adaptation and use of new knowledge; 2. the interactive learning that occurs when organisations engage in generation, diffusion, adaptation and use of new knowledge, and the way this leads to new products and processes – i.e. innovation; and 3. the institutions that govern how these interactions and processes take place. The reason it is believed that the framework is relevant to INRM is that both see innovation as an inherently complex process undertaken by a network of actors. Both also recognise innovation as a social process, involving interactive 'learning by doing' in which innovations and the institutions (norms, expectations, ways of organising) co-evolve. As a result innovation, including rural innovation, is an inherently unpredictable, non-linear process. This conclusion has profound implications for all types of evaluation, considered below.

Evaluation Appropriate for INRM

The term evaluation covers a huge area of enquiry and can fulfil many purposes. Patton (1997) identifies three main uses for evaluation findings which are: 1. judge merit or worth; 2. generate knowledge; and 3. improve projects and programmes. Traditionally, evaluation carried out in both national and international agricultural research has focused on 1 and 2, that is judging merit and generating knowledge. Cost–benefit analysis, audits, showing accountability to donors and quality control are all activities that fall under the former while extrapolating principles about what work, theory

Table 14.1. The shifts and expanded options in development practice, including evaluation practice, implied by an Innovation Systems perspective (Watts *et al.*, 2003).

Evaluation	From	Expanded to include
Paradigm of and for	• Things	• People
Orientation and power	• Top-down	• Bottom-up
Key words	• Planning	• Participation
Modes/approaches	• Standardised	• Diverse
	• Linear	• Complex
	• Reductionist	• Systems
Conditions	• Controlled	• Uncontrolled (able)
	• Stable	• Dynamic
	• Predictable	• Unpredictable
Research mode	• Experimental	• Constructivist
Learning	• *Ex post*	• Continuous
Roles	• Teacher	• Facilitator
	• Supervisor	• Coach
	• External evaluator	• Evaluation facilitator
Outcomes	• Products and infrastructure	• Processes and capability
Valued behaviours	• Rigourous/objective	• Critical self-reflection
Dominant professions	• Agricultural scientists and economists	• All
Patterns of change	• Predetermined/prescriptive	• Evolutionary
Characteristic management tools	• Logframes and external review	• Action research, participatory review and reflection
Main purpose of evaluation	• Accountability and control	• Learning and improvement
Accountability to	• Donors and peers	• All stakeholders, especially the poor
Vision of capacity development	• Build capacity of others	• Develop own capacity
Treatment of failure	• Buried or punished	• Valued as a learning opportunity
Consequences of failure	• Cataclysmic	• Continuous programme readjustment

building and policy making all result from the latter. While these types of evaluation are still necessary for INRM, much more emphasis needs to be placed on evaluation aimed at improving projects and programmes. This type of evaluation focuses on stimulating learning about what is working and what is not, and as a result helps improve the management of projects and programmes. In INRM, this evaluation needs to serve the learning needs of all the stakeholders involved, from farmers to researchers. Traditionally, the learning from evaluations has been assimilated by the agricultural economists who made these evaluations, and the information written up in journals that are inaccessible to non-specialists.

As well as having many uses, evaluation can occur at different stages in the project cycle, and beyond. In the past, evaluation in agricultural research has focused on *ex ante* impact assessment to set priorities, and *ex post* impact

of the success criteria chosen was 'participating farmers make changes that improve the technology for them, they continue with these improvements and promote and pass them on to others'. The indicators included percentages of: 1. farmers who had made modifications; 2. had kept them; and 3. had passed them on to others.

In general, criteria, indicators and the impact pathway itself can change during a project, based on learning. Getting stakeholders together to agree on the impact pathway helps create a common understanding of what the project is trying to achieve, and this makes achieving impact more likely. All stakeholders should also be involved in designing the monitoring system and collecting data that serves their information needs. However, all information required cannot be collected through participatory approaches (Campbell *et al.*, 2001) and other extractive methods, such as structured questionnaires, are sometimes needed.

IPE shares many similarities with Outcome Mapping, developed over the last 5 years by the Canadian International Development Research Centre (IDRC) (Earl *et al.*, 2001). In Outcome Mapping, the outcomes are changes in people's behaviour. Outcome Mapping is based on individual projects and organisations documenting their contribution to developmental change, rather than attempting to quantify their impact in terms of rate of return to investment. IDRC sees the quantification of impact as detrimental to learning and adaptive management because the drive to claim credit interferes with the creation of knowledge. Instead, Outcome Mapping argues that donors should make recipients accountable for demonstrating that they are progressing towards impact and improving effectiveness, not for developmental impact itself, which in any case nearly always occurs well after a project has finished. Hence, in Outcome Mapping there is a change in emphasis in evaluation on helping to improve, rather than prove, on helping to understand rather than to report, and on creating knowledge rather than taking credit. In this shift to accountability for learning, impact assessment ceases to be an attempt to attribute and quantify based on often inappropriate economic models, and becomes more like making a legal case, built on evidence from many sources. Douthwaite *et al.* (2003a) make a similar argument, which, interestingly draws on the experience of GTZ in Germany, who, like IDRC in Canada, is a project implementer. Douthwaite *et al.* (2003a) argue that plausible *ex post* impact assessment needs to describe the innovation processes that took place and therefore good M&E information is a prerequisite.

Ex post impact assessment

Based on the arguments in the last section it is believed that the emphasis for *ex post* impact assessment should be placed on: 1. the processes of knowledge generation and diffusion; 2. the creation of organisational capabilities, i.e. the collective ability to develop appropriate solutions to identified problems; and 3. the emergence and evolution of innovation networks (Guba and Lincoln, 1989). However, donors will still need to demonstrate to their own constituencies that money spent has contributed to development. It is argued that *ex post* impact assessment for INRM needs to be different from conventional impact assessment of agricultural research that is largely based on the use of inappropriate economic models (Hall *et al.*, 2002). These approaches attempt to relate changes in impact indicators to research investments. Ekboir (2003) states that this is valid only if an implicit assumption is true: that the link between indicators and investments dominates all other relationships that influence the impact indicators. Ekboir (2003) goes on to say that this is only true for minor improvements along stable technological paths, such as breeding improved germplasm for commercial irrigated production systems. Such an assumption is not likely to be valid for much of INRM research. Hence, rather than try to attribute impact using 'heroic' assumptions, *ex post* impact assessment in INRM should focus on establishing which development changes (e.g. poverty alleviation) have taken place, and building a case based on a variety of different information sources which show that the project made a contribution. Box 14.1 gives an example of the unpredictability, time-lags and interactions of stakeholders in a rural innovation process. In this example, because zero tillage interacted with traditional seed improvement research, macroeconomic policies, commercial policies of herbicide producers and an institutional innovation (the farmers' associations), it is impossible to say what percentage of the impact can be attributed to research, which is what conventional impact assessment attempts to do.

Box 14.1. Real-life problems in attribution of impact (from Ekboir and Parellada, 2002).

Argentina, Brazil, Paraguay and Uruguay have enjoyed a six-fold increase in the production of grains since the 1970s. This increase came about as a result of farmers adopting three different technologies: soybeans in the late-1960s, zero tillage in the 1990s, and improved cereals and oilseeds germplasm since the early 1970s. The adoption was triggered not only by the availability of new technologies but also by public policy changes and private firms' commercial strategies. The impact of technologies, policies and commercial strategies cannot be separated because without zero tillage, the impact of improved germplasm would have been very small, since zero tillage was necessary to stop soil erosion and improve water management. At the same time, new and improved germplasm increased the profitability of zero tillage, fostering adoption. But adoption of zero tillage only became technically feasible with the development of glyphosate and economically feasible when it became substantially cheaper in the early 1990s.[4] Finally, the liberalisation policies introduced in the late 1980s and early 1990s forced farmers to look for new technologies in order to reduce costs.

The zero tillage innovation itself was developed despite terracing being identified by the overwhelming majority of researchers in the late 1960s as the most promising solution to the problems of soil erosion caused by soybean cultivation. Zero-tillage systems were eventually developed by a network of agents. This included agrochemical companies, a few public-sector researchers, farmers and agricultural machinery manufacturers. A key component of zero tillage's success was promotion by associations of farmers that also included researchers and private companies. These associations were created in the late 1980s with support from agrochemical companies.

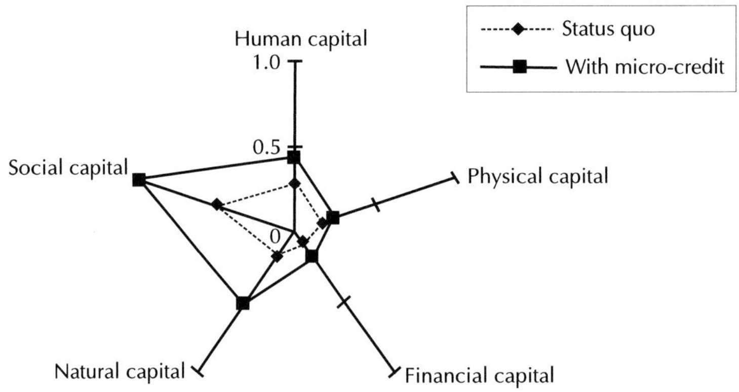

Fig. 14.2. A radar plot showing the effect of a micro-credit scheme on the five livelihood capitals in Chivi, Zimbabwe (Campbell *et al.*, 2001).

To build plausible impact cases, INRM needs to quantify and describe verifiable developmental changes to which it has contributed. These impacts can occur at a variety of spatial and temporal scales and can be context-specific. Campbell *et al.* (2001) suggest an approach based on the use of criteria and indicators, which can be selected with the help of the 'impact pathway' or 'outcome map'. Campbell *et al.* (2001) suggest that the SLF can also guide indicator selection because with the recognition of five capital asset types SLF helps avoid disciplinary bias. Moreover, SLF has been vigorously debated in the literature and is widely understood. However, each of these capital assets may require measurement of several variables, which makes it difficult in practice to identify few proxies that can be monitored over time.

Campbell *et al.* (2001) suggest five different approaches to amalgamating indicators to give an integrated account of change. These are: 1. simple additive indices; 2. combining indicators derived using principal component analysis; 3. two-dimensional plots of variables derived by principal component analysis; 4. radar plots of changes in the five livelihood capitals; and 5. the use of canonical correlation to combine indicators across scales. Depending on the approach used, combining indicators within and across each of the capital assets can create several practical problems. Campbell *et al.* (2001) discuss the pros and cons in the application of the different approaches for aggregating indicators and give examples for each of these approaches. We illustrate here only the application of the radar diagram approach. Figure 14.2 shows a radar plot of the impact of micro-credit schemes on the five capitals in Chivi district in South Zimbabwe. Campbell *et al.* (2001) indicate that the data were generated from a decision support system where the impacts seem to have been simulated with and without the micro-credit scheme. For each of the capital assets, a proxy variable was selected: 1. *physical capital*, percentage of households with 'improved roofing' (income generated from activities sponsored by the micro-credit scheme are often used to improve household assets); 2. *financial capital*, percentage of households achieving a 'high' level of savings; 3. *natural capital*, percentage of households taking measures to improve the fertility of their fields; 4. *social capital*, percentage of households adhering to community-based rules and 5. *human capital*, percentage of committees exposed to, and practising, improved methods of organisation. The radar plot is very effective at quickly communicating that micro-credit is strongly correlated with improvements in social capital, followed by natural capital, and rather less on financial, physical and human capital. Clearly, an assessment that looked only at the effect of micro-credit on financial capital, which on the face of it would appear reasonable, would miss a large part of the impact. However, it will be useful to note that attribution of the changes shown in the radar diagram to the credit intervention cannot be made unless the experiment has a proper counterfactual. Simulation models (as was done for this example) or statistical techniques can be used to test the attribution problem.

Campbell *et al.* (2001) state that simulation modelling is a particularly important tool for impact assessment in INRM because it can help predict outcomes in the complex systems in which INRM works. Complex adaptive systems theory helps to put some bounds on the predictive powers of simulation modelling in INRM by establishing that complex adaptive processes evolve by the interaction of trends and random events, subject to the initial conditions. Processes evolve through a succession of many small variations interrupted by rare catastrophic mutations. The mutations can be triggered by small changes in any variable and then spread through the system. Even though it is possible to model the probability distribution of the changes, it is impossible to predict whether the next change will be small or catastrophic. Even though limited predictability of major trends is possible, random events may derail these predictions. Additional information can reduce, but not eliminate, the uncertainty which increases with the time horizon considered (Dixit and Pindyck, 1994).

However, irrespective of the accuracy of predictions made, simulation modelling is an important learning tool (Twomlow *et al.*, 2003). It provides a suitable framework by which to understand the consequences of changes in the components of a system in both the long and short terms, on a range of scales. Moreover, simulation modelling can be applied in a participatory mode by using it to generate a number of likely scenarios that can provide useful discussion points between researchers and farmers. Simulation modelling can also provide an effective and efficient framework for extrapolating research findings and the understanding of system processes to other sites and management conditions (Foti *et al.*, 2002).

Ex ante impact assessment and priority setting

One of the main reasons for carrying out *ex ante* impact assessment has been to guide priority setting. The ISs recognition of the indeterminate and complex nature of innovation suggests that *ex ante* impact assessment can only recognise technological trends once they have emerged (Rycroft and Kash, 1999). While most of the returns to research will come from research on existing technological trends, these returns will eventually fall unless new trends emerge. *Ex ante* impact assessment can only give some estimates for simple projects along established research and market lines. But even in these cases, the intrinsically random nature of the process means that *ex ante* projections of impact will probably be wrong and should only be used for priority setting with caution. Greater emphasis should be given to two complementary approaches. Firstly, researchers must be allowed to spend some of their time exploring new areas and ideas beyond those prescribed by *ex ante* impact assessment. Knowledge-management literature suggests this should be as much as 20% (von Krogh *et al.*, 2000). Secondly, a research institution can build a consensus with its major stakeholders on strategic areas where its resources should be concentrated using technology foresight methods. According to Georghiou (1996) technology foresight involves 'systematic attempts to look into the longer-term future of science, the economy, the environment and society with a view to identifying the emerging generic technologies and underpinning areas of strategic research likely to yield the greatest economic and social benefits'. Technology foresight approaches include the Delphi method and scenario building. The Delphi method is a technique used to arrive at a group position on an issue under investigation and consists of a series of repeated interrogations, usually by means of questionnaires, of a group of individuals whose opinions or judgments are of interest. After the initial interrogation of each individual, each subsequent interrogation is accompanied by information usually presented anonymously about the preceding round of replies. The individual is thus encouraged to reconsider and, if appropriate, to change his/her previous reply in the light of replies provided by other members of the group. After two or three rounds, the group position is determined by averaging (Ziglio, 1996). Scenario building is often used in industry by companies like Shell to develop a number of possible situations and then work back from those futures to establish how credible they are, and how the organisation would respond or change if they came true (van der Heijden, 1996).

Even though particular outcomes cannot be predicted with certainty, it is possible to identify factors that will, with high probability, affect the chances of success or failure. Among these factors, probably the three most important are: 1. the information flows within individual institutions; 2. information flows within the innovation network; and 3. the patterns of collaboration among agents. Institutions with more horizontal information flows are able to adapt faster to changing environments and to identify earlier emerging commercial and technological opportunities (von Krogh *et al.*, 2000). Strong information flows enable each agent to understand the capabilities and needs of other agents and what they are doing. Collaboration patterns determine the collective capabilities of the network (Dosi *et al.*, 2000). Close collaboration brings together the capabilities of the individual agents and helps to fuse them into collective capabilities. In this way, the network can undertake more complex and extensive activities.

Once research projects have begun, the M&E described above can help to modify priorities and identify new areas of research. Early identification of farmer adoption/non-adoption and modification allows the research process to be adapted and allows new priority areas for research to be set. For example, M&E carried out by the International Crops Research Institute for the Semi-Arid Tropics (ICRISAT) in Malawi and Zimbabwe found that limited access to inorganic fertilisers and improved legume seeds meant that there was little adoption/adaptation of soil fertility management interventions (Dimes *et al.*, 2004; Twomlow *et al.*, 2004). This helped to focus research on to short-term solutions that carry little risk or require only limited investment, and those that require enabling environments to be developed, thus encouraging households to make a major change in the way they allocate the resources they are willing to invest.

Evaluation of scientists

The INRM paradigm and ISs view have profound implications for the evaluation of NRM scientists. Given the dynamic and unpredictable nature of innovation and the difficulties of attributing impact, scientist evaluation should focus on their contribution to achieving the outcomes specified in Outcome Mapping or Impact Pathway Analysis rather than on achieving development impact itself. The production of research outputs, such as publications, varietal releases, methodologies and tools, are necessary but not sufficient for achieving research outcomes. Researchers should also be assessed in relation to external qualitative assessments of research programmes. A third area of assessment should be in relation to behaviour known to foster innovation, such as participation in innovation networks, collaboration with colleagues, and knowledge sharing (Huffman and Just, 2000). These assessments should form part of an incentive scheme that also includes enforcement of quality standards and adequate salaries and funding.

Conclusions

In this chapter it has been shown that INRM is the result of an evolution of learning from experience that began with FSR in the early 1970s. INRM is an approach to research and development that builds the capacity of farmers and other natural resource managers to manage change in sustainable ways. The evolution of thinking in INRM has mirrored similar advances in the understanding of research, development and innovation processes, one of which is the ISs framework from the fields of evolutionary and institutional economics. Both INRM and the ISs view acknowledge that rural innovation is an inherently indeterminate and complex process, involving the interactions and co-learning of a network of actors, of which farmers and researchers are just two. The ISs view has some important implications for the evaluation for INRM. The focus of evaluation needs to shift from being about accountability and public awareness to supporting learning and adaptive management of all the stakeholders involved in a project. Specifically, more emphasis should be placed in the use of evaluation to improve, rather than prove, on helping to understand rather than to report, and on creating knowledge rather than taking credit. In this shift towards accountability for learning, *ex post* impact assessment ceases to be an attempt to quantify an intervention's impact based on inappropriate economic models. Instead it becomes a rational argument, built like a legal case using evidence from many sources that an intervention contributed to developmental impact. The overall developmental impacts, for example, reduction in poverty, should be quantified but not as an intervention's contribution to that impact, unless the link between the intervention and the impact dominates all others.

In this chapter it is argued that a key source of the evidence needed for impact assessment is the monitoring and evaluation carried out within the project cycle, which also provides the real-time information necessary to facilitate the adaptive management of all stakeholders necessary for successful INRM. To be most effective M&E should be based on a shared view amongst the stakeholders of the outcomes they expect the project to contribute, and how these outcomes contribute to larger-scale developmental impact. This shared view should be recorded as an 'outcome map' or 'impact pathway' that then helps frame the M&E, and the selection of criteria and indicators. Good M&E will identify and describe incipient processes of knowledge generation and diffusion, the emergence and evolution of innovation networks, and the creation of organisational capabilities. The job of the impact assessor at some time in the future is to convincingly show how these incipient processes and capabilities grew and contributed to wider-scale development changes that occurred in the project area. In this chapter a number of methods of measuring, describing and understanding these development changes including the SLF, simulation modelling and various approaches of combining indicators to give an integrative picture have been reviewed.

Finally, evaluation appropriate for INRM is very different from the conventional evaluation practice in many IARCs and NARES. Whether INRM-type evaluation becomes more common will depend largely on donors making IARCs and NARES accountable, not for impact in unrealistically short time-periods, but accountable for learning, adapting and achieving outcomes that are known to contribute to development. The signs are positive. IDRC, GTZ and DFID have started to make the change, not just for INRM but for all types of integrated development projects. The CGIAR Institutional Learning and Change Initiative, supported by IFAD, The Rockefeller Foundation and GTZ and BMZ, is recommending evaluation techniques that support learning and change, and are fully consistent with those outlined in this chapter.

Endnotes

[1] INRM is assumed to include all efforts in integrated genetic resource management. As such, at ICRISAT, INRM is now referred to as IGNRM to make this linkage more explicit.

[2] The IITA Benchmark Approach conducts research in a characterised benchmark area that contains within it farming system dynamics and diversity that is representative of a portion of a wider agroecological zone. The benchmark areas are characterised in terms of population density and access to markets (Douthwaite *et al.*, 2003b).

[3] Often called the Transfer of Technology model (Chambers and Jiggins, 1986) or the Pipeline model (Clark, 1995).

[4] Glyphosate is a broad-spectrum herbicide released commercially by Monsanto in the early 1970s.

References

Biggs, S.D. (1990) A multiple source model of innovation of agricultural research and technology promotion. *World Development* 18 (11), 1481–1499.

Campbell, B., Sayer, J.A., Frost, P., Vermeulen, S., Ruiz-Perez, M., Cunningham, A. and Ravi, P. (2001) Assessing the performance of natural resource systems. *Conservation Ecology* 5(2), 22. http://www.consecol.org/vol5/iss2/art22 index.html

CGIAR/TAC (Consultative Group on International Agricultural Research/Technical Advisory Committee) (1998) *Third System Review of the Consultative Group on International Agricultural Research (CGIAR).* CGIAR System Review Secretariat, Food and Agriculture Organization of the United Nations (FAO), Rome, Italy.

Chambers, R. and Jiggins, J. (1986) Agricultural research for resource poor farmers: a parsimonious paradigm. *Discussion Paper 220.* Institute of Development Studies (IDS), University of Sussex, Brighton, UK, 44 pp.

CIFOR (Center for International Forestry Research) (1999) Integrated natural resource management: the Bilderberg Consensus. *Presented at a workshop on Integrated Natural Resources Management, 3–5 September 1999, Oosterbeek, The Netherlands.* http://www.inrm.cgiar.org/documents/bb_meeting.htm (accessed 19 April 2003).

Clark, N. (1995) Interactive nature of knowledge systems: some implications for the third world. *Science and Policy* 22, 249–258.

Dimes, J., Muza, L., Malunga, G. and Snapp, S. (2004) Trade-offs between investments in nitrogen and weeding: On-farm experimentation and simulation analysis in Malawi and Zimbabwe. In: Friesen, D.K. and Palmer A.F.E. (eds) *Integrated Approaches to Higher Maize Productivity in the New Millennium: Proceedings of the Seventh Eastern and Southern Africa Regional Maize Conference, 5–11 February, 2002, Nairobi, Kenya.* Centro Internacional de Mejormiento de Maíz y Trigo, Mexico, D.F. Mexico, and Kenya Agricultural Research Institute, Nairobi, Kenya, pp. 452–456.

Dixit, A.K. and Pindyck, R.S. (1994) *Investment under Uncertainty.* Princeton University Press, Princeton, New Jersey, 476 pp.

Dosi, G., Nelson, R.R. and Winter, S.G. (2000) Introduction: the nature and dynamics of organizational capabilities. In: Dosi, G., Nelson, R.R. and Winter, S.G. (eds) *The Nature and Dynamics of Organizational Capabilities.* Oxford University Press, New York, 408 pp.

Douthwaite, B., Kuby, T., van de Fliert, E. and Schulz, S. (2003a) Impact pathway evaluation: an approach for achieving and attributing impact in complex systems. *Agricultural Systems* 78(2), 243–265.

Douthwaite, B., Baker, D., Weise, S., Gockowski, J., Manyong, V.M. and Keatinge, J.D.H. (2003b) An impact assessment of IITA's benchmark area approach. *IITA Impact Series Publication.* International Institute of Tropical Agriculture (IITA), Ibadan, Nigeria.

Earl, S., Carden, F. and Smutylo, T. (2001) *Outcome Mapping: Building Learning and Reflection into Development Programs.* International Development Research Centre (IDRC), Ottawa, Canada, 140 pp.

Ekboir, J. (2003) Why impact analysis should not be used for research evaluation and what the alternatives are. *Agricultural Systems* 78(2), 166–184.

Ekboir, J. and Parellada, G. (2002) Public–private interactions and technology policy in zero-tillage innovation processes – Argentina. In: Byerlee, D. and Echeverría, R. (eds) *Agricultural Research Policy in an Era of Privatization: Experiences from the Developing World.* CAB International, Wallingford, UK, 320 pp.

Foti, R., Rusike, J. and Dimes, J. (2002) Risk diversification opportunities through legumes in smallholder farming systems in the semi-arid areas of Zimbabwe. In: Waddington, S.R. (ed.) *Grain Legumes and Green Manures for Soil Fertility in Southern Africa: Taking Stock of Progress. Proceedings of a Conference held 8–11 October 2002, Leopard Rock Hotel, Vumba, Zimbabwe.* Soil Fert Net and Centro Internacional de Mejoramiento de Maíz y Trigo–Zimbabwe, Harare, Zimbabwe.

Freeman, C. (1987) *Technology and Economic Performance: Lessons from Japan.* Pinter, London, UK.

Fresco, L.O. and Kroonenberg, S.B. (1992) Time and spatial scales in ecological sustainability. *Land Use Policy* 9(3), 155–168.

Funnel, P. (2000) Developing and using a program theory matrix for program evaluation and performance monitoring. In: Rogers, P.J., Petrsino, A., Hasci, T.A. and Huebner, T. (eds) *Program Theory in Evaluation. Challenges and Opportunities. New Directions for Evaluation 87.* John Wiley and Sons, Baltimore, Maryland.

Gardner, G. and Halweil, B. (2000) Overfed and underfed: the global epidemic of malnutrition. *Worldwatch Paper 150.* Worldwatch Institute, Washington, DC.

Georghiou, L. (1996) The UK technology foresight programme. *Futures* 28(4), 359–377.

Guba, E.G. and Lincoln, Y.S. (1989) *Fourth Generation Evaluation.* Sage Publications, London, UK, 300 pp.

Hagmann, J. and Chuma, E. (2002) Enhancing the adaptive capacity of the resource users in natural resource management. *Agricultural Systems* 73, 23–39.

Hall, A.J., Rasheed Suliaman, V., Clark, N.G. and Yoganand, B. (2002) From measuring impact to learning institutional lessons: an innovation systems perspective on improving the management of international agricultural research. *Agricultural Systems* 78, 213–241.

Hall, A.J., Yoganand, B. and Crouch, J.H. (2003a) The evolving culture of science in the CGIAR: concepts for building a new architecture of innovation in agribiotechnology. In: Hall, A.J., Yoganand, B., Rasheed Suliaman, V. and Clark, N.G. (eds) *Post-harvest Innovations in Innovation: Reflections on Partnership and Learning.* Crop Post-Harvest Programme, South Asia, Patancheru, India and NR International, Aylesford, UK, pp. 94–122.

Hall, A.J., Yoganand, B., Sulaiman, R.V. and Clark, N.G. (eds) (2003b) *Post-harvest Innovations in Innovation: Reflections on partnership and Learning.* Crop Post-Harvest Programme, South Asia, Patancheru, India, and NR International, Aylesford, UK, 168 pp.

van der Heijden, K. (1996) *Scenarios: The Art of Strategic Conversation.* John Wiley and Sons, Chichester, UK, 305 pp.

Holling, C.S., Berkes, F. and Folke, C. (1998) Science, sustainability and resource management. In: Berkes, F. and Folke, C. (eds) *Linking Social and Ecological Systems. Management Practices and Social Mechanisms for Building Resilience.* Cambridge University Press, Cambridge, UK, 476 pp.

Huffman, W.E. and Just, R. (2000) Setting efficient incentives for agricultural research: lessons from principal-agent theory. *American Journal of Agricultural Economics* 82(4), 828–841.

von Krogh, G., Ichijo, K. and Nonaka, I. (2000) *Enabling Knowledge Creation: How to Unlock the Mystery of Tacit Knowledge and Release the Power of Innovation*. Oxford University Press, New York, 292 pp.

Mackay, R. and Horton, D. (2003) Institutional learning and change in the CGIAR. Summary record of the Workshop held at IFPRI, Washington, DC, 4–6 February 2003. *IFPRI Impact Assessment Discussion Paper Brief No. 18,* International Food Policy Research Institute (IFPRI), Washington, DC.

Malecki, E.J. (2003) Digital development in rural areas: potentials and pitfalls. *Journal of Rural Studies* 19(2), 201–214.

McCown, R.C. (2001) Farming systems research and practice. *Proceedings of the 10th Australian Agronomy Conference, Hobart, Australia, 28 January–1 February 2001.* http://www.regional.org.au/au/asa/2001/plenery/4/mccown.htm#P2_45 (Accessed 15 December 2003).

Nelson, R.R. and Winter, S.G. (1983) *An Evolutionary Theory of Economic Change.* The Belknap Press of Harvard University Press, Cambridge, Massachusetts, 452 pp.

Nicolis, G. and Prigogine, I. (1989) *Exploring Complexity.* W.H. Freeman and Co., New York, 328 pp.

van Noordwijk, M., Tomich, T.P. and Verbist, B. (2001) Negotiation support models for integrated natural resource management in tropical forest margins. *Conservation Ecology* 5(2), 21. http://www.consecol.org/vol5/iss2/art21

Patton, M.Q. (1997) *Utilization-Focused Evaluation.* Third Edition. Sage Publications, Thousand Oaks, California, 431 pp.

Population Reference Bureau (2001) *2001 World Population Data Sheet.* Population Reference Bureau, Washington, DC. http://www.prb.org/Content/NavigationMenu/Other_reports/2000-2002/sheet1.html

Ravnborg, H.M. (1992) The CGIAR in transition: implications for the poor, sustainability and the national research systems. *Network Paper no. 31.* Agricultural Administration (Research and Extension) Network, Overseas Development Institute (ODI), London, UK.

Röling, N. (1988) *Extension Science: Information Systems in Agricultural Development.* Cambridge University Press, Cambridge, UK.

Rycroft, R.W. and Kash, D.E. (1999) *The Complexity Challenge: Technological Innovation for the 21st Century*. Science, Technology and the International Political Economy Series. Cassell, New York, 303 pp.

Sayer, J. and Campbell, B. (2001) Research to integrate productivity enhancement, environmental protection, and human development. *Conservation Ecology* 5(2), 32. http://www.consecol.org/vol5/iss2/art32

Scoones, I. (1998) Sustainable rural livelihoods: a framework for analysis. *IDS Working Paper 72,* University of Sussex, Brighton, UK.

Springer-Heinze, A., Hartwich, F., Henderson, J.S., Horton, D. and Minde, I. (2003) Impact pathway analysis: an approach to strengthening the impact orientation of agricultural research. *Agricultural Systems* 78(2), 267–385.

Twomlow, S.J., Dimes, J., Delve, R., Thornton, P., Carberry, P. and Probert, M. (2003) Simulation modeling and participatory research: Closing the gap between research and management in smallholder farming systems. Paper presented to the *2nd Forum for Agricultural Research in Africa Plenary, 19–20 May 2003, Dakar, Senegal.*

Twomlow, S.J., Rusike, J. and Snapp, S.S. (2004) Biophysical or economic performance which reflects farmer choice of legume 'Best Bets' in Malawi. In: Friesen, D.K. and Palmer A.F.E. (eds) *Integrated Approaches to Higher Maize Productivity in the New Millennium: Proceedings of the Seventh Eastern and Southern Africa Regional Maize Conference, 5–11 February 2002, Nairobi, Kenya.* Centro Internacional de Mejormiento de Maíz y Trigo, Mexico, D.F. Mexico, and Kenya Agricultural Research Institute, Nairobi, Kenya, pp. 480–486.

Watts, J., Mackay, J.R., Horton, D., Hall, A., Douthwaite, B., Chambers, R. and Acosta, A. (2003) Institutional learning and change: An introduction. *ISNAR Discussion Paper Nos. 3–10.* International Service for National Agricultural Research Systems (ISNAR), The Hague, The Netherlands.

World Bank (2000) *The World Development Report 2000/2001.* Oxford University Press, New York.

Ziglio, E. (1996) The delphi method and its contribution to decision making. In: Alder, M. and Ziglio, E. (eds) *Gazing into the oracle: The Delphi Method and its Application to Social Policy an Public Health.* Jessica Kingsley Publishers Ltd, London, UK, pp. 3–33.

15 NRM Impact Assessment in the CGIAR: Meeting the Challenge and Implications for CGIAR Centres

T.G. Kelley[1] and H.M. Gregersen[2]

[1] *Consultative Group on International Agricultural Research (CGIAR) Science Council Secretariat, Food and Agriculture Organization of the United Nations (FAO), Rome, Italy*
[2] *CGIAR Science Council, Solvang, California, USA*

Introduction

Natural resources management (NRM) research is both a growing and a changing part of the Consultative Group on International Agricultural Research (CGIAR) portfolio of research. Yet, to date, there is little convincing evidence that such research is having a significant impact in terms of the CGIAR goals that are related to sustainable poverty alleviation and food security. At the same time, there is increasing emphasis – particularly since the early 1990s – on showing impact from agricultural research. Donors would like to see demonstrated linkages between the research they fund and improvements in the livelihoods of the poor.

A clear distinction should be made between the different components of the overall assessment and evaluation process of research projects and programmes. There is *ex ante* impact assessment (IA), used primarily for planning and priority setting; there is monitoring and evaluation, for assessing the quality and relevance of on-going research and generating essential feedback for project managers; and there is *ex post* IA, which assesses changes in selected indicators that can be attributed to specific research-related interventions primarily for accountability purposes but also as an input in forward planning of new research programmes. A distinction can and should also be made between studies that examine adoption (a component of *ex post* IA) and those that analyse the direct and indirect effects of adopted research in terms of achieving the major objectives sought, i.e. *ex post* IA studies that have as their goal the measurement of impact on sustainable poverty alleviation and food security in the case of the CGIAR. Reaching the latter

 Natural Resource Management in Agriculture: Methods for Assessing Economic and Environmental Impacts (eds B. Shiferaw, H.A. Freeman and S.M. Swinton)

goal is the focus of this chapter, in which the CGIAR is colloquially referred to as the System.

The chapter is organised as follows:

- An investigation of investment trends in the CGIAR, showing that NRM-related research and policy research are a growing part of the portfolio. NRM research, as used here, encompasses research on land, water, and biodiversity resources management and is typically focused on producing knowledge that results in technologies, information, and methods/processes that enhance the productivity of ecosystem resources in a sustainable manner. The primary clients are farmers, communities and policy makers
- An exploration of some of the definitional issues related specifically to NRM and the move within the CGIAR towards integrated natural resources management (INRM) that involves a much broader and more complex conceptual base
- An assessment of the challenges ahead in terms of assessing the impacts of NRM research at both the centre and the System level within the CGIAR
- Some conclusions and recommendations.

NRM Environment-related Research Growing in Relation to Other Research

Although overall financial contributions have been fairly stable during the last 10 years – growing at an average annual rate of 0.6% in nominal terms between 1994 and 2001 and declining by 1.6% in real terms (World Bank, 2003) – over this period, some significant patterns of investments have emerged.

Table 15.1 shows CGIAR investment shares by undertakings/activities from 1994 to 2002. CGIAR investments in 'Increasing productivity' have fallen from 47% of the total in 1994 to 34% in 2002. Within this main activity, the sub-activity 'Germplasm enhancement and breeding' investments fell from 23% in 1994 to 18% in 2001, while the sub-activity 'Production systems development and management' fell from 24% to 17%. The two largest components within the Production systems sub-activity saw their investments shares fall the most: Cropping systems from 16% to 9% and Livestock systems from 6% to 5%. At the same time, investments in Tree systems fluctuated around 3% while investments in Fish systems actually rose. Between 1994 and 2002 CGIAR investments in 'Protecting the environment' rose from 15% to 19% and Improving policies from 10% to 15%. This trend in CGIAR investment away from productivity-enhancing activities, for which there are proven impacts on poverty, raises some questions about the current direction and focus of the CGIAR (World Bank, 2003).

Although the CGIAR activity 'Protecting the environment' is one of the fastest-growing areas of research activity within the CGIAR, it is also an area for which there is, as yet, only limited documented impact. As noted in the recent Operations Evaluation Department (OED) Overview of the CGIAR

Table 15.1. CGIAR research agenda investments by undertaking/activity, 1994–2001.

	1994		1995		1996		1997		1998		1999		2000		2001		2002[a]	
CGIAR activities	US$	%	US$	%	US$	%	US$	%	US$	%	US$	%	US$	%	US$	%	US$	%
Increasing productivity (of which)	124.3	47	134.4	47	129.1	40	133.1	40	124.3	37	117.3	34	119.7	36	123.3	35	125.4	34
Germplasm enhancement and breeding	61.9	23	64	22	58.8	18	63.7	19	60	18	61.2	18	61.8	18	64.1	18	NA[b]	NA
Production systems development and management	62.4	24	70.5	25	70.2	22	69.4	21	64.3	19	56.1	16	57.9	18	59.3	17	NA	NA
• Cropping systems	41.6	16	38.5	13	40.5	12	35.1	11	32.7	10	29.3	8	32.1	10	32.7	9	NA	NA
• Livestock systems	15.7	6	21.1	7	18.4	6	18.7	6	19.7	6	15.6	4	13.8	4	16.7	5	NA	NA
• Tree systems	3.9	1	8.9	3	9.2	3	14.2	4	10.4	3	9.3	3	8.3	3	7.9	2	NA	NA
• Fish systems	1.2	0.5	1.9	1	2.2	1	1.4	0.4	1.5	0.4	1.9	0.5	3.7	1	1.9	1	NA	NA
Protecting the environment	40.1	15	45.3	16	53.7	17	57.4	17	64.5	19	67.9	20	60.4	18	67.2	19	66.5	18
Saving biodiversity	22.6	9	28.5	10	34.6	11	35.3	11	37.2	11	36.2	10	34.8	10	34.2	10	36.9	10
Improving policies	26	10	25.2	9	38.9	12	37.3	11	39.9	12	46.8	13	48	14	49	14	55.4	15
Strengthening national agricultural research systems (NARS)	51.7	20	52.6	18	68.7	21	70.2	21	70.9	21	78.6	23	74.6	22	81.1	23	84.9	23
Total	264.7	101	286	100	325	101	333.3	100	336.8	100	346.8	100	337.5	100	354.8	101	369	100

Source: World Bank, CGIAR Financial Reports 1994–2002

[a] 2002 figures are taken from the CGIAR Annual Report for 2002, CGIAR (2002).

[b] NA=Data not available.

Report (World Bank, 2003) NRM research in the CGIAR is 'under-evaluated' and requires more accountability. The use of the term 'under-evaluation' relates to four distinct areas:

1. Productivity/efficiency of resource use
2. Science quality
3. Comparative advantage
4. Impacts on the ground.

This OED assessment raises questions not only about the shift in priorities and investments by the System over time, from crop germplasm improvement to NRM research, but is critically important as the CGIAR contemplates whether to adopt four new Challenge Programmes, all of which have strong NRM and INRM dimensions (Interim Science Council, 2002).

One of the major recommendations from the OED's analysis is the need for the CGIAR to give more prominence to basic plant breeding and germplasm improvement and to reshaping NRM research in order to focus tightly on productivity enhancement and sustainable use of natural resources (NRs) for the benefit of developing countries. The latter part of this recommendation – the need to focus more on the productivity dimensions of NRM – reflects a growing awareness among CGIAR stakeholders of an increasing trend towards conducting research on environmental protection/environmental services, for which little impact has thus far been demonstrated, at the expense of contributing to productivity enhancement for which a considerable amount of documented impact exists. Some also question the CGIAR's comparative advantage in NRM research with a strong environment conservation focus. For a detailed discussion of NRM research in the CGIAR see Barrett (2002).

It is difficult to be precise about the cumulative level of investments in NRM research activities for the System since its inception, principally because the CGIAR Activity definitions have changed over time, and those definitions encompassed different aspects of NRM research. For example, of the five principal CGIAR Activities used for classification purposes between 1992 and 2001, two – 'Protecting the environment' and 'Increasing productivity through production systems development and management' – captured different aspects of NRM research. The CGIAR investment allocated to `Protecting the environment' amounted to almost US$500 million (in full cost terms) between 1992 and 2001 – based on an average investment share of 16.5%. Over the same period, investments in 'Production systems development and management' accounted for roughly US$630 million (averaging 21% of the total investment). Certainly not all of this research can be defined strictly as on NRM, but these figures offer some indication of the significant level of investment in NRM – ever since 1992.

Table 15.2 shows investments by the CGIAR across the 16 CGIAR centres in the System from 1994 through 2002. Many of the major commodity centres and the eco-regional centres – those centres that are strongly focused on productivity/enhancement – have seen their investment levels reduced significantly, both in nominal and real terms, consistent with the trend towards lower investment in crop germplasm and increasing productivity.

Table 15.2. Financial allocations to CGIAR centres, 1994–2002 (in US$ million).

Centre[a]	1994	1995	1996	1997	1998	1999	2000	2001	2002	2003[b]
CIAT	33.6	31.5	31.0	31.7	32.1	28.7	29.5	29.7	32.3	35.2
CIFOR	6.1	9.0	9.0	10.6	11.3	11.5	12.6	12.6	11.7	13.9
CIMMYT	31.2	31.7	28.9	28.6	30.1	33.8	39.0	40.7	41.3	39.0
CIP	22.8	23.2	24.2	23.4	22.2	20.0	20.2	19.7	19.2	19.4
ICARDA	19.0	19.3	21.1	22.3	25.2	19.5	23.4	21.3	24.3	26.7
ICLARM	6.6	7.8	9.6	9.0	10.6	14.2	10.4	13.1	12.3	17.4
ICRAF	17.0	16.9	17.4	21.8	20.4	20.6	20.7	22.9	21.8	26.6
ICRISAT	30.8	30.0	29.7	27.7	26.5	21.2	23.3	23.9	24.7	23.5
IFPRI	13.8	13.8	16.0	18.2	20.1	20.8	21.2	22.5	22.7	25.5
IITA	33.4	31.4	31.2	30.8	29.2	30.7	30.1	35.3	32.6	38.5
ILRI	29.1	29.6	28.3	26.1	24.6	26.6	26.5	28.2	27.5	28.7
IPGRI	14.5	12.6	16.4	18.8	21.2	20.1	21.5	23.1	25.6	29.4
IRRI	39.8	38.1	38.3	35.4	34.8	32.5	32.6	32.6	33.4	29.3
ISNAR	10.4	11.3	10.7	9.9	9.6	8.2	8.2	8.1	8.9	10.6
IWMI	8.9	10.2	10.0	10.0	9.4	8.8	8.9	11.4	20.7	21.5
WARDA	8.1	9.7	8.7	8.6	10.0	10.8	9.4	9.7	9.8	10.3
Total	325.1	326.1	330.5	332.9	337.3	328.0	338.5	355.0	369.0	396.0

Source: World Bank, CGIAR Financial Reports, 1994–2003

[a] CGIAR Consultative Group on International Agricultural Research
CIAT Centro Internacional de Agricultura Tropical
CIFOR Center for International Forestry Research
CIMMYT Centro Internacional de Mejoramiento de Maíz y Trigo
CIP Centro Internacional de la Papa
ICARDA International Center for Agricultural Research in the Dry Areas
ICLARM World Fish Centre
ICRAF World Agroforestry Centre
ICRISAT International Crops Research Institute for the Semi-Arid Tropics
IFPRI International Food Policy Research Institute
IITA International Institute of Tropical Agriculture
ILRI International Livestock Research Institute
IPGRI International Plant Genetic Resources Institute
IRRI International Rice Research Institute
ISNAR International Service for National Agricultural Research
IWMI International Water Management Institute
WARDA Africa Rice Center
[b] Estimated.

For example, the International Rice Research Institute's (IRRI's) budget fell from US$38.7 million (1994–1996 average) to US$31.8 million (2001–2003 average), the Centro Internacional de la Papa's (CIP's) from US$23.4 to US$19.4 million, the International Livestock Research Institute's (ILRI's) from US$29.0 to US$28.1 million, and the International Crops Research Institute for the Semi-Arid Tropics's (ICRISAT's) from US$30.2 to US$24.0 million. The major exception to this trend is the Centro Internacíonal de Mejoramiento de Maíz y Trigo (CIMMYT) – its budget rose from US$30.6 to US$40.3 million, although they are currently experiencing some reduction and downsizing. When viewed in real terms, i.e. after adjusting for inflation, the impacts of these reduced resources are even more significant. The centres that expanded during this period were usually those focusing on NRM (particularly environmental protection) and policy. Thus, the International Water Management Institute's (IWMI's) budget rose from US$9.5 (1996–1998) to US$17.9 million (2001–2003), the World Fish Centre's (ICLARM's) from US$8.0 to US$14.3 million, the World Agroforestry Centre's (ICRAF's) from US$17.1 to US$23.8 million, the Center for International Forestry Research's (CIFOR's) from US$8.0 to US$12.7 million, the International Food Policy Research Institute's (IFPRI's) from US$14.5 to US$23.6 million and the International Plant Genetic Resources Institute's (IPGRI's) from US$14.5 to US$26.0 million.

Given the relatively significant and growing investment in NRM research within the CGIAR and the almost total lack of evidence of research impact, it is essential that the CGIAR gear up rapidly and with significant effort to document the impacts of the past investments in NRM-related research to provide information that can be used in shaping future investment in research.

From NRM to INRM

Traditional NRM research in the CGIAR tended to be more narrowly defined and included such agronomy-related themes such as soil and nutrient management, irrigation and land-cover management, water harvesting and so on. It had a strong emphasis on maintaining or increasing resource productivity. Indeed, it also aimed to complement the germplasm improvement research to exploit the benefits of new cultivars.

More recently, there has been a growing interest in the CGIAR in INRM research. This is a broader research paradigm that emphasises the nexus of productivity enhancement–environmental protection–human development as a multiple research objective across different time and spatial scales, from field plot to landscape levels (Sayer and Campbell, 2001; Turkelboom *et al.*, 2003). This paradigm runs parallel to the integrated watershed management paradigm that has been in use for many years (TAC, 1997, 2001b; Brooks *et al.*, 2003).

In recognising the complexity of the integrated people/productivity/protection nexus, INRM research is oriented toward enhancing adaptive capacity by: 1. incorporating more participatory approaches; 2. embracing key principles related to multiscale analyses and interventions; and 3. the use of a variety of new and improved tools such as: systems analysis, geographic information systems (GIS), and other information and communication technologies. Integration provides the key: across scales, components, stakeholders and disciplines. Invariably, INRM must concern itself with sociopolitical, economic, and ecological variables (Campbell *et al.*, 2001). Clearly, this represents a significant departure from traditional NRM research that simply aimed to maintain ('maintenance research') or raise productivity of resource use in a sustainable manner, that is, over the long term.

Because INRM is fundamentally different – it is more development-oriented, attempts to catalyse change, focuses on the (non-linear) 'process' of change, and is not top-down – IA cannot use the static linear models of commodity crops – the traditional economic surplus or econometric approaches (CGIAR, 2000). In addressing INRM IA, the participants at the inter-centre INRM Workshop, held in Penang in August 2000, noted that INRM-based methods are more like continuous assessment that includes regular feedback to improve performance (CGIAR, 2000). Therefore, IA methodologies should employ a highly adaptive research approach and emphasise 'process' at least as much as results. Still, after some time, one should be able to assess the contribution of the research product to resilience, and resilience directly affects vulnerability – a key component of poverty.

There are many welcome features to the new INRM paradigm that address a range of highly important, here to fore neglected topics and dimensions of NRM research in the CGIAR. But there are also concerns, particularly related to the highly conceptual nature of the definition of INRM and, thus, the problems introduced in attempting to do specific, quantitative *ex post* IA. Among the concerns related to assessment and evaluation are issues related to:

- Cost effectiveness: highly participatory (farmers and stakeholders), process-oriented, with strong and continuous monitoring and evaluation (M&E) assessments – all positive developments, but the cost implications therein are high, especially if attempted for every project
- Learning-by-doing, strong participatory approaches, and creating empowerment are all important, and the process does matter, but are these goals in their own right, or are they means for achieving more basic goals related to the welfare of people? If they are the goals, then this is a departure from the previous science-based focus of the CGIAR; and the question arises as to how progress/impact is to be measured
- Scaling-up from site specificity. Will there be a need to define a new set of 'processes' for each site or at each level of scaling up? Which aspects are generic and which need to be defined anew each time?
- What is the appropriate role of the CGIAR here – how far should it move towards the development end of the spectrum? Will impact still be defined in terms of producing international public goods, a stated objective of the CGIAR?
- How will INRM achieve 'lasting impacts on people and the environment across relatively large areas and within reasonable time frames' (CGIAR, 2000) where it has not yet been possible to demonstrate impacts even for the more limited and structured NRM research? What special aspects of INRM render it more conducive to/effective at scaling up?

In short, while the INRM research concept is certainly more comprehensive and more process-oriented than the conventional and more limited NRM research, it does raise questions about the ability to measure and assess *ex post* impacts in the traditional sense of the term. One of the fundamental issues that should be debated soon is the nature of the impacts that need to be measured for INRM research. While the use of IA as a learning tool for those doing research is quite clear in the case of INRM, it is not as clear exactly what INRM thinking is about the other major use, namely, as a means of justifying and guiding investment by the investors and donors in the CGIAR. If these investors and donors are looking for more conventional quantitative measures, or at least indications of INRM investment impact, then the IA challenges are substantial. This conclusion also holds true for the more-focused, narrowly defined NRM research.

In addressing these challenges, first priority for *ex post* IA should be given to the older, already completed NRM research projects, for three reasons: 1. the impacts tend to be more tractable in measurement since there generally is a single, focused goal (productivity enhancement); 2. they provide the retrospective view necessary to measure *ex post* impacts, i.e. allowing for significant research and adoption lag periods to have elapsed; and 3. the lessons learned from such assessment – both methodological and outcome-based lessons – will be a valuable input in developing acceptable means for measuring impacts of the broader, more-diffuse INRM research.

Ex post Impact Assessment of CGIAR NRM Research: the Challenges

As mentioned above, with the growing trend towards greater investment in NRM-related activities and centres, there is also growing need to demonstrate impact. Many investors and donors need evidence of impact from their investments in order to secure additional resources.

Insufficient evidence of NRM research impact

As mentioned earlier, there is a growing concern with the lack of evidence of the impacts of CGIAR NRM research. This concern was expressed most recently in a meta-evaluation of the CGIAR by the World Bank's OED (World Bank, 2003). This concern is now explored in greater detail and some possible explanations offered.

Pingali (2001) provides an overview of some of the important IA (*ex ante* and *ex post*) work within the CGIAR since its inception. Research related to crop germplasm improvement (CGI) effects clearly dominates the literature. Generally speaking, *ex post* impacts for CGI research by CGIAR centres are relatively well documented. The recent Standing Panel on Impact Assessment (SPIA)-commissioned CGI IA, by involving all eight CGIAR commodity centres, documents the significant contribution made by the CGIAR to improving agricultural productivity through germplasm improvement (Evenson and Gollin, 2003). This contribution includes:

- Poverty reduction (higher incomes, more employment and lower food prices)
- Large land savings [Nelson and Maredia (2002) estimate a land savings equivalence between 130 and 320 million ha from CGIAR research]
- High rates of return on commodity improvement research investments.

Pingali's review of the literature reveals relatively few 'crop management and improved input use' and other NRM-related CGIAR impact studies to-date, a finding that corroborates an earlier review by Byerlee and Pingali (1994). In the following discussion it should be noted that the impact of research on NRM is concerned with two main effects: productivity enhancement and environmental and natural resource protection or the sustainability dimension.

The SPIA-commissioned benefit–cost meta-analysis of CGIAR investment

This analysis (Raitzer, 2003) systematically reviews and evaluates IA studies of economic benefits derived from CGIAR innovations (known 'success stories'), so as to produce a range of plausible and highly credible benefit–cost ratios for the entire investment in the CGIAR (since 1972). The SPIA consultant spent considerable time reviewing the evidence and literature searching for documented large-scale CGIAR research impacts, where estimated benefits were at least US$50 million. Results show a notable absence of big success stories for NRM – at least not among those documented – with the exception of biocontrol or integrated pest management (IPM) research (e.g. Bokonon-Ganta *et al.*, 2001; Zeddies *et al.*, 2001) that is not usually classified under NRM *per se*. Thus, the documented evidence of the economic impact of NRM research in the CGIAR is virtually nil, at least when considering moderate to large-scale effects.[1]

Alston et al. *(2000) meta-analysis*

This survey of the rates of returns for all types of agricultural research, including smaller-scale studies, found very few NRM-related studies (less than 4% of the total studies reviewed). Hence, unlike the situation in CGI, for which large-scale adoption of yield-enhancing CGIAR-derived varieties has been documented for a range of CGIAR crops, there are as yet few examples of successful (widely adopted) CGIAR-generated improved NRM technologies for which demonstrable impact has been measured and assessed. Further, the NRM IAs included in the Alston *et al.* study showed significantly lower average rates of return than the CGI-related IAs.

External Programme and Management Reviews (EPMRs) of the CGIAR centres

These reviews have evaluated NRM research components in the centres and the evidence they provide is not always positive in terms of the effectiveness of such research. On the contrary, with the exception of the CIP EPMR in 2002, recent EPMR reports have been quite specific in their criticism of the quality of science, achievements and on-the-ground impacts from the NRM research programmes.

Each of the above points to a similar conclusion: there is little documented evidence of impact – economic or otherwise – of CGIAR research on NRM and related topics. Some of the recent studies that do attempt to document impact, like those of ILRI's *ex post* IA series, show limited impact or even negative rates of return on investments (Elabasha *et al.*, 1999; Rutherford *et al.*, 2001). Other NRM 'impact' reports are anecdotal or early (limited) adoption studies (e.g. ICRISAT's 'A Rainbow Painted on the Last Frontier' and 'Joining Hands to Halt Erosion' type stories). Some are quite general and conjectural (claims difficult to substantiate one way or other), or they tend to focus on descriptions of potential or probable impact, e.g. the International Board for Soil Research and Management (IBSRAM) impact report by Maglianao (1998). It seems that most NRM impact type studies give much stronger emphasis to evaluation of adoption and constraint issues, i.e. they have a strong learning component, rather than focusing on documenting the impact *per se*, as has been done in the CGI studies.[2]

Why the lack of documented evidence of impact?

Why are there so few documented success stories of NRM research in the CGIAR, studies that go beyond anecdotal evidence and selective small-scale case study results? Some of the more plausible reasons might include the following.

Lack of sustained critical mass investment

No doubt the lack of evidence partly reflects a lack of sustained emphasis on NRM research over the last few decades. But this can be only a partial explanation. While CGIAR investments in CGI research have been much larger than for NRM research, the absolute levels of investments by the centres in NRM research and its earlier precedents, for example, farming systems research, are still considerable, as indicated earlier. The levels indicated certainly qualify this type of research.

It should not be forgotten that research in soil and water management and in cropping/farming systems in general represented a significant part of many CGIAR centres' research agenda during the CGIAR's first two decades, and these were typically focused on productivity-enhancing aspects of NRM. Major investments were made in such areas as broadbed-and-furrow management, minimum and zero tillage systems, alley cropping, watershed management and other soil and water management related research. To date, far too little of this has been assessed in terms of impact – whether in terms of improvements in resource productivity, in enhancing the environment, or in terms of quality.

Inappropriate methods

NRM IA has lagged behind assessment of the impacts of germplasm improvement and certain technology developments. Approaches are needed that capture environmental services and other (non-crop yield) gains due to such NRM/INRM research as maintenance and loss reduction, risk reduction, quality improvement, reduction of negative environmental externalities, and compatibility with off-farm labour schedules. Certainly, lack of appropriate methods has constrained efforts to document impact from NRM (Izac, 1998). Economic surplus methods for measuring and attributing the impact for CGI research may often not be appropriate in the case of NRM research. While this may apply to some of the current efforts in process-oriented INRM, it does not explain the lack of NRM impact assessments for research focused mainly on productivity improvements – the lion's share of NRM efforts before the mid-1990s and a significant portion of it afterwards.

When addressing NRM research impacts, a whole range of other issues needs to be considered. Markets are largely missing for the environmental services provided. Different valuation methods exist, all of which are highly imperfect and tricky to use, and hence there is a need for a range of values reflecting different perspectives and valuation methods. Externalities are spread over different scales and hence difficult to capture as each level needs to be done with different tools. The time dimension is crucial and hence the choice of discounting key. There are also important problems of resilience and irreversibilities that need to be taken into account in constructing counterfactual scenarios. For these reasons, designing control groups for NRM treatments is particularly difficult because of the spatial and temporal dimensions involved.

The difficulty in measuring and attributing impact of NRM/INRM research is of a significantly higher order than for CGI research (Izac, 1998). The issues relate particularly to: complexity issues (in scale, in time), non-linearity (causality), the economic and non-economic dimensions, operation-indicator issues, higher costs, more disciplines involved, longer time lags, attribution problems, and difficulty in extrapolation. The problem is confounded because some of the gains and impacts from CGI investments that have been supported by improved crop and soil management derived through NRM research are also hidden and cannot be recognised.[3] This is a measurement/allocation problem, but without some evidence it remains conjectural or anecdotal at best. There is a need to develop means to measure and subsequently document the key role improved resource management has played in realising on-the-ground impacts.

Given the level of investment in NRM research in the CGIAR to date – most of it focused on productivity improvements – we should not ignore productivity impacts using the conventional market model. Underpinning this is the core issue of efficiency of resource use. Virtually, all sustainable paths to poverty alleviation are derived directly or indirectly through increased productivity.

Notwithstanding the present need for new methods and approaches to measure the more-complex and less-tangible effects of NRM research, it remains the case that even simple impact measures, such as adoption and use of NRM products, are still scarce and hence complexity itself may not be the primary reason for a lack of documented impact in NRM research in the CGIAR.

Lack of impact per se

It must be recognised that, as in the case of other types of research, some NRM research in the CGIAR has failed to generate the appropriate technologies or institutional arrangements that adequately address the needs of poor farmers and communities. Looking through centre annual reports from the late 1970s through to the early 1990s, one can appreciate the range of NRM research-related activities in which the CGIAR has been involved. Some of these early efforts focused on: water harvesting, broadbed-and-furrow management, erosion control through contour bunding, zero tillage in Africa, use of green cover crops and/or mulching, ley farming, alley cropping, and better management of crop stover. While most of these are commendable in terms of the science applied, they ultimately appear not to have generated sufficient (sustainable) wide-scale adoption among farmers. In such cases, where it is evident that adoption is lacking, there is little incentive to assess impact. Thus, lack of impact *per se* could be a major reason behind the lack of evidence of impact. This is in no way an indictment of the quality of research conducted – not all research can be expected to result in a proven, adopted technology – nor does it overlook the fact that some technologies have indeed been adopted. In some cases, lack of an effective delivery mechanism could explain low adoption, although this reason may be used more frequently than is justified.

One hypothesis to explain why NRM research may not have had more impact is that the information or technology generated through the research is not, in itself, sufficient to catalyse wide-scale adoption. Its use and adoption is contingent on a great many other pre-conditions. Relative to germplasm improvement, NRM improvements require many more actors to get impacts on the ground, such as extension, policy, institutions, organised farmers and communities. Because it is often location-specific, and the CGIAR has not yet developed adequate links with many of these actors at the local level, it is inherently difficult to generate impacts.

To the extent that for either technical, economic, or social reasons, research-led technologies have not been adopted, it might be useful to distinguish between NRM research focused on individual farmer-based decision-making (more technology-focused) vs. that focused on group/community-based decision-making (more rules/institution-focused, technology less important). With respect to the individual farmer, the attractiveness of new technology depends primarily on expected profitability/risk levels and additional labour or other inputs required with the 'improved NRM technologies'. Perhaps not appreciated sufficiently is the fact that many farmers in developing countries are looking for ways to reduce their labour input in agriculture, and not to increase it, or to have other opportunities that are more profitable or less risky, or give them higher utility, e.g. investments in children's education. An opportunity-cost assessment approach would pick up that aspect. Some of ICRISAT's research on the non-use of fertilisers in southern Africa shows this to be the case (Rusike *et al.*, 2003).

With respect to NRM research focused on group- or community-based decision making, the emphasis on such key issues as property rights and the need for community action has resulted in a number of promising success stories, as brought out in, e.g. the Systemwide Collective Action and Property Rights (CAPRi) external review (Interim Science Council, 2002). Here, the major constraint is scaling up, or scaling out. Without that ability, the investment is usually not cost-effective. Indeed, this was one of the major conclusions reached at the Agroforestry Dissemination Workshop held at ICRAF in September 1999: 'The developing world has no shortage of successful "pilot" schemes and projects that have sought to address the problems of poverty, food security and environmental degradation. There are too few cases where these successful pilots have led to widespread impact on a sustainable basis' (Cooper and Denning, 2000). Exacerbating scaling problems is the fact that extension funding has fallen significantly in recent years, and the greater the complexity of technological adoption, the greater the need for extension (Douthwaite *et al.*, 2001). Thus, this lack of impact may not be attributable to research itself. The entire impact pathway needs to be considered, and that includes the dissemination and adoption processes. Essential here is acquiring a better understanding of how resource management changes take place under different sets of agricultural policies and economic and social environments. This sets the stage for more effective targeting of technology and greater impact.

Meeting the challenge: NRM impact assessment at the System level

Although the need to document NRM *ex post* impact assessments was highlighted at the SPIA-sponsored IA Workshop held at the Food and Agriculture Organization of the United Nations (FAO) in May 2000 (TAC, 2001a), sufficient resources have not been available to embark on any System-level assessment of NRM impacts. In 2003 SPIA initiated three main activities aimed at a better understanding of the impacts of the CGIAR's work in NRM. These are:

1. Developing improved methods for assessing NRM impacts
2. Working with centres to develop empirical evidence of impacts from centre activities
3. Developing through CGIAR Science Council (SC) evaluations, some empirical evidence of impacts from System-wide activities.

Activity 1. Development of improved methods for assessing NRM impacts

The need to develop appropriate methods has been highlighted. SPIA has commissioned a background paper on current NRM IA methodology and will host a workshop inviting experts and key stakeholders of the System to discuss the output of this paper and to help develop strategic guidelines for conducting *ex post* IA for NRM/INR research. At the same time, the Intercentre Working System on INRM is preparing a parallel paper reflecting the centres' collective judgement on the best way to move forward and illustrating the variety of past NRM IA work. In bringing together the best currently available methods for assessing returns from NRM research, SPIA will act as a catalyst in developing new approaches and clarifying objectives and purposes for NRM IA. It bears recalling that 'the purpose of any impact study must be well articulated to guide choices as to stage, product emphasis, geographic scope, precision of measurement, and other parameters' (Anderson, 1997). This would include assessment of the relative contributions of different types of research, such as CGI and NRM research in an overall on-the-ground *ex post* IA.

Activity 2. Empirical evidence of impacts from centre activities

This activity involves a set of six case study assessments of the impacts of selected centre NRM projects/activities. SPIA/SC is providing some resources and oversight for selected centres to undertake credible, comparable empirical assessments of the impacts of selected NRM projects/programmes in the context of the CGIAR mission and goals. The interim Science Council (iSC) and Intercentre Working Group on INRM have also completed a series of mini case studies illustrating the approaches to, and results from, INRM research in the CGIAR.

Activity 3. Empirical evidence of impacts from System-wide activities

This involves an assessment of the impacts associated with one of the longest running System-wide programmes that focus primarily on NRM activities – the Alternatives to Slash and Burn (ASB). In order to ensure effectiveness and efficiency in the use of CGIAR funds, the IA will be carried out jointly with a more traditional iSC type of programme evaluation. ASB already has produced a draft report on impact pathways associated with its programme.

In addition to the above, there is a clear need for an inventory of NRM and related research activities in the CGIAR, with some measure of the levels of investments therein. This, of course, will require that centres come to grips with the definitional issues raised above and that the CGIAR develops a set of uniform definitions and boundaries on such issues.

Implications for CGIAR Centre NRM Impact Assessment Activities

Every centre ought to have some highly convincing impact stories that reflect its major research thrusts. Commodity- and eco-regional-oriented centres have concentrated most of their *ex post* IA effort on documenting impacts of their germplasm improvement work. These have been convincing stories. Clear recognition of this is evident from the numerous King Baudouin Awards (for excellence in research), virtually all of these awarded for work related to widespread impacts of improved cultivars developed by the centres and their partners.

At the same time, these centres have had difficulty in documenting rigorously and on a large-scale the impact from their research in NRM and related areas. All of these commodity and eco-regional centres have had considerable activity in, and a strong commitment to, NRM research. This provides a strong incentive for these centres to play a leadership role in demonstrating the poverty impacts from CGIAR investments in NRM research. Databases exist, for example, the ICRISAT adoption database, that will allow preliminary assessments of the adoption and assumed benefits from NRM research-derived technologies.

The challenge here is great: firstly, in developing a framework for understanding, identifying, and measuring poverty alleviation impacts from NRM research; and secondly in actually measuring and documenting the impact from a few selected NRM research-related projects/areas of work.

With respect to identifying an overall framework and methods, the SPIA/IFPRI poverty impacts study provides an excellent platform on which to build. This project has just completed seven individual case studies on CGI and NRM work. The study uses a livelihood framework, combines quantitative and qualitative analyses, and economics and social indicators to trace the effects of research on different groups of people (Adato and Meinzen-Dick, 2002). In short, it represents a good model for NRM research IA, although scaling up still represents a major challenge and will need to be considered carefully.

With respect to the selection of specific NRM research topics and case studies, one approach [that used by SPIA at the System level in the Benefit–Cost Meta-Analysis (Raitzer, 2003)] would be to compare for a given centre the total cost of all NRM investments to date against the aggregated benefits from one or two large-scale centre NRM success stories. For a major donor to any one of these centres, or for a new donor considering major support to NRM research, a legitimate question would be 'how effective have investments in NRM research been at this or any other centre, or how effective is it likely to be – and what's the evidence, i.e. the impacts that have been achieved and are likely to be achieved?' NRM scientists are clearly the best placed to specify which investments are likely to have generated the highest payoffs.

Aside from the accountability motivation, a close examination of the benefits to date from NRM research activities would provide highly useful reflection as to what has and what has not worked, and insight as to why, i.e. draw on the learning function associated with IA. Quantitative economic assessment based on productivity improvements could be combined with qualitative assessments of the nature and scope of environmental improvements. If well documented, the latter would be particularly useful.

Conclusions and Recommendations

Both for accountability purposes and to help shore up funding in the CGIAR, *ex post* IAs of CGIAR research activities are essential. Donors continue to ask for evidence that the research they fund has impact on the livelihoods of the poor.

CGIAR-centre and CGIAR Activity-level investment trends show a clear shift in emphasis away from commodity improvement and NRM research for productivity enhancement towards policy and NRM/INRM research focused on environmental services. The latter are not under so much pressure to document successes, but this could well change in the near future. At present, there is insufficient evidence of the impact of these investments in research.

There is concern among some stakeholders of the CGIAR that investment in environmental protection and environmental services research – for which there is insufficient impact information – is increasing at the expense of efforts aimed at enhancing productivity through more traditional NRM production systems research and through CGI – for which a substantial amount of impact has been convincingly documented.

The recommendations flowing from this chapter are as follows:

- It is essential that the CGIAR gear up rapidly to document large-scale (beyond 'pilot' or site-specific cases) *ex post* impacts of both the more traditional NRM and the newer INRM research to demonstrate the essential international public good (IPG) element required to justify CGIAR investment
- Priority for *ex post* IA should be given to the more traditional productivity-enhancing NRM research projects because these are likely to be more tractable in measurement. Many of these simply have a longer track record, meaning that significant research and adoption lag periods have elapsed – a prerequisite for measuring impact in CGIAR goal terms. The lessons learned from such assessments – both methodological and outcome-based – will be a valuable input in developing acceptable means for measuring impacts of INRM research. Information about impact of this kind will be helpful (and may be decisive) in underpinning arguments in support of NRM and INRM research. Such information will help bring sustained funding for the long-term processes of discovery that help achieve global food security (Anderson, 1997). To the extent that impacts from NRM research are not forthcoming, diagnosing why is of critical importance for resource allocation decisions and the design of future research programmes
- These CGIAR centres with longer histories and more experience in NRM and FSR research (ICRISAT, ICARDA, IITA and CIAT) should play a leading role in documenting impacts from their more-successful NRM research-derived technologies. Drawing on the promising conceptual and analytical frameworks being used within the CGIAR, and utilising adoption and information databases (formal and informal), these centres are in an ideal position to initiate and take the lead in *ex post* IAs of NRM research.

Acknowledgements

Comments and suggestions for improvement on an earlier draft from Derek Byerlee, Elias Fereres and David Raitzer are gratefully acknowledged. The views expressed are those of the authors [the Secretary and Chair for the Standing Panel on Impact Assessment (SPIA) of the CGIAR's Science Council] and do not necessarily represent those of SPIA or the Science Council.

Endnotes

[1] SPIA, however, commissioned a study to examine the indirect NRM impacts from CGIAR research – particularly the positive and negative environmental impacts from CGI. The Environmental Impact Study by Maredia and Pingali (2001) found that most of the evidence about the negative environmental impacts of genetic resources (GR) technology is anecdotal, and it is difficult to establish causality in any event. Yes, there have been adverse environmental consequences such as pesticide overuse in rice and fertiliser contamination of groundwater from wheat production, but this is not a direct consequence of improved germplasm. The Environmental Impact Study by Nelson and Maredia (forthcoming) estimates the land saved equivalent from agricultural research, again, as an indirect (positive) impact on the environment.

[2] This is not to imply that anecdotal evidence is not genuine, but that much more remains to be done in rigorously documenting on a larger scale some of the promising work and early achievements of the CGIAR centres in the area of NRM, e.g. CIFOR's research and impact on changes in public policies related to management of forests, the increased use and benefit from trees and woody shrubs as a result of ICRAF's work, better irrigation management resulting from IRRI's and IWMI's efforts, the fisheries work at ICLARM influencing policies and community action for conservation of fish stocks, higher productivity and land conservation through ILRI's research on range improvement, the rice–wheat initiative (zero tillage) led by CIMMYT, and so on. These are all promising developments and worthy of additional resources by which solid impacts on poverty can be demonstrated.

[3] For example, Bell *et al.* (1995) attempted to measure the genetic and agronomic contributions to increasing wheat yields in northwest Mexico. They estimated a 28% yield gain due to genetic factors, a 48% yield gain attributed to increasing N fertiliser and the remaining 24% was attributed to 'other factors' – possibly including increasing P fertiliser, among others. These results, while isolating the pure genetic contribution, do not in themselves establish a contribution from NRM research.

References

Adato, M. and Meinzen-Dick, R. (2002) Assessing the impact of agricultural research on poverty using the sustainable livelihoods framework. *IFPRI Discussion Paper no. 89.* International Food Policy Research Institute (IFPRI), Washington, DC, 46 pp.

Alston, J.M., Chan-Kang, C., Marra, M.C., Pardey, P.G. and Wyatt, T.J. (2000) *A meta-analysis of rates of return to agricultural R&D:* Ex pede Herculem? International Food Policy Research Institute (IFPRI), Washington, DC, 148 pp.

Anderson, J.R. (1997) *On Grappling with the Impact of Agricultural Research.* Prepared for International Centers Week, 31 October 1997, Washington, DC, 13 pp. (unpublished).

Barrett, C.B. (2002) Natural resources management research in the CGIAR: A meta-evaluation. Thematic Working Paper for the Operations Evaluation Department (OED) Report *'The CGIAR at 31: An Independent Meta-Evaluation of the CGIAR,'* World Bank, Washington, DC, 141 pp.

Bell, M.A., Fischer, R.A., Byerlee, D. and Sayre, K. (1995) Genetic and agronomic contributions to yield gains: a case study for wheat. *Field Crops Research* 44, 55–65.

Bokonon-Ganta, A.H., DeGroote, A.H. and Neuenschwander, P. (2001) Socio-economic impact of biological control of mango mealybug in Benin. *Agriculture, Ecosystems and Environment* 1905, 1–12.

Brooks, K., Folliott, P., Gregersen, H. and DeBano, L. (2003) *Hydrology and the Management of Watersheds*, 3rd edn. Iowa State University Press, Ames, Iowa, 574 pp.

Byerlee, D. and Pingali, P. (1994) Rice research in Asia: fulfilments and frustrations. Presented at the *XXII Conference of the International Association of Agricultural Economists in Harare, Zimbabwe, 22–29 August 1994* (mimeo).

Campbell, B., Sayer, J.A., Frost, P., Vermeulen, S., Ruiz Pérez, M., Cunningham, A. and Prabhu, R. (2001) Assessing the performance of natural resource systems. *Conservation Ecology* 5(2), 22.

CGIAR (Consultative Group on International Agricultural Research) (2000) *Integrated Natural Resource Management in the CGIAR: A Brief Report of the Integrated Natural Resource Management (INRM) Workshop, Penang, Malaysia, 21–25 August 2000.* CGIAR Secretariat, World Bank, Washington, DC, 50 pp.

CGIAR (Consultative Group on International Agricultural Research) (2002) *Annual Report 2002. Research for Sustainable Development: From Words to Action.* Consultative Group on International Agricultural Research (CGIAR), World Bank, Washington, DC, 56 pp.

Cooper, P.J.M. and Denning, G.L. (2000) *Scaling up the Impact of Agroforestry Research: Report of the Agroforestry Dissemination Workshop. World Agroforestry Centre (ICRAF), Nairobi, Kenya, 14–15 September 1999.* ICRAF, Nairobi, Kenya, 43 pp.

Douthwaite, B., Keatinge, J.D.H. and Park, J.R. (2001) Why promising technologies fail: The neglected role of user innovation during adoption. *Research Policy* 30, 819–836.

Elbasha, E., Thornton, P.K. and Tarawali, G. (1999) An *ex post* economic assessment of planted forages in West Africa. *ILRI Impact Assessment Series no. 2.* International Livestock Research Institute (ILRI), Nairobi, Kenya, 61 pp.

Evenson, R.E. and Gollen, D. (2003) *Crop Variety Improvement and its Effect on Productivity: The Impact of International Agricultural Research.* CAB International, Wallingford, UK, 522 pp.

Interim Science Council (iSC) (2002) *Report of the First External Review of the Systemwide Programme on Collective Action and Property Rights (CAPRi). Report presented at the Annual General Meeting, Manila, The Philippines, 1 November 2002.* iSC Secretariat, Food and Agriculture Organization of the United Nations (FAO), Rome, Italy, 43 pp.

Izac, A.-M.N. (1998) *Assessing the Impact of Research in Natural Resources Management.* Synthesis of a Workshop, 27–29 April 1998, Nairobi, Kenya. World Agroforestry Centre (ICRAF), Nairobi, Kenya, 38 pp.

Maglianao, A.R. (1998) IBSRAM's impact: Making the difference in sustainable land management research. *Issues in Sustainable Land Management no. 4.* International Board for Soil Research and Management (IBSRAM), Bangkok, Thailand, 43 pp.

Maredia, M. and Pingali, P. (2001) *Environmental Impacts of Productivity-Enhancing Crop Research: A Critical Review. Report prepared on behalf of the CGIAR Standing Panel on Impact Assessment.* Science Council Secretariat, Food and Agriculture Organization of the United Nations (FAO), Rome, Italy, 36 pp.

Nelson, M. and Maredia, M. (2002) *Environmental Impacts of the CGIAR: An Assessment.* Report prepared on behalf of the CGIAR Standing Panel on Impact Assessment. Science Council Secretariat, Food and Agriculture Organization of the United Nations (FAO), Rome, Italy, 46 pp. (unpublished).

Pingali, P.L. (2001) In: Feldman, M.P. (ed.) *Milestones in Impact Assessment Research in the CGIAR, 1970–1999, With an Annotated Bibliography of Impact Assessment Studies Conducted in the CGIAR, 1970–1999.* Standing Panel on Impact Assessment, Technical Advisory Committee (TAC) of the Consultative Group on International Agricultural Research (CGIAR), Mexico, DF, Mexico, 37 pp.

Raitzer, D. (2003) *Benefit–Cost Meta-Analysis of Investment in the Consultative Group on International Agricultural Research (First Draft). Report prepared on behalf of CGIAR Standing Panel on Impact Assessment.* Science Council Secretariat, Food and Agriculture Organization of the United Nations (FAO), Rome, Italy, 45 pp.

Rusike, J., Dimes, J. and Twomlow, S. (2003) Risk-return tradeoffs of smallholder investments in improved soil fertility management technologies in the semi-arid areas of Zimbabwe. Presented at the *International Association of Agricultural Economists (IAAE) Mini-Symposium on Soil Fertility and Food Security for the Poor in Southern Africa: Technical, Policy and Institutional Challenge, Durban South Africa, August 16–22, 2003* (unpublished).

Rutherford, A.S., Odero, A.N. and Kruska, R.L. (2001) The role of the broadbed maker plough in Ethiopian farming systems: an *ex post* impact assessment study. *ILRI Impact Assessment Series no. 7.* International Livestock Research Institute (ILRI), Nairobi, Kenya, 157 pp.

Sayer, J.A. and Campbell, B. (2001) Research to integrate productivity enhancement, environmental protection, and human development. *Conservation Ecology* 5(2), 32.

TAC (Technical Advisory Committee of the Consultative Group on International Agricultural Research) (1997) *Priorities and Strategies for Soil and Water Aspects of Natural Resources Management Research in the CGIAR.* TAC Secretariat, Food and Agriculture Organization of the United Nations (FAO), Rome, Italy, 110 pp.

TAC (Technical Advisory Committee of the Consultative Group on International Agricultural Research) (2001a) *The Future of Impact Assessment in the CGIAR: Needs, Constraints and Options. Proceedings of a workshop organized by the Standing Panel on Impact Assessment of the Technical Advisory Committee, 3–5 May 2000, FAO, Rome.* TAC Secretariat, Food and Agriculture Organization of the United Nations (FAO), Rome, Italy.

TAC (Technical Advisory Committee of the Consultative Group on International Agricultural Research). (2001b) *NRM Research in the CGIAR: A Framework for Programme Design and Evaluation. Presented at the Annual General Meeting of the Consultative Group on International Agricultural Research (CGIAR) in Washington DC, 30–31 October 2001.* TAC Secretariat, Food and Agriculture Organization of the United Nations (FAO), Rome, Italy.

Turkelboom, F., La Rovere, R., Hagmann, J., El-Khatib, R. and Jazeh, K. (2003) *Putting INRM into Action: Workshop Documentation. 4th INRM Workshop, International Center for Agricultural Research in the Dryland Areas (ICARDA), Aleppo, Syria, 16–19 September, 2002.* Consultative Group on International Agricultural Research (CGIAR) Task Force on Integrated Natural Resource Management (INRM), ICARDA, Aleppo, Syria.

World Bank (1994–2002) *CGIAR Financial Reports. 1994–2002.* Consultative Group on International Agricultural Research (CGIAR) Secretariat, World Bank, Washington, DC.

World Bank (2003) *The CGIAR at 31: An Independent Meta-Evaluation of the CGIAR. Volume 1: Overview Report.* Operations Evaluation Department (OED), World Bank, Washington, DC, 46 pp.

Zeddies, J., Schaab, R.P., Neuenschwander, P. and Herren, H.R. (2001) Economics of biological control of cassava mealybug in Africa. *Agricultural Economics* 24, 209–219.

16 Towards Comprehensive Approaches in Assessing NRM Impacts: What We Know and What We Need to Know

S.M. Swinton[1], B. Shiferaw[2] and H.A. Freeman[3]

[1] *Michigan State University, East Lansing, Michigan, USA*
[2] *International Crops Research Institute for the Semi-Arid Tropics (ICRISAT), Nairobi, Kenya*
[3] *International Livestock Research Institute (ILRI), Nairobi, Kenya*

Introduction

The deliberate use of ecosystems by mankind to meet food, feed, industrial, and social and environmental needs inevitably alters the natural ecosystem functions and services. While flux is inherent to ecological systems and their evolution, the natural resource base is currently facing unprecedented human pressure due to population growth and rising consumer demand that follows rising incomes. This human pressure creates a growing need to improve the productivity of existing natural resources and to counter processes that deplete their productive capacity. Governments around the world have responded to the degradation of the natural resource base with projects aimed at sustaining productivity levels and environmental quality. The rising proportion of research funds directed at natural resource management (NRM) at the Consultative Group for International Agricultural Research (CGIAR) is indicative of strong interest (Kelley and Gregersen, Chapter 15, this volume). The increase in funding brings with it an increased need for accountability, *ergo* the urgency of improving impact assessment of NRM investments.

The introduction to this volume summarised the special difficulties in measuring the impacts of agricultural technologies that are designed to enhance the sustainability of natural resources needed for human survival. The technologies themselves are diverse; they range from genetic improvements that allow crops to grow in inhospitable places to conservation practices that reduce soil loss and water pollution. Although a few NRM innovations boost farm revenues (e.g. via enhanced yields), most of the benefits to resource users come from cost-savings, reduced vulnerability to risk (e.g. yield stability) and the avoidance of declining productivity. Examples of such technologies include biologically based soil fertility management, soil and water conservation, water harvesting, integrated pest management, water-saving irrigation, minimum tillage, agroforestry and forest management, rangeland management, and biodiversity conservation. These innovations when adopted provide significant positive environmental and sustainability benefits both on-site and off-site.

The special characteristics of NRM technologies mean that a balanced economic impact assessment must be able to measure environmental and sustainability impacts above and beyond what would have occurred in their absence, a task that has often been ignored in impact assessments heretofore (Nelson and Maredia, 1999). As outlined in the introductory chapter, comprehensive NRM impact assessments pose special problems for establishing the counterfactual, measuring environmental effects, placing a value on those effects, and integrating the final results into a unifying framework.

This book has focused squarely on addressing the methodological challenges for evaluating the impacts of NRM. The preceding 15 chapters have presented and discussed the key issues, challenges, indicators, and valuation and evaluation methods. The sections that dealt with methodological advances were further enriched through case studies that illustrate how impact evaluations can integrate economic and environmental impacts. As agricultural research and development enters a new era through harnessing biotechnology and integrating genetic and resource management, diverging perspectives are emerging on how future impact assessments need to be carried out. The book has highlighted some of these views and outlined areas for future research.

This concluding chapter synthesizes the conceptual, methodological and empirical issues for evaluating the impacts of NRM technology and policy interventions. The intention is to highlight the salient features raised across the chapters and offer some insights on the key lessons, policy conclusions, knowledge gaps, and areas that need further research.

What We Know: The State of the Art in NRM Impact Assessment

Substantial experience has now been gained in applying economic impact assessment methods to productivity-enhancing agricultural research. Measuring changes in economic surplus associated with improvements in agricultural technologies is the most commonly used method in evaluating social net gains from research investments. Alston *et al.* (1995) and Maredia *et al.* (2000) provide a good review of best practices for *ex post* impact evaluation of the economic impacts of agricultural research programs. Despite extensive work on environmental valuation and benefit–cost analysis, there is a dearth of literature on methods for valuation of ecosystem services from NRM technology. Likewise, there are few empirical studies of the social impacts of NRM. Recent years have witnessed a gradual shift in the evaluation literature towards looking at the non-productivity related environmental and sustainability impacts of crop and resource management interventions (e.g. Traxler and Byerlee, 1992; Pingali *et al.*, 1994; Gumtang *et al.*, 1999; Gupta and Abrol, 2000; Pretty *et al.*, 2000). Improved methods are now being developed for comprehensive evaluation of the economic welfare impacts of agricultural interventions – including the productivity and environmental costs and benefits.

Beyond market-based assessments

A recent survey of 1100 agricultural research impact assessment studies found that only 11 included environmental impacts (Alston *et al.*, 1998). In the face of rising expenditures on NRM projects, the fact that 99% of past impact studies relied on measures of economic efficiency alone highlights the need for better assessment of impacts related to sustainability and environmental quality. Because many NRM problems involve economic externalities and/or public goods, neither the problems nor the impacts of NRM technologies designed to solve them are readily measured in markets. It is now widely accepted that impact assessment of NRM interventions should look beyond conventional market-based techniques. Non-market valuation methods, now widely in use in the developed countries for assessment of environmental impacts, can be tested and adapted for evaluating the non-marketed impacts of agricultural and NRM practices. One major challenge is how to measure or find indicators for the dynamic and multidimensional impacts of NRM technologies in agriculture.

Measurement problems

In order to assign economic values to changes in the flow of ecosystem goods and services, the essential first step is to understand how the new interventions affect the quality or quantity of the resource in question and how that translates into changes in goods and services that people value. Changes in ecological functions and processes may be very gradual and take a long time to manifest. Moreover, the dynamic, interdependent nature of ecosystems makes it hard to measure a clear cause–effect relationship from an NRM technology intervention. A basic hurdle in measurement and quantification of biophysical changes therefore has been the incomplete understanding of how NRM practices affect ecosystem health and sustainability.

However incomplete, human knowledge about ecosystems is growing. Long-term experimentation in selected systems has provided useful information about system dynamics and how crop and resource management interventions affect agricultural productivity and resource conditions. The need for such experimentation is even stronger in locations where variability of production conditions is high and data from a short time-series will fail to capture the underlying variation. Simulation models that emulate soil, water, nutrient and crop interactions are now widely used. If properly validated using site-specific biophysical and climatic conditions, such models can be very helpful in evaluating the impacts of multiple changes. This is particularly the case for the integrated interventions of what has now come to be known as integrated natural resource management (INRM). Satellite imagery and geographic information systems are becoming useful tools for monitoring the spatial and temporal dynamics of changes in patterns of land use, vegetation cover, drought stress, surface water, water logging and land degradation. These tools are also gaining importance in yield forecasting and assessment of production risks.

A combination of these scientific advances is making it possible for biophysical scientists and agro-ecologists to estimate physical, chemical and biological changes in agro-ecosystems associated with NRM interventions. When such changes can be understood or predicted, certain measurable indicators can be developed to quantify the magnitude of change associated with a given intervention. Indicators may be developed through experimentation and proper monitoring of changes over a sufficient period of time or through the application of exploratory and predictive simulation models. Chapters 3, 4 and 5 in this volume examined specialised indicators of soil quality, water quantity and quality, and changes in other agro-ecosystem services.

For impact assessment purposes, the most useful indicators of ecosystem functions and services show impacts within 3–5 years of an NRM intervention. For soils, Pathak *et al.* (Chapter 3, this volume) find that biological indicators like soil respiration, microbial biomass, and C and N mineralisation are useful, as are physical indicators such as nutrient runoff and soil loss. Relative to changes in soil quality indicators, changes in surface and groundwater quantity and quality can be observed in a relatively short period of time (Sahrawat *et al.*, Chapter 4, this volume). In order to measure NRM impacts on agro-biodiversity, Wani *et al.* (Chapter 5, this volume) suggest the following indicators for observation within 3–5 years: the index of surface percentage of crops, crop agro-biodiversity factor, and surface variability factors. Changes in biodiversity indicators related to genetic variability, species diversity or richness require longer periods to become visible, indicating the need for long-term follow up and monitoring. Changes in the level of carbon sequestered in soils and vegetation may require even long periods, making simulation modelling a promising approach for predictive purposes.

Valuation problems

When public funds have been invested in developing environmental services and measurable indicators of those services have been identified, a natural question is how to value changes in their status (as a step toward measuring return on investment). The value of a given resource or environmental service is measured in terms of trade-offs that consumers face with or without the change. The techniques for eliciting this information depend on the kind of markets at hand. For marketable goods and services, observed market behaviour can be used. Two such methods, applied by Drechsel *et al.* to valuing soil fertility changes, are the calculation of replacement cost and the value of a productivity change (Chapter 9, this volume). Even when a natural resource service is not traded, so long as there exist marketed substitutes, the behaviour observed in markets for the substitute can be used for valuation of changes in quality or quantity. However, markets for factor inputs (e.g. land and labour) in developing countries are often imperfect, limiting the usefulness of market prices in valuation studies. Even when markets function well, NRM technologies may not generate goods and services that are traded in markets. As Shiferaw *et al.* (Chapter 2) show, the social benefits associated with changes in NRM are typically non-marketed, ruling out the use of actual markets to measure the economic values of changes in natural resource service flows due to NRM technologies. However, techniques exist for estimation of non-use values and indirect use values that are not traded in markets. In particular, contingent valuation and similar non-market valuation techniques need to be tested and developed for application to NRM impacts in agriculture. Although benefit transfer methods have been proposed to reduce the cost of estimating non-market values, they are of limited relevance when economic and ecological conditions differ markedly between the original location and the one where the values would be applied.

Attribution Problems

Establishing a cause–effect relationship between NRM programme interventions, intermediate outcomes and developmental or environmental impacts can be challenging. First, *ex post* impact assessments often rely on scanty cross-sectional adoption data, making it difficult for the impact evaluator to see the full picture of technology dissemination. Second, crop and resource management research often is not embodied in an observable physical entity that farmers can adopt or reject. The improved management practices are knowledge-intensive techniques transmitted as a recommendation or as a cognitive framework regarding such topics as pest management or soil conservation. Among the multiple sources of such information, it may be difficult to attribute changes in management practices to any given source (Traxler and Byerlee, 1992). Third, in contrast to crop improvement research, NRM research frequently involves multiple interactions, multiple stakeholders, and participatory processes. These characteristics pose formidable complications to the attribution of project impacts to a given research or development intervention (Freeman *et al.*, Chapter 1, this volume). Douthwaite *et al.* (Chapter 14, this volume) discuss the rationale for qualitative, step-wise and adaptive monitoring and evaluation methods for understanding the innovation process and how adoption begets outcomes that in turn beget impacts.

Beyond attribution, NRM impact analyses must measure impacts against the counterfactual case of what would have occurred in the absence of the NRM intervention – whereas scientific experiments typically include a control treatment as a baseline against which to judge other intervention effects. Although social programmes are often practically (or ethically) constrained from including a true control treatment, impact assessments must still characterise and try to measure the counterfactual case.

Various quasi-experimental approaches are suitable alternatives (Cook and Campbell, 1979). Baseline data are essential for reliable estimates of the changes attributable to the NRM intervention. One practical approach is the double-difference comparison. This method involves comparing relative changes in performance indicators before and after the NRM intervention between participants and non-participants. Careful research design and statistical analysis can help control for selection bias and other attribution problems (Pender, Chapter 6, this volume).

Integration of Resource and Environmental Impacts into Economic Impact Assessment

The economic surplus (ES) framework is the most desirable approach for summarising the economic welfare impacts of agricultural research investments. The classic ES approach measures the shift in a product supply curve resulting from technological change. The supply shift triggers changes in consumer surplus and producer surplus. Although benefit–cost analysis has been applied to a number of NRM projects, there have been scarcely any attempts to apply the ES approach (Alston *et al.*, 1995; Swinton, Chapter 7, this volume). The ease with which resource and environmental impacts can be integrated using this framework depends on the type of NRM intervention. Because non-market environmental or health effects often are not directly tied to agricultural output, productivity and environmental impacts must be calculated separately. Estimating ES for environmental impacts will require a simulated or surrogate market in which the marginal willingness to pay (WTP) (demand) curve can be estimated separately. Further research is needed to define the conditions under which the total ES may be measured as a sum of the economic surplus from productivity changes in the marketed commodity plus the estimated economic surplus from the simulated markets for environmental and health services. NRM may also change the quality of the products, which may induce a shift in consumer demand as well as in producer supply. Impact evaluation in this case will require measurement of the supply as well as the demand shifts (Swinton, Chapter 7, this volume).

At present, however, the suggested methods for integrating quantitative estimates of both marketed productivity impacts and non-marketed environmental impacts are untested. The current state of the art is exemplified by Bantilan *et al.* (Chapter 11, this volume), which combines an estimate of economic surplus based on marketed productivity changes with an inventory of environmental benefits and costs. The authors conduct a qualitative assessment of the environmental benefits vs. costs, concluding that environmental net benefits are positive. Based on this result, they infer that the market-based net benefits estimated from productivity enhancement alone are a lower bound for the true combined net benefits from both productivity and environmental dimensions.

Two broad classes of empirical methods are used to estimate changes in ES. When past data are available about the performance of NRM interventions, econometric regression methods can be used for several important purposes. First, econometrics is widely used to test the potential effects of NRM changes on productivity (Pender, Chapter 6, this volume). When data from a sufficiently large sample is available, econometric methods are useful in testing whether investments in specific crop and resource management practices had significant effects on productivity or on the quality of the resource base (Pender, Chapter 6; Kerr and Chung, Chapter 10, this volume). Careful econometric analysis can substantially reduce the problems of attribution. Second, econometric inverse demand models are used to estimate the price elasticity of demand for marketed (and non-marketed) products. Third, econometric models can identify the factors determining both: the likelihood of adoption of an NRM innovation; and the degree of NRM used by those who have adopted.

When sufficient data are not available for econometric estimation, an alternative useful approach for estimating the magnitude and form of production and environmental effects is bioeconomic modelling (Kruseman and Bade, 1998; Barbier and Bergeron, 2001; Okumu *et al.*, 2002; Holden and Shiferaw, 2004). Using mathematical relationships, bioeconomic models link economic behavioural objectives with key ecological and production processes that determine biophysical outcomes (Oriade and Dillon, 1997). As discussed by Holden (Chapter 8, this volume) such integration allows the analysis of efficiency, distributional and sustainability impacts of proposed technology and/or policy interventions (Ruben *et al.*, 2001). The approach can also be used to measure the impact of these interventions *ex post*. A household-scale example is the impact analysis of soil and water conservation technologies (Shiferaw and Holden, Chapter 12, this volume). At the regional scale, computable general equilibrium (CGE) models become very useful to capture the economy-wide impacts of technology and policy interventions (Holden and Lofgren, Chapter 13, this volume). CGE models are particularly suited for assessing price effects and distributional issues associated with technical and policy interventions.

What We Need to Know – Areas for Future Research

Despite recent progress in developing methods for evaluating the impacts of productivity enhancing technologies on the one hand and for measuring natural resource service flows and their value on the other, these advances have not been unified in NRM impact studies. With very few exceptions, NRM impact evaluations have failed to incorporate the non-productivity related impacts (resource and environmental service flows) into economic impact assessments. This volume has brought together some of the methodological tools that can be used to integrate the sustainability impacts with the productivity impacts of agricultural NRM interventions. But the state of the art does not yet permit us to advocate 'best practices' for comprehensive evaluation of NRM impacts. Several knowledge gaps first beg the attention of researchers.

How does NRM affect ecosystem functions and services?

Our understanding of the impacts of human interventions on ecosystem functions and services at different scales and how this affects productivity, sustainability and environmental outcomes is still inadequate. The concept of 'natural resource management' itself is very broad, ranging from crop and livestock management practices to strategies for managing natural resources such as soils, water, biodiversity, fish and forests. Agricultural activities may have important externalities, such as global warming. Improved NRM enhances the provision of essential ecosystem services that reduce such negative environmental externalities. How different types of NRM interventions affect the flow of ecosystem services at different spatial and temporal scales is, however, not clearly understood. While there are several reports on the environmental impacts of intensive agricultural activities (e.g. the Green Revolution), there are few empirical examples for crops other than wheat and rice (Maredia and Pingali, 2001). The limited evidence and insufficient understanding of the key links between agricultural activities and how NRM would regulate this link, prevent quantification and measurement of key outcomes and potential impacts on human welfare. As Altieri has argued, 'what is lacking ... is the explicit description of the scientific basis of NRM and of methods to increase our understanding of the structure and dynamics of agricultural and natural resource ecosystems and providing guidelines to their productive and sustainable management' (Altieri, 2002, p. 7). Such understanding is a key first step in enhancing attribution of certain environmental outcomes to NRM interventions. Progress toward better definition of agro-ecosystem functions and services is urgently needed. Simulation modelling offers an increasingly valid and cost-effective tool for understanding the biophysical dynamics of NRM interventions.

Indicators of ecosystem performance

To the extent that agricultural natural resource functions *are* understood, the measurement of their status and service flows remains too costly for practical impact assessment purposes. Inexpensive but reliable indicators continue to be needed. A core set of environmental and sustainability indicators would allow researchers to check for deviation from trend by gathering time-series data for regular ecosystem monitoring, not to mention establishing the counterfactual to NRM interventions. Some preliminary steps have been taken by the Heinz Foundation (http://www.heinzctr.org/ecosystems/index.htm), which started to monitor the state of US ecosystems in 1999. Their efforts offer useful criteria for consideration, although their indicators obviously need adaptation to the developing country settings of most NRM projects.

How to enhance attribution of impacts?

More systematic thinking is needed about how to measure the dissemination of knowledge-based technologies that are not embodied in improved tools or germplasm. Knowledge-based innovations appear less well suited to the reduced form input demand approaches that economists have used for embodied technologies like improved seeds. Better indicators for ecosystem performance measurement can help. So too can direct approaches to measuring farmers' knowledge and attitudes and how they affect the choice of management practices. The knowledge–attitudes–practices (KAP) model from epidemiology may be a start, as the explicit measurement of changes in knowledge and attitudes of a treatment group compared with a control can confirm attribution to project interventions. Indeed, explicit attribution becomes doubly important – albeit doubly complicated – when NRM technologies are introduced in tandem with genetic technologies or a newly supportive public policy. Careful adherence to sound impact assessment methods (especially the double-difference method) and strict adherence to avoid or measure selection bias among beneficiaries will have to be joined to closer scrutiny of knowledge and attitudes.

Can we properly value non-market ecosystem services?

Even when we can understand and measure cost-effectively the resource and environmental service flows from NRM interventions, shortcomings in our ability to measure the welfare impacts of these changes can impede accurate assessments. The reviews in this volume have identified several techniques used for valuation of non-market outcomes in the developed world (Shiferaw *et al.*, Chapter 2). Valuation methods for non-marketed ecosystem services (e.g. carbon sequestration in soil or biodiversity preservation) need to be tested and refined.

Many methods for measuring WTP for environmental services presuppose that consumers directly demand the service in question. Yet many agricultural NRM services do not fit that description. Few consumers would pay for the presence of *Rhizobium* bacteria in soil, yet the nitrogen-fixing services that they perform provide plant nutrition and, if carefully timed, may reduce nitrate leaching into drinking water supplies. In short, the demand for the services of *Rhizobium* bacteria is indirect, not direct. As such, it is analogous to the demand for other agricultural inputs. Two key factors differ, however. First, whereas conventional derived input demand arises solely from market prices and factor endowments, part of the derived demand for NRM services originates in a direct demand for health that affects the non-market valuation of exposure to reduced drinking water quality. Second, that same health component involves externalities to neighbours of the producer, a stakeholder group whose members' utility is not included in an indirect demand function based upon the marketed agricultural product. Moving from theory to practice in measuring indirect WTP for environmental and health services will be complex. A major desirable innovation is to find lower cost – yet accurate – ways to estimate downward-sloping inverse demand curves as a basis for estimating elasticities of demand for non-market environmental and health services.

For policy purposes, a simpler approach than measuring WTP is to measure farmers' willingness to *accept* compensation for the non-marketed health and environmental services that they provide. This will be a compensating surplus measure for farmers to provide essential services to society. Such measures will require more bioeconomic modelling in order to estimate the opportunity costs implicit in providing cost-increasing health and environmental services.

Can the economic surplus approach be extended for integrated assessment?

Despite its strengths, the economic surplus approach has been criticised on several counts. For purposes of NRM impact assessment, the most serious of these is its failure to account for environmental impacts that are external to functioning markets. One of the goals of NRM is to reduce the undesirable on-site and off-site externalities associated with agricultural production. Two chapters in this volume (Swinton, Chapter 7 and Bantilan *et al.*, Chapter 11) have discussed the ways to extend the economic surplus approach towards comprehensive evaluation of productivity and environmental impacts. However, progress in this area has been hampered by measurement problems, the high cost of WTP estimation, and the difficulties of mixing values assessed from different market settings (e.g. real markets and hypothetical ones). Initial efforts to integrate productivity and environmental impacts in a comprehensive assessment should focus on simple cases where price elasticities of demand can readily be estimated. Serious thinking is needed on how to combine productivity and environmental effects in computing a single, comprehensive measure of impact from NRM interventions.

Alternatively, the economic surplus approach to productivity impact assessment may be supplemented by qualitative information. Some audiences uncomfortable with the demanding assumptions required for many WTP estimation studies may consider these methods more valid. The participatory methods for interdisciplinary analysis of adoption pathways, processes and outcomes may also contribute to participant empowerment that can enhance impacts, whether or not they enhance impact assessment *per se*.

What do we know about economy-wide impacts?

In addition to direct effects associated with supply shifts, agricultural productivity interventions also generate indirect economic effects through product and factor market linkages. The overall effect of technical change from research and development (R&D) interventions hence depends on system-wide growth and multiplier effects induced through input use, factor markets and production linkages (Maredia *et al.*, 2000). For comprehensive evaluation of large-scale NRM impacts, it would be useful to include these general equilibrium or economy-wide effects. While this can be done using a CGE model (Holden and Lofgren, Chapter 13, this volume), there is limited experience in developing CGE models that incorporate environmental and sustainability impacts. In situations where substantial impacts occur from both general equilibrium market effects and sustainability effects, it could be very rewarding to develop and employ such methods. Standard CGE models (Lofgren *et al.*, 2002) are now being developed for many developing countries, and these models deserve research into possibilities for adaptation to evaluate NRM technology and policy impacts.

Simple steps toward better impact assessments

Advance planning can greatly improve the quality of NRM impact assessments. The classic principles of quasi-experimentation remain relevant: to compare affected and unaffected groups before and after the program intervention, taking care not to bias results due to non-random selection of participants (Cook and Campbell, 1979). The few NRM impact assessments available have had adequate baseline data only on the productivity dimension, not on environmental and health dimensions. In some instances, this is because intended environmental and health outcomes had not been clearly specified at project outset.

For effective impact assessments, baseline data on all intended outcomes dimensions is necessary. Acquiring such data calls for projects *before implementation begins:* 1. to specify clearly the intended outcomes; 2. to choose acceptable indicators of important outcome dimensions; 3. to identify comparable, paired groups inside and outside the intervention area; and 4. to budget for and to conduct baseline studies on the intended outcomes and related variables for the paired groups within and without the NRM intervention zone. It goes without saying that planning and budget are also needed for one or more follow-up studies to measure progress toward the intended outcomes – again, among comparable households both affected and unaffected by the NRM programme. Ensuring that appropriate baseline and follow-up data are collected is not only possible; it will also greatly facilitate advances in the methodological areas listed above.

Institutionalising NRM impact assessment

If it is to affect institutional decision making, NRM impact evaluation needs to be integrated into programme planning in research and development institutions. An institutional learning cycle from programme planning to implementation to impact analysis and back to programme planning can help to ensure that lessons are learned and plans modified systematically. Such a process is more likely to prevent uncorrected flaws from turning well-conceived programmes into vaunted 'failures'.

NRM impact assessment can be conducted internally within R&D institutions or subcontracted to external evaluators. But there are compelling reasons for R&D institutions to institutionalise IA capacity 'in-house' if they are to be effective in influencing internal programme planning. Some R&D institutions have institutionalised impact assessment within an independent economics or social science programme. However, given the trend toward replacing disciplinary research areas with thematic research areas, a promising approach is to institutionalise IA capacity into a specialised impact assessment unit that reports directly to senior management. Staff in this unit should be drawn from both the social and the biophysical sciences, in order to provide comprehensive analysis of the multi-dimensional and non-monetary impacts of NRM interventions. Such a unit can provide intellectual leadership for all IA studies and can provide a platform for integrating the results from impact assessment studies into institutional learning and research planning for the purpose of enhancing future impacts.

An independent IA unit can be effective at forging strategic alliances between research institutes, development partners, and advanced research institutes. The chapters in this volume demonstrate how comprehensive assessment of NRM impacts can emerge from strategic partnerships between university-based researchers (with a comparative advantage in development of theories and methods for assessing NRM impacts) combined with researchers in R&D institutions (with comparative advantages in empirical applications of these methods, synthesis of experience, and scaling-up results).

Research managers also need to think carefully about how much to invest in impact assessment. A standing IA unit can be expensive, and R&D institutions exist primarily to generate impacts, not to measure them. Yet in a world where many institutions claim to generate impacts and compete for funds to sustain their efforts, a competitive advantage can be built from the institutional capabilities to perform high-quality impact assessments and to adapt programme planning systematically based upon the lessons learned. Building such capabilities will require a modest proportion of core funds on a continuing basis, with the understanding that the IA unit will help to attract competitive funds through collaboration with thematic units on project design.

Conclusions

Reducing poverty and ensuring livelihood security for the millions of impoverished people whose subsistence depends on agriculture will not be possible without judicious management of the productive resource base. But failure to demonstrate desired impacts could undermine current R&D efforts in developing and disseminating new innovations that provide dual productivity and sustainability benefits. Although many NRM interventions do not provide direct short-term net economic benefits to producers, they do generate non-marketed ecosystem goods and services that are essential for sustaining agricultural productivity and environmental quality. Failure to incorporate the value of environmental and health outcomes of agricultural NRM investments will lead to bias and likely underestimation of their social net benefits, followed by underinvestment from the standpoint of social welfare. The mirror image of such misallocation of R&D resources is equally troubling, for it entails overinvestment in agricultural programmes that may cause environmental and health damage.

Methods for comprehensive economic impact assessment that would integrate productivity, environmental and sustainability impacts are only just beginning to emerge. This volume has assembled recent methodological advances from this nascent area. It has critiqued the methodological *status quo*, and sought to define new horizons for experimentation to refine current practices and to develop second-generation methods that address existing and emerging challenges. The key challenges relate to measurement, indicators, valuation and attribution of impacts.

As we look into the future, NRM is entering a new era. With the emerging recognition that participatory NRM projects can empower individuals and communities, empowerment is shifting from being an unintended to an intended benefit. As it becomes an explicitly intended outcome of integrated NRM projects, empowerment begs the same needs for measurement, attribution and valuation that have challenged assessments of environmental and health dimensions of NRM interventions. Likewise, INRM projects typically prioritise poverty alleviation, making measures of income distribution effects another newly important dimension of NRM impact assessment.

In addition to application of new methods from environmental and resource economics, future NRM impact assessments have much to gain from employing a mix of quantitative and qualitative approaches. This can enrich interpretation and communication of outcomes and assist in their attribution. Qualitative methods can be especially helpful at elucidating *how* outcomes came to be. Such process understanding has particular value for unanticipated outcomes, with an eye to ensuring that desirable ones can be replicated and undesirable ones avoided in future.

Participatory impact assessments by NRM project beneficiaries may also enhance the empowerment outcome and associated impacts. However, the role of such participatory assessments should be recognised as a self-monitoring activity that is part of the project effort, not a true impact assessment of what would have occurred without the project. Accurate impact assessment, even of environmental and empowerment dimensions, must adhere to the basic principles of: 1. freedom from participant selection bias; 2. before vs. after comparisons aided by baseline information; and 3. 'with vs. without' measures of what the intervention accomplished.

The needs for improved methods for economic and social impact assessment are matched by needs for improved understanding of ecosystem performance. Following Altieri's call for increased 'understanding of the structure and dynamics of agricultural and natural resource ecosystems and providing guidelines to their productive and sustainable management' (Altieri, 2002, p. 7), the need for close future interdisciplinary collaboration is clear. Forging strong linkages and effective dialogue among ecologists, economists, and other social scientists is a *sine qua non* for future advances in scientifically sound natural resource management interventions and for thorough and balanced evaluations of their impacts.

References

Alston, M.J., Norton, W.G. and Pardey, P.G. (1995) *Science Under Scarcity. Principles and Practice for Agricultural Research Evaluation and Priority Setting.* Cornell University Press, Ithaca, New York, and London, UK, 585 pp.

Alston, J.M., Marra, M.C., Pardey, P.G. and Wyatt, T.J. (1998) Research returns redux: A meta analysis of returns to agricultural R&D. *EPTD Discussion Paper No. 38.* International Food Policy Research Institute (IFPRI), Washington, DC, 44 pp.

Altieri, M. (2002) Agroecology: the science of natural resource management for poor farmers in marginal environments. *Agriculture, Ecosystems and Environment* 93, 1–24.

Barbier, B. and Bergeron, G. (2001) Natural resource management in the hillsides of Honduras. Bioeconomic modeling at the microwatershed level. *IFPRI Research Report No. 123.* International Food Policy Research Institute (IFPRI), Washington, DC, 59 pp.

Cook, T.D. and Campbell, D.T. (1979) *Quasi-Experimentation: Design and Analysis Issues for Field Settings.* Houghton Mifflin, Boston, Massachusetts. 405 pp.

Gumtang, R.J., Pampolino, M.F., Tuong, T.P. and Bucao, C. (1999) Groundwater dynamics and quality under intensive cropping systems. *Experimental Agriculture* 35, 153–166.

Gupta, R.K. and Abrol, I.P. (2000) Salinity build-up and changes in the rice-wheat system of the Indo-Gangetic plains. *Experimental Agriculture* 63(2), 273–284.

Holden, S.T. and Shiferaw, B. (2004) Land degradation, drought and food security in a less-favoured area in the Ethiopian highlands: a bioeconomic model with market imperfections. *Agricultural Economics* 30(1), 31–49.

Kruseman, G. and Bade, J. (1998) Agrarian policies for sustainable land use: bio-economic modeling to assess the effectiveness of policy instruments. *Agricultural Systems* 58, 465–481.

Lofgren, H., Harris, R.L. and Robinson, S. (2002) A Standard Computable General Equilibrium (CGE) Model in GAMS. *Microcomputers in Policy Research.* Volume 5. International Food Policy Research Institute (IFPRI), Washington, DC, 69 pp.

Maredia, M. and Pingali, P. (2001) *Environmental Impacts of Productivity-Enhancing Crop Research: A Critical Review.* Standing Panel on Impact Assessment (SPIA), Food and Agriculture Organization of the United Nations (FAO), Rome, Italy, 36 pp. http://www.sciencecouncil.cgiar.org/publications/ispubs.htm.

Maredia, M., Byerlee, D. and Anderson, J. (2000) *Ex post* evaluation of economic impacts of agricultural research: A tour of good practice. *Paper presented at the workshop on the Future of Impact Assessment in the CGIAR: Needs, Constraints and Options. Standing Panel on Impact Assessment (SPIA), May 3-5, 2000, Rome, Italy,* 39 pp. (unpublished) http://www.cgiar.org/who/wwa_spiameet.html.

Nelson, M. and Maredia, M. (1999) Environmental impacts of the CGIAR: An initial assessment. *Paper presented initially at the International Centers Week, 25–29 October 1999, Washington, DC. Standing Panel for Impact Assessment. Food and Agriculture Organization of the United Nations (FAO), Rome, Italy,* 71 pp. (unpublished) http://www.sciencecouncil.cgiar.org/publications/ispubs.htm.

Okumu, B.N., Jabbar, M.A., Coman, D. and Russel, N. (2002) A bioeconomic model of integrated crop–livestock farming systems: the case of Ginchi watershed in Ethiopia. In: Barrett, C.B., Place, F. and Aboud, A. (eds) *Natural Resource Management in African Agriculture.* CAB International, Wallingford, UK, pp. 235–249.

Oriade, C. and Dillon, C.A. (1997) Developments in biophysical and bioeconomic simulation of agricultural systems: A review. *Agricultural Economics* 17, 45–58.

Pingali, P.L., Marquez, C.B. and Palis, F.G. (1994) Pesticides and Philippine rice farmer health: A medical and economic analysis. *American Journal of Agricultural Economics* 76(3), 587–592.

Pretty, J.N., Brett, C., Gee, D., Hine, R.E., Mason, C.F., Morison, J.I.L., Raven, H., Rayment, M.D. and van de Bijl, G. (2000) An assessment of the total external costs of UK agriculture. *Agricultural Systems* 65(2), 113–136.

Ruben, R., Kuyvenhoven, A. and Kruseman, G. (2001) Bioeconomic models for ecoregional development: Policy instruments for sustainable intensification. In: Lee, D.R. and Barrett, C.B. (eds) *Tradeoffs or Synergies? Agricultural Intensification, Economic Development and the Environment.* CAB International, Wallingford, UK, pp. 115–134.

Traxler, G. and Byerlee, D. (1992) Economic returns to crop management research in a post-green revolution setting. *American Journal of Agricultural Economics* 74(3), 573–582.

Index

Page numbers in **bold** refer to illustrations and tables